# Advances in Intelligent Systems and Computing

## Volume 316

**Series editor**

Janusz Kacprzyk, Polish Academy of Sciences, Warsaw, Poland
e-mail: kacprzyk@ibspan.waw.pl

## About this Series

The series "Advances in Intelligent Systems and Computing" contains publications on theory, applications, and design methods of Intelligent Systems and Intelligent Computing. Virtually all disciplines such as engineering, natural sciences, computer and information science, ICT, economics, business, e-commerce, environment, healthcare, life science are covered. The list of topics spans all the areas of modern intelligent systems and computing.

The publications within "Advances in Intelligent Systems and Computing" are primarily textbooks and proceedings of important conferences, symposia and congresses. They cover significant recent developments in the field, both of a foundational and applicable character. An important characteristic feature of the series is the short publication time and world-wide distribution. This permits a rapid and broad dissemination of research results.

## Advisory Board

Chairman

Nikhil R. Pal, Indian Statistical Institute, Kolkata, India
e-mail: nikhil@isical.ac.in

Members

Rafael Bello, Universidad Central "Marta Abreu" de Las Villas, Santa Clara, Cuba
e-mail: rbellop@uclv.edu.cu

Emilio S. Corchado, University of Salamanca, Salamanca, Spain
e-mail: escorchado@usal.es

Hani Hagras, University of Essex, Colchester, UK
e-mail: hani@essex.ac.uk

László T. Kóczy, Széchenyi István University, Győr, Hungary
e-mail: koczy@sze.hu

Vladik Kreinovich, University of Texas at El Paso, El Paso, USA
e-mail: vladik@utep.edu

Chin-Teng Lin, National Chiao Tung University, Hsinchu, Taiwan
e-mail: ctlin@mail.nctu.edu.tw

Jie Lu, University of Technology, Sydney, Australia
e-mail: Jie.Lu@uts.edu.au

Patricia Melin, Tijuana Institute of Technology, Tijuana, Mexico
e-mail: epmelin@hafsamx.org

Nadia Nedjah, State University of Rio de Janeiro, Rio de Janeiro, Brazil
e-mail: nadia@eng.uerj.br

Ngoc Thanh Nguyen, Wroclaw University of Technology, Wroclaw, Poland
e-mail: Ngoc-Thanh.Nguyen@pwr.edu.pl

Jun Wang, The Chinese University of Hong Kong, Shatin, Hong Kong
e-mail: jwang@mae.cuhk.edu.hk

More information about this series at http://www.springer.com/series/11156

Peter Sinčák · Pitoyo Hartono
Mária Virčíková · Ján Vaščák
Rudolf Jakša
Editors

# Emergent Trends in Robotics and Intelligent Systems

Where Is the Role of Intelligent Technologies in the Next Generation of Robots?

 Springer

*Editors*
Peter Sinčák
Technical University of Kosice
Kosice, Slovakia

Pitoyo Hartono
School of Information Science
  and Technology
Chukyo University
Toyota, Japan

Mária Virčíková
Technical University of Kosice
Kosice, Slovakia

Ján Vaščák
Technical University of Kosice
Kosice, Slovakia

Rudolf Jakša
Technical University of Kosice
Kosice, Slovakia

ISSN 2194-5357          ISSN 2194-5365    (electronic)
ISBN 978-3-319-10782-0    ISBN 978-3-319-10783-7    (eBook)
DOI 10.1007/978-3-319-10783-7

Library of Congress Control Number: 2014948175

Springer Cham Heidelberg New York Dordrecht London

© Springer International Publishing Switzerland 2015
This work is subject to copyright. All rights are reserved by the Publisher, whether the whole or part of the material is concerned, specifically the rights of translation, reprinting, reuse of illustrations, recitation, broadcasting, reproduction on microfilms or in any other physical way, and transmission or information storage and retrieval, electronic adaptation, computer software, or by similar or dissimilar methodology now known or hereafter developed. Exempted from this legal reservation are brief excerpts in connection with reviews or scholarly analysis or material supplied specifically for the purpose of being entered and executed on a computer system, for exclusive use by the purchaser of the work. Duplication of this publication or parts thereof is permitted only under the provisions of the Copyright Law of the Publisher's location, in its current version, and permission for use must always be obtained from Springer. Permissions for use may be obtained through RightsLink at the Copyright Clearance Center. Violations are liable to prosecution under the respective Copyright Law.
The use of general descriptive names, registered names, trademarks, service marks, etc. in this publication does not imply, even in the absence of a specific statement, that such names are exempt from the relevant protective laws and regulations and therefore free for general use.
While the advice and information in this book are believed to be true and accurate at the date of publication, neither the authors nor the editors nor the publisher can accept any legal responsibility for any errors or omissions that may be made. The publisher makes no warranty, express or implied, with respect to the material contained herein.

Printed on acid-free paper

Springer is part of Springer Science+Business Media (www.springer.com)

# Preface

# Science Is an Integral Part of Culture

"Science is an integral part of culture.
It's not this foreign thing, done by an arcane priesthood.
It's one of the glories of the human intellectual tradition."

Stephen Jay Gould, Harward University
Expert in History of Science

Research, level of education, knowledge and outlook of people determine the society and the cultural aspects of the community. Technical University of Kosice in Slovakia with over 15000 students and approximately 2000 staff members form a very important community in the city of Kosice. There is a number of research groups at the University environment and therefore all of them influence the educational and complex cultural level of city of Kosice. The Creative thinking is an imminent part of science and therefore we do believe that science is a part of Culture and that the level of research give impact to the level of culture in the community. The good example can be confirmed with the fact that a real research can lead to international collaborations and gathering of international visitors to the city. From this point of view Symposium on Emergence Technology in Artificial Intelligence and Robotics is a nice example where more than 15 people from all over the world take part in the event as plenary speakers. The importance of the symposium is underlined with the fact that His Excellency the Japanese Ambassador in the Slovak Republic took a patronage over this event. Also EU-Japan Fest Foundation and Ministry of Education of Slovak Republic contributed to this research and cultural event as a part of European Capitol of Culture - 2013.

Intelligent Robotics is a part of the growingly important Artificial Intelligence and Robotics field. This domain is now under rapid development, shifting from academic attention to commercial domain where worldwide business communities are becoming aware of promising financial profit in Intelligent Robotics in close and distance future. Robotics is now also expanding from industrial to non-industrial environment (homes, public places, hospitals, cultural events, art, social events etc.) thanks to computer technology and also to fast and essential changes in human computer interaction.

The Ministry of Culture of the Slovak Republic supported a world programming contest in the domain of social robotics. The goal was to design an ideal social companion. The proclamation of the results will take place during Symposium on Emergence Technology in Artificial Intelligence and Robotics.

Many studies confirm that Intelligent Robotics will be a large domain supporting transformation of employments from simple low knowledge jobs to high tech job positions. Intelligent Robotics will bring machines into our everyday life and these non-living artifacts will co-create a culture of the mankind. Intelligent Robotics will have an impact to human culture and therefore the importance of this meeting within the European Capitol of Culture - 2013 framework is absolutely consistent and we can proclaim that science is a part of cultural life in Slovak Republic.

**Prof. Peter Sinčák**
Technical University of Košice, Slovakia

**Prof. Pitoyo Hartono**
Chukyo University, Nagoya, Japan

# Contents

## Part II: Intelligent Systems

# 70th Anniversary of Publication: Warren McCulloch & Walter Pitts - A Logical Calculus of the Ideas Immanent in Nervous Activity

Jiří Pospíchal and Vladimír Kvasnička

Institute of Applied Informatics, Faculty of Informatics and Information Technologies,
Slovak Technical University, Bratislava, Slovakia
{pospichal,kvasnicka}@fiit.stuba.sk

**Abstract.** In 1943, a paper by Warren McCulloch & Walter Pitts [6] entitled *"A logical calculus of the ideas immanent to nervous activity"* was published, which is now considered as one of the seminal papers that initiated the formation of artificial intelligence and cognitive science. In this paper concepts of logical (threshold) neurons and neural networks were introduced. It was proved that an arbitrary Boolean function may be represented by a feedforward (acyclic) neural network composed of threshold neurons, i.e. this type of neural network is a universal approximator in the domain of Boolean functions. The present paper recalls the core achievements of this paper and puts it into perspective from the point of view of further achievements based on their approach. Particularly, S. Kleene [5] and M. Minsky [7] extended this theory by their study of relationships between neural networks and finite state machines. The present paper is not a standard research article where new ideas or approaches would be presented. However, the 70th anniversary of publication of the McCulloch and Pitts paper should be sufficiently important to recall this core event in computer science and artificial intelligence. In particular, the main concept of their paper opened unexpected ways to study processes in the human brain. Their approach offers a way to treat a core philosophical mind/body problem in such a way that the brain is considered as a neural network and the mind is interpreted as a product of its functional properties.

## 1    Introduction and Basic Concepts

Logical neurons and neural networks were initially introduced in Warren McCulloch´s and Walter Pitts´s paper [6]. This paper demonstrated that neural networks are universal approximators for a domain of Boolean functions; i.e. an arbitrary Boolean function can be represented by a feedforward neural network composed of threshold neurons. We have to mention from the very beginning that this work is very difficult to read; its mathematical-logical part was probably written by Walter Pitts, who was in both sciences a total autodidact. Thanks to logician S. Kleene [5] and computer scientist M. Minsky [7], this work has been "translated" at the end of the fifties into a form using standard language of contemporary logic and mathematics and its important ideas became generally available and accepted.

© Springer International Publishing Switzerland 2015
P. Sinčák et al. (eds.), *Emergent Trends in Robotics and Intelligent Systems*,
Advances in Intelligent Systems and Computing 316, DOI: 10.1007/978-3-319-10783-7_1

An elementary unit of neural networks is threshold (logical) neuron of McCulloch and Pitts. It has two binary values (i.e. either state 1 or state 0). It may be interpreted as a simple electrical device - relay. Let us postulate that a dendritic system of threshold neuron is composed of excitation inputs (described by binary variables $x_1$, $x_2$, ..., $x_n$ which amplify an output response) and inhibition inputs (described by binary variables $x_{n+1}$, $x_{n+2}$, ..., $x_m$ which are weakening an output response), see Fig. 1.

An activity of threshold neuron is set to one if the difference between a sum of excitation input activities and a sum of inhibition activities is greater than or equal to the threshold coefficient $\vartheta$, otherwise it is set to zero.

$$y = \begin{cases} 1 & (x_1 + ... + x_n - x_{1+n} - ... - x_m \geq \vartheta) \\ 0 & (x_1 + ... + x_n - x_{1+n} - ... - x_m < \vartheta) \end{cases} \tag{1}$$

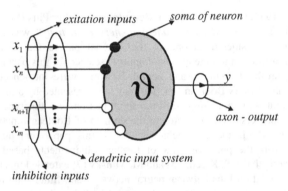

**Fig. 1.** Diagrammatic visualization of McCulloch and Pitts neuron which is composed of the dendritic system for information input (excitation or inhibition) activities and axon for information output. A body of neuron is called the soma, it is specified by a threshold coefficient $\vartheta$.

If we introduce a simple step function

$$s(\xi) = \begin{cases} 1 \, (\text{if } \xi \geq 0) \\ 0 \, (\text{otherwise}) \end{cases} \tag{2a}$$

then an output activity may be expressed as follows:

$$y = s\left( \underbrace{x_1 + ... + x_n - x_{1+n} - ... - x_m - \vartheta}_{\xi} \right) \tag{2b}$$

An entity $\xi$ is called the internal potential. Simple implementations of elementary Boolean functions of disjunctions, conjunctions, implication and negation are presented in Fig. 2.

Let us note that the above mentioned simple principles (1-2) "all or none" for neurons have been introduced in the late twenties and early thirties of the former century by English physician and electro-physiologist Sir Edgar Adrian [1] when he studied output neural activities by making use of very modern (for that time) electronic equipment based on electron-tube amplifiers and cathode-ray tubes for a visualization of measurements.

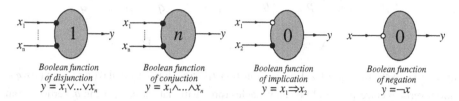

**Fig. 2.** Four different implementations of threshold neurons which specify Boolean functions of disjunction, conjunction, implication and negation, respectively. Excitatory connections are terminated by a black dot whereas inhibition connections by open circles.

In the original paper [6] McCulloch and Pitts have discussed a possibility that inhibition is absolute, i.e. any active inhibitory connection forces the neuron into the inactive state (with zero output state). The paper itself shows that this form of inhibition is not necessary and that "subtractive inhibition" based on formulae (1-2) gives the same results.

## 2    Boolean Functions

Each Boolean function [12, 13] is represented by a syntactic tree (derivation tree) which represents a way of its recurrent building, going bottom up, initiated by Boolean variables and then terminated (at the root of tree) by a composed Boolean function (formula of propositional logic), see Fig. 3, diagram A. Syntactic tree is a very important notion for a construction of its subformulae, each vertex of tree specifies sub formulae of the given formula: lowest placed vertices are assigned to trivial subformulae p and q, forthcoming two vertices are assigned subformulae $p \Rightarrow q$ and $p \wedge q$, highest placed vertex – root of the tree – is represented by the given formula $(p \Rightarrow q) \Rightarrow (p \wedge q)$.

We see that for an arbitrary Boolean function we may simply construct a neural network which simulates functional value of the Boolean function, see Fig. 3, where this process is outlined for formula $(p \Rightarrow q) \Rightarrow (p \wedge q)$. It means that these results may be summarized in a form of a theorem.

**Theorem 1.** Each Boolean function, represented by a syntactic tree, can be alternatively expressed in a form of neural network composed of logical neurons that correspond to connectives from the given formula.

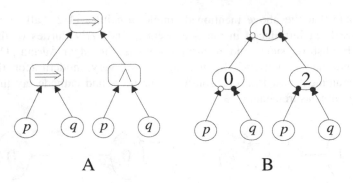

**Fig. 3.** (A) Syntactic tree of a Boolean function (propositional formula) $(p \Rightarrow q) \Rightarrow (p \wedge q)$. Bottom vertices correspond to Boolean variables (propositional variable) $p$ and $q$, vertices from the next levels are assigned to connectives implication and conjunction, respectively. An evaluation of the syntactic tree runs bottom up. (B) Neural network composed of logical neurons of connectives which appear in a given vertex of the syntactic tree of diagram A. We see that between syntactic tree and neural network there exists a very close one-to-one correspondence, their topologies are identical, they differ only in vertices. Figuratively speaking, we may say that a neural network representing a Boolean function $\varphi$ can be constructed from its syntactic tree by direct substitution of its vertices by proper logical neurons.

This theorem belongs to basic results of the seminal paper by McCulloch and Pitts [6]. It claims that an arbitrary Boolean function represented by a syntactic tree may be expressed in a form of neural network composed of simple logical neurons that are assigned to logical connectives from the tree. It means that neural networks with logical neurons are endowed with an interesting property that these networks have a property of universal approximator in a domain of Boolean functions. The above outlined constructive approach based on existence of syntactic tree for each Boolean function is capable of accurate simulation of any given Boolean function.

The architecture of neural network based on the syntactic tree which is assigned to an arbitrary Boolean function may be substantially simplified to the so-called 3-layer neural network composed of

(1) a layer of input neurons (which copy input activities, they are not computational units),
(2) a layer of hidden neurons and
(3) a layer of output neurons;

where neurons from two juxtaposed layers are connected by all possible ways by connections. This architecture is minimalistic and could not be further simplified. We demonstrate a constructive way of how to construct such a neural network for an arbitrary Boolean function.

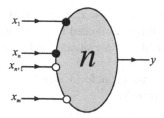

**Fig. 4.** A logical neuron for simulation of an arbitrary conjunctive clause which is composed of propositional variables or their negations that are mutually connected by conjunctions, $y = x_1 \wedge ... \wedge x_n \wedge \neg x_{n+1} \wedge ... \wedge \neg x_m$

Applying simple generalization of the concept of logical neuron, we may immediately show that a single logical neuron is capable of simulating a conjunctive clause $x_1 \wedge ... \wedge x_n \wedge \neg x_{n+1} \wedge ... \wedge \neg x_m$.

In the theory of Boolean functions [12, 13], a very important theorem is proved that each Boolean function may be equivalently written in a form of disjunctive normal form

$$\varphi = \bigvee_{\substack{\tau \\ (val_\tau(\varphi)=1)}} x_1^{(\tau)} \wedge x_2^{(\tau)} \wedge ... \wedge x_n^{(\tau)} \tag{3}$$

where

$$x_i^{(\tau)} = \begin{cases} x_i & \left(\text{if } val_\tau(x_i) = 1\right) \\ \neg x_i & \left(\text{if } val_\tau(x_i) = 0\right) \end{cases} \tag{4}$$

A final form of the Boolean function (3) is outlined in Fig. 4. Results of this illustrative example may be summarized in a form of the following theorem.

**Theorem 2.** An arbitrary Boolean function f can be simulated by a 3-layer neural network.

We have to note that, according to the theorem 2, the 3-layer neural networks composed of logical neurons are a universal computational device for a domain of Boolean functions; each Boolean function may be represented by this "neural device" called the neural network. This fundamental result of McCulloch`s and Pitts`s paper [6] preceded modern result, after which 3-layer feed-forward neural network with a continuous activation function is a universal approximator of continuous functions specified by a table of functional values [13].

We may question what kind of Boolean functions is a single logical neuron capable to classify correctly? According to Minsky and Papert, this question may be solved relatively quickly by geometric interpretation of computations running in logical neuron [8]. In fact, logical neuron divides input spaces into two half spaces by a hyperplane $w_1x_1 + w_2x_2 + ... + w_nx_n = \vartheta$ for weight coefficients $w_i = 0, \pm 1$. Then, we say that a Boolean function $f(x_1, x_2, ..., x_n)$ is *linearly separable*, if and only if there exists such a hyperplane $w_1x_1 + w_2x_2 + ...+ w_nx_n = \vartheta$ which separates a space of input activities in such a way that objects evaluated by 0 are situated in the first part of the space, whereas objects evaluated by 1 are situated in the second part of the space.

**Theorem 3.** Logical neurons are capable to *simulate* correctly *only those Boolean functions that are linearly separable.*

A classical example of a Boolean function which is not linearly separable is a logical connective "exclusive disjunction" which may be formally specified as a negation of a connective of equivalence, $(x \oplus y) \Leftrightarrow \neg(x \equiv y)$, in computer-science literature this connective is usually called the XOR Boolean function, $\varphi_{XOR}(x, y) = x \oplus y$. Applying the technique from the first part of this chapter, we may construct a neural network which simulates this inseparable Boolean function. From its functional values we may directly construct its equivalent form composed of two clauses

$$\varphi_{XOR}(x_1, x_2) = (\neg x_1 \wedge x_2) \vee (x_1 \wedge \neg x_2) \tag{5}$$

Then this Boolean function is simulated by the neural network displayed in Fig. 5.

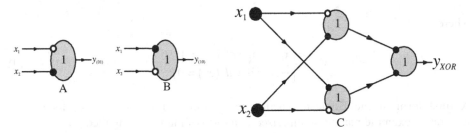

**Fig. 5.** Diagrams A and B simulate single conjunctive clauses from (5). Diagram C represents 3-layer neural network which hidden neurons are taken from diagrams A and B, respectively. An output neuron corresponds to a disjunctive connective.

## 3    Formal Specification of Neural Networks

From our previous discussion it follows that a concept of neural network [13] belongs to fundamental notions of artificial intelligence (not only those networks that are composed of logical neurons). Neural network is defined as an ordered triple

$$\mathcal{N} = (G, w, \vartheta) \tag{6}$$

where $G$ is a connected oriented graph, $w$ is a matrix of weight coefficients and $\vartheta$ is a vector of threshold coefficients.

Until now, we did not use time information in an explicit form. We postulate that time t is a discrete entity and is represented by natural integers. Activities of neurons in time $t$ are represented by a vector $x(t)$, in the time $t = 0$ a vector $x(0)$ specifies initial activities of a given neural network. Relation for an activity of the ith neuron in time $t$ is specified by

$$x_i^{(t)} = s\left(\sum_j w_{ij} x_j^{(t-1)} - \vartheta_i\right)$$    (7)

where summation runs over all neurons that are predecessors of the ith neuron, activities of these neurons are taken in the time t-1. Neural network $\mathcal{N}$ may be understood as a function which maps an activity vector $x^{(t-1)}$ in the time $t$-1 onto an activity vector $x^{(t)}$ in the time t , $x^{(t)} = F\left(x^{(t-1)}; \mathcal{N}\right)$, where the function F contains the specification $N$ of the given network as a parameter.

## 4    Finite State Machine (Automaton)

A finite state machine [4, 5, 7] works in discrete time events 1, 2,..., $t$, $t+1$,... . It contains two tapes of input symbols and output symbols, respectively, where output symbols are determined by input symbols and internal states $s$ of the machine

$$state_{t+1} = f\left(state_t, input\ symbol_t\right)$$    (8a)

$$output\ symbol_{t+1} = g\left(state_t, input\ symbol_t\right)$$    (8b)

where functions $f$ and $g$ specify the given machine and are considered as its basic specification:

1. Transition function $f$ determines the next state; this is fully specified by an actual state and an input symbol,
2. Output function $g$ determines the output symbol, this is fully specified by an actual state and an input symbol.

**Definition 1.** A finite state machine (with an output, called alternatively the Mealy automaton) is defined by an ordered 6-tuple $M = (S, I, O, f, g, s_{ini})$, where $S = \{s_1,...,s_m\}$ is a finite set of internal states, $I = \{i_1, i_2,...,i_n\}$ is a finite state of input symbols, $O = \{o_1, o_2,...,o_p\}$ is a finite set of output symbols, $f : S \times I \rightarrow S$ is a transition function, $g : S \times I \rightarrow O$ is an output function, and $s_{ini} \in S$ is an initial state.

Transition and output functions may be used for a construction of a model of a finite state machine, see Fig. 6.

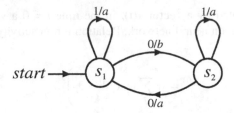

**Fig. 6.** An example of finite state machine composed of two states, $S = \{s_1, s_2\}$, two input symbols, $I = \{0,1\}$, two output symbols, $O = \{a,b\}$ and an initial state $s_1$

Finite state machines are determined as a mapping of input string of symbols onto output string of symbols

$$G\left(\underbrace{100111010...}_{input\ string\ x}; f, g\right) = \underbrace{\square\ abaaaabaa...}_{output\ string\ y}.$$

where the symbol $\square$ in an output string means an "empty token", symbols of output string are shifted by one-time step with respect to the input string. A mapping $G$ is composed of functions $f$ and $g$ which specify a "topology" of the finite state machine.

**Theorem 4 [5, 8].** Each neural network can be represented by an equivalent finite state machine with output.

Existing proof of this theorem is simple and constructive, we can construct for a given neural network single elements from the definition 1, $M = (S, I, O, f, g, s_{ini})$.

For a given neural network we unambiguously specify a finite state machine which is equivalent to the given neural network. This means that any neural network may be represented by an equivalent finite state machine.

A proof of inverse theorem with respect to theorem 4 (i.e. each finite state machine may be represented by an equivalent neural network) is not a trivial one, the first who proved this inverse form was Minsky in 1967 in his famous book "Computation: Finite and Infinite Machines" [7] by making use of a very sophisticated constructive approach. For a given finite state machine, an equivalent neural network can be constructed.

**Theorem 5 [7].** Each finite state machine with output (i.e. the Mealy automaton) can be represented by an equivalent recurrent neural network.

To summarize our results, we have demonstrated that neural networks composed of logical neurons are powerful computational devices: (1) feedforward neural networks represented by the acyclic graph are universal approximators of Boolean functions and (2) there is a property of mutual equivalency between finite state machines and neural networks. An arbitrary finite state machine may be simulated by a recurrent neural network and, conversely, an arbitrary neural network (feedforward of recurrent) may be simulated by a finite state machine. Further relation between finite automata and neural networks was further studied by many authors, e.g. Noga et al. [10] and Hsien a Honavar [3].

## 5    Conclusions

McCulloch and Pitts's paper is very ostensibly "neural" in the sense that it used an approach for specification of neuron activities based on a simple rule all-or-none. However, McCulloch–Pitts neural networks are heavily simplified and idealized when compared to the then known properties of neurons and neural networks. Their theory did not offer testable predictions or explanations for observable neural phenomena. It was removed from what neurophysiologists could do in their labs. This may be why neuroscientists largely ignored McCulloch's and Pitts's theory. For this scientific community, its main power is not in a capability to produce verifiable hypothesis but in a fact that such extremely simple neural theory offers arguments of basal character for a discussion about "philosophical" problems concerning a brain and mind relationship. It cannot be expected that a further "sophistication" of this theory (e.g. the rule "all-or-none" [1, 2] is substituted by another more realistic rule or "spiking" neurons are used, etc.) will negatively influence general results deduced from the model.

One example of seminal papers influenced by results of McCulloch and Pitts was the work of well-known John von Neumann [9] who is known as a creator of the so-called "von Neumann computer architecture", which was outlined in his famous 1945 technical report. He mentioned that various mechanical or electrical devices have been used as elements in existing digital computing devices. It is worth mentioning that the neurons are definitely elements in the above sense. From the early 1940s the McCulloch–Pitts neuron was considered by many non-neuroscientists to be the most appropriate way to approach neural computation, largely because the work of McCulloch and Pitts was so well known.

McCulloch's and Pitts's views – that neural nets perform computations (in the sense of computability theory) and that neural computations explain mental phenomena – permanently belong to the mainstream theory of brain and mind. It may be time to rethink the extent to which those views are justified in light of current knowledge of neural mechanisms. The philosophical impact of the paper of McCulloch and Pitts is broadly discussed in many works oriented to the famous problem of connections between mind and brain [2, 11, 12].

**Acknowledgments.** This chapter was supported by Grant Agency VEGA SR No. 1/0553/12 and 1/0458/13.

## References

1. Adrian, E.D.: The Basis of Sensation. The Action of the Sense Organs. Norton & Company, New York (1928)
2. Boden, M.: Mind As Machine: A History of Cognitive Science, vol. I, II. Oxford University Press, Oxford (2006)

3. Chun-Hsien, C., Honavar, V.: Neural network automata. In: Proc. of World Congress on Neural Networks, vol. 4, pp. 470–477 (1994)
4. Hopcroft, J.E., Motwani, R., Ullman, J.D.: Introduction to Automata Theory, Languages, and Computation. Pearson Education, New York (2000)
5. Kleene, S.C.: Representation of events in nerve nets and finite automata. In: Shannon, C.E., McCarthy, J. (eds.) Automata Studies. Annals of Mathematics Studies, vol. 34, pp. 3–41 (1956)
6. McCulloch, W.S., Pitts, W.H.: A Logical Calculus of the Ideas Immanent in nervous Activity. Bulletin of Mathematical Biophysics 5, 115–133 (1943)
7. Minsky, M.L.: Computation. Finite and Infinite Machines. Prentice-Hall, Englewood Cliffs (1967)
8. Minsky, M., Papert, S.: Perceptrons. An Introduction to Computational Geometry. MIT Press, Cambridge (1969)
9. von Neumann, J.: First Draft of a Report on the EDVAC (1945), http://qss.stanford.edu/~godfrey/vonNeumann/vnedvac.pdf (retrieved October 1, 2012)
10. Noga, A., Dewdney, A.K., Ott, T.J.: Efficient simulation of finite automata by neural nets. J. ACM 38(1991), 495–514 (1991)
11. Piccinini, G.: The First Computational Theory of Mind and Brine: A Close Look at Mcculloch and Pitts, Logical Calculus of Ideas Immanent in Nervous Activity. Synthese 141, 175–215 (2004)
12. Quine, W.V.O.: Mathematical Logic. Harvard University Press, Cambridge (1981)
13. Rojas, R.: Neural Networks. A Systematic Introduction. Springer, Berlin (1996)
14. Searle, J.: Mind: a brief introduction. Oxford University Press, New York (2004)

# Part I
# Robotics

# Theoretical Analysis of Recent Changes and Expectations in Intelligent Robotics

Peter Sinčák, Daniel Lorenčík, Mária Virčíková, and Ján Gamec

Center for Intelligent Technologies, Department of Cybernetics and Artificial Intelligence,
Faculty of Electrical Engineering and Informatics, Technical University of Kosice, Slovakia,
The European Union Letna 9, 04001 Košice, Slovakia
{peter.sincak,daniel.lorencik,maria.vircikova,
jan.gamec.2}@tuke.sk

**Abstract.** This paper deals with the current trend towards moving from industrial robots to the service or social robotics era. The time when robots populated the environment 'alone' is over and networked robotics and the acquisition and understanding of crowd-sourcing is fully supported by the trends in Computer technology.

Cloud Robotics is a new phenomenon supported by Cloud Computing and the main challenges question how these new trends change the tools of Artificial Intelligence and the forms of its contributions to human behavior simulation. The challenging question within this domain is the quality of Human-Robot and Robot-Robot interactions in a general environment or industrial scenario.

The role of emotions seems to be increasingly important and the impact of synthetic robotic emotions on humans is essential to human performance and productivity or entertainment in everyday life. The paper also emphasizes the importance of tele-monitoring, linked with tele-operation, as an important part of the knowledge acquired in cloud robotics and crowd sourcing. The paper draws together certain theoretical predictions on the future of intelligent robotics domains.

## 1 Introduction

The first glimmer of robotics began when the mankind came up with mechanical copies of biological organisms like animals, humans, etc. History proves these examples began with the Archytas of Tarentum [1] who, in 350 B.C., constructed the 'artificial mechanical Dove' which used some form of compressed air to be able to fly rather high. History is very rich with similar, yet more advanced mechanical constructors, including Leonardo da Vinci in the 15th century who designed a mechanical machine that looked like a soldier or a knight with armored protection. He also had certain engineering ideas of machines similar to helicopters and hang-gliders.

In 1505, he published the work and his drawings confirmed his ideas. Since that time, there is no verification whether it was actually constructed, but it was a very

© Springer International Publishing Switzerland 2015
P. Sinčák et al. (eds.), *Emergent Trends in Robotics and Intelligent Systems*,
Advances in Intelligent Systems and Computing 316, DOI: 10.1007/978-3-319-10783-7_2

good inspiration for technical motivations toward later inventions and practical contribution for mankind.

Later, in the early 18th century in Grenoble, Jacques de Vaucanson had an extreme passion for automation and mechanical systems. Thus, in 1738, he designed an automated Flute and Tambourine Player and a mechanical DUCK that was able to imitate the behavior of real ducks; all these inventions used to entertain local Royalty and higher society [2].

Very important step toward computers was made in 1801, when Joseph Jacquard invented punch cards and used them to control a textile machine. Later, this technology was used and supported by the work of Charles Babbage and much later by IBM for their first computers.

Tele-operations advanced dramatically in the field of robotics in 1898. In this year, Nikola Tesla utilized radio waves to control a boat and confirmed the importance of radio-waves in the remote operation of objects at a distance from human control. This philosophical approach is very important even today when and all aspects of tele-operations are not yet fully solved. Tesla's role [3] is not as appreciated as much as it should be and the importance of this innovative man is enormous.

**Fig. 1.** Vaucanson's design of musical Automata and mechanical Duck. Tesla's tele-operated ship.

The introduction of the word 'Robot' came from Central Europe when a Czech writer Karel Capek and his brother Jozef used it in Karel's 1921 play 'R.U.R.,' an acronym for 'Rossuum's Universal Robots.' [4] Robot derives from the Czech word 'Robota' which means 'compulsory job'. Just a few years later, in 1926, Fritz Lang made the movie 'Metropolis' in which 'Maria,' a female robot, played a role.

The convergence of the development of robotics and computers is evident, thus, the contributions of Allan Turing and John von Neumann are extremely important – first with Turing's tests of computer/robot intelligence and also with von Neumann's architecture for computer hardware. Later, in 1940, Issac Asimov in his series 'A Strange Playfellow' called Robbie an artificial person. In 1950 'I, Robot', a popular novel, was written. Asimov is well known for his popularization of the term 'Robotics' as well as for claiming three robotics laws (later adding a Zero-th Law) that are very important in present times. The community of lawyers is beginning to consider the implications of robots and the significance of their existence, ownership and autonomy.

The Rockefeller Foundation set aside a special grant to fund an Artificial Intelligence seminar for selected people in 1956. In the same year, John McCarty and Marvin Minski established an Artificial Intelligence Lab at MIT, but later McCarty moved to Stanford and Minsky dominated the MIT AI Lab. The Stanford AI Lab, coordinated by McCarty, became an important AI research destination in the early 1960.

In 1961, the MH-1 Mechanical Hand was developed at MIT and, in 1962, a new robot made by the UNIMATE Company was used on General Motors Assembly Lines to replace repetitive and laborious operations. Following on in 1966, the Stanford Research Institute (SRI) established the Shakey robot with a degree of intelligence and sensors. In the same year, MIT, led by Joseph Weizenbaum, began the project ELIZA for dialog engineering that could be fully used in robot-human communications.

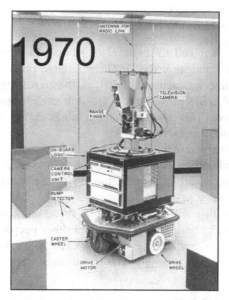

**Fig. 2.** The Shakey Robot from Stanford University, constructed in 1970 [5]

## 1.1    Computer Technology versus Robotics

The development of Computer technology was very influential to the development of robotics. The relationship between computers and robotics anchors the basic understanding of whether all processing for robots will be done on-board the robot or follow a remote-brain approach as promoted at the University of Tokyo in the early 90s, providing full 'off-board processing'. This is a crucial question in autonomous systems that include the distribution of computer power for both 'on and off board' robots.

Computer trends clearly favor Cloud technology [6] and its advantages seem to outweigh such disadvantages and negatives of Cloud computing as security as well as others, currently being solved. The importance of wireless technology is enormous both in information and command transmission as well as in powered wireless transmission, critical in many areas, including robotics. In general, it is believed that wireless connectivity will rapidly improve around the globe in the sense of its speed and reliability.

All of the above technologies of cloud and wireless environments fully support the idea of progress in manual tele-operations and an almost autonomous use of robots where humans serve supervisory roles. Tele monitoring and tele-control are not yet solved by the current available technology if we consider the globe as an opportunity for robot operation and human/robot collaboration. This goal must include a crowd sourcing know-how and learning process in general. Learning procedures will change, influenced by computer technologies, big data solutions and the increased speed of computers in general.

## 1.2    Human-Computer/Robot Interaction

The previous problem is linked to the level of Human-Computer Interaction (HCI) and Natural Computer Interaction (NCI). There is huge progress in speech (mainly English) human machine interaction as well as the movement detection achieved mainly by the NATHAL/KINECT project [7]. Multimodal interface is the future of human-robot interaction. This also allows emotions to be incorporated between human – computer – robot interactions.

Much cross-disciplinary research on affective computing, emotional technology and emphatic computing has been investigated and a number of results from psychological emotional model implementation to NCI are underway. This interaction seems to be more and more psychologically and socially plausible, therefore, we are heading toward robots that will be companions or co-workers with humans. That means that humans can be replaced by robots and other people can work with them (robots). An interesting example is the early BAXTER project that would cooperate with and/or replace humans at assembly lines in industrial environments.

## 1.3    Legal Issues and Robotics

Legal Issues and Robotics [8] is a completely new area, but legal issues connected with technology are pretty well-known cases in history. What is important to understand is that this is the main legal issue concerning the reliability of a machine guaranteed by the machine/robot producer. The problems concerning more robots among humans will be very similar to having more cars in society, but this will become more and more complicated if mankind allows robots without a proper owner's and producer's guarantee. The problem is who decides whether and how robots can harm people. These are very complicated questions that can slow down the practical use of robots in everyday life.

## 2    Cloud Computing and Its Impact on Robotics

Cloud computing is very important to present trends in computing. It saves money, makes computing a service that is fully scalable and provides complete solutions for users who don't care about the technicalities of the task. But the problem must be specified and the requirements of the task must be defined.

Cloud Computing and Cloud Robotics are linked with the basic technology as eternally linked systems (ELS). Cloud Robotics [9] was first mentioned by James Kuffner from CMU and all efforts have been focused on creating a framework for Cloud Robotics with various types of robots, regardless of their inputs and abilities to acquire knowledge. Good examples of such projects are the RoboEarth and similar projects of that type.

The major question is if Cloud Robotics will change core approaches to Artificial Intelligence, with learning being the major and key challenge. Crowd Source learning, data and knowledge acquisition lead to Big Data and Data Discovery issues important to robotic systems.

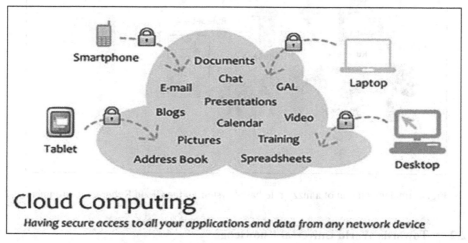

**Fig. 3.** Basic Concept of the Cloud Computing Paradigm which is becoming more and more financially advantageous over classical computing approaches [10]

### 2.1    Cloud Robotics and Crowd Source Knowledge Acquisition

Cloud Robotics has an important impact on knowledge acquisition and still more crowd based approaches are being used to build a global knowledge base [11], used and sourced by robots. A good example of such an approach is building a large fuzzy rule database. Certainly it is not an easy approach, since we have a twin-structure database. One part is the assumption portion of a rule and the second is the consequence portion. Both must be sequentially and incrementally consistent with all rules to build a large database.

If numerous machines and/or people can build knowledge bases in the form of IF-THEN rules, then numerous machines can use them. So we are applying well-known tools of Artificial or Computational Intelligence in the Cloud Computing / Robotic environment and a key question is how this technology will influence the core methods and approaches to AI and Computational Intelligence.

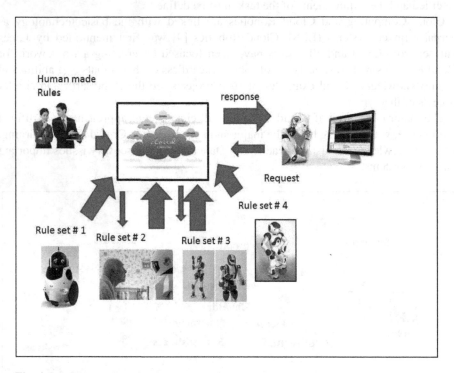

**Fig. 4.** Implementation of a fuzzy rule-based system in the Cloud Robotics environment

## 2.2   Towards World Universal Knowledge

Universal World Knowledge has been the dream of many generations and was featured in a number of sci-fi movies and analyzed from the technological point of view [12]. In the movie 'I, Robot', a VIKI computer (Virtual Interactive Kinetic Intelligence) played the role of the World Universal Knowledge. There is a number of projects related to Cloud Robotics and a universally fused knowledge base, as in project RoboEarth [ ] and a number of others. The concept of Cloud Robotics varies. One approach is that there will be virtual robots in the cloud and once the personalization of user preferences is done and a personal user switch on the robot profile is downloaded from the cloud, an actual robot for personalized human computer interaction will be accomplished.

## 2.3    AI Bricks as Important Parts of Cloud Robotics

The software agents for Cloud robotics with modular concepts can be useful tools for the future of crowd-based artificial intelligence powered by many people. The Agent is a virtual model of a robot or some particularly well and clearly defined procedures based on Artificial Intelligence (AI bricks) used by other agents. The scope of the problem is device and problem dependent. The concept of cloud robotics is under rapid development and concentrated technological innovation will exert a major influence on Cloud Robotics technology including wireless tele-robotics, learning and other important issues related to robotics [13].

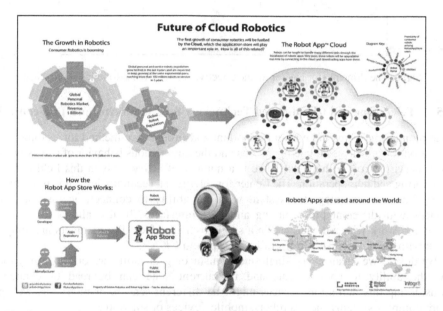

**Fig. 5.** Basic concept of Cloud Robotics, designed by Grishin Robotics, Ltd. of New York

## 2.4    Evaluating the Contribution of Cloud Robotics

There is extensive discussion whether Cloud Robotics is mostly a hype or a breakthrough technological revolution. That poses the question of how we measure the contribution of cloud technology to robotics as well as the efficiency of knowledge fusion and its utilization of the integrated knowledge by numbers of robots or robot-like entities. The basic parameters to observe are time, incremental stages, replication and impact of so-called Crowd Learning Systems on the knowledge base for robots.

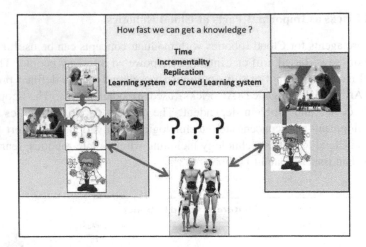

**Fig. 6.** The Principle for measuring the effectiveness of Cloud Robotics approaches

### 2.5    Tele-Scope and Cloud-Based Technology for Monitoring and Tele-control

Tele-monitoring seems to be a very important and progressive method for leading tele-operation towards assisted tele-operations and the autonomous behavior of robots under the supervisory role of humans. There is a number of active tools in this field of tele-monitoring and tele-operation. The Center for Intelligent Technologies is one of them.

Telescope is an integrated system with the ability to connect, access and join devices with different programming and user interfaces. It also allows to control, share and work with a device within a single user interface. Programmers are able to work with the device using uniform JavaScript and C# API.

By device we mean every single mechanism or gear (software or hardware) that has the ability to communicate and its current state can be read or changed. Considering this definition, devices might be different sensors or motors as well as more complex systems such as robots, mobile devices or software.

As a result, Telescope systems can communicate with a device but also provide communication between devices. For instance, controlling of a motor based on data from a speed controller can be simply achieved using this feature. The main features of Telescope systems are:

- System availability anywhere. This means both geographically and technically, with availability on PC and mobile devices using a web browser
- System availability within different web browsers
- System runs in real-time using WebSocket technology[1]
- Scalability, meaning the system's ability to deal with large amounts of connections. For this purpose we used the Redis database system[2] and its cluster mode.

---

[1] http://www.websocket.org/

[2] http://www.redis.io/

**System Telescope Consists of 2 Larger Parts, as May Be Seen (Fig. 7):**

- Back-office service – Event server:

The essential part of back-end is the Event server which is designed to communicate and cooperate with other event servers. The server uses Python language and WebSocket technology used for communication. On the other hand, faster data storage and communication is achieved by using a Redis NoSQL Server that creates a cluster and mediates the communication between user and device based on a unique identifier.

- Front-office service – Telescope user interface,

Telescope is a WEB-based system covered by web service available for PCs, mobile devices (smartphones iOS, Android) and smart TVs. After connecting to the Telescope Web Service[3], CloudFlare[4] technology responds to user requests and guarantees the availability of static content anywhere in the world. Dynamic content is provided by PHP server which allows the users to login. Subsequently, JavaScript code takes control over communication with the server that processes the events and the actual data from the device is displayed.

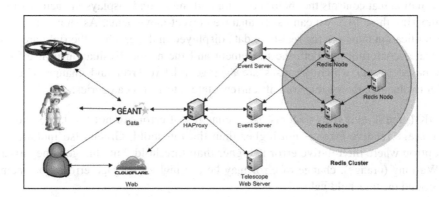

**Fig. 7.** Telescope system architecture overview

Overall communication within the Telescope system is joint in a single node, handled by a powerful HAProxy proxy server[5], and directs communication to the dedicated nodes.

---

[3] www.telescopesystem.com
[4] www.cloudflare.com
[5] http://haproxy.1wt.eu/

## 2.6     Monitoring and Analyzing Real-Time Data Using Telescope System

As mentioned before, the Telescope system uses WebSocket technology that provides access to real-time data. For example, we chose robot NAO[6] from a French company Aldebaran Robotics.

The basic requirement for connecting the robot to the Telescope is a wrapper program that translates communication between the robot and Event server. Telescope system also offers a security mode with user authorization using a unique ID and allowing users to share devices among each other.

Back to the NAO, the first thing to do is authorize yourself in the Telescope system using your personal ID. Only after that the system will provide a list of devices available to your account and relevant information such as online status, name or identifying number. The main requirement for monitoring data from the robot is the ability to retrieve data from the robot in real time, which can be achieved in 2 different ways. One is to regularly request specific data from a robot in a loop. The second is to wait for a value to be changed and handle each event.

We chose the first option, so the program will request data from the robot in specific time sequences. In addition, NAO robot supports 2 different return values from one motor – an actuator value and a sensor value. The first represents information that controls the motor behavior and the second displays the actual value. Comparing these two, we can easily analyze correct movements. As shown in Fig.8, the application monitors the real-time data displayed in the graph in the right area and evaluates conformity of both the movement and the motor. Evaluation in itself is a rule-based system which premises are average relative error and change of error. After evaluation, the system sorts its current state into 1 of 3 categories:

- OK (green), everything seems okay, changes of error are not too large and the average relative error is not higher than the threshold. Green also includes an option where the relative error is higher than threshold, but change is negative
- Warning (orange), change of error may be too high or average error is balancing around the threshold value
- Error (red), means a serious problem

This diagnostic system is based on an approach that connects NAO from everywhere on the planet to everywhere on the planet without the necessity of having a public IP address for a NAO robot. This improves the security and portability of the system. This is also a Cloud-ready solution, tending to become a software as a service provided through the warranty and post-warranty periods as an aftercare policy of the producer or robot owner.

---

[6] https://community.aldebaran-robotics.com/nao/

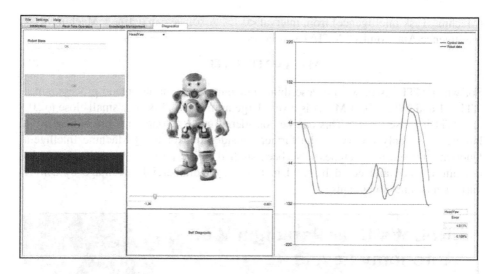

**Fig. 8.** NAO diagnostic system using Telescope system

# 3    Finding a Way towards Learning Machines

There in a number of approaches for creating an intelligent Machine. Learning is essential for this process and depends on learning goals. Learning tools can be divided into 3 categories:

1. Learning from data – mainly based on neural networks
2. Learning from humans in the form of experience – rules – Fuzzy Inference Machine
3. Learning by demonstration, using Kinect or other approaches for learning from human manual tele-operation and dividing data into functional blocks that can be reused in other tasks given by humans to robots.

Figure 8 shows the understanding of the U.S. IROBOT Company of creating an intelligent machine using tele-operation and engaging learning procedure into the process. The application potential of this approach is rather large, since a number of autonomous machines are expected to appear on the market in the near future.

One of the major problems related to Intelligent Machines are the difficulties in measuring the autonomy of the system. For the most part, there is no universal approach to this problem. We can state that it is a task or mission oriented approach and therefore we write it as:

$$GTI = HTI + MTI \tag{1}$$

Where

GTI (Global Task Intelligence) is always value 1, is a sum of Human Intelligence „HTI (Human Task Intelligence) from interval <0,1> and Machine Intelligence  MTI

(Machine Task Intelligence) from interval <0,1>. Also, we can define a Machine Task Intelligence Autonomy"(MTIA) as follows:

$$\text{MTIA} = \text{MTI} / \text{HTI} \tag{2}$$

So, when MTIA is 0, we are describing a manual, fully human-made process, since HTI is 1 and MTI is 0. If MTIA is a very large number, HTI is very small close to „0" and MTI is close to „1". This can be considered as an autonomous mission where a human is the only observer. The further consideration of MIQ Machine Intelligent Quotients related to machines have been studied in the past []. The MIQ should be domain-oriented and could be used in future for commercial advantage by selling various machines to humans.

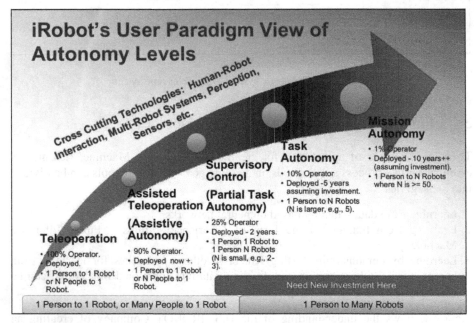

**Fig. 9.** The approach of Tele-operation toward Mission Autonomy operation [14]

# 4    Emotions and Robots–A Part of the Human Robot Interaction

## 4.1    Theory of RIEM (Robotic Integral Emotional Models)

The proposed theory of RIEM comes from social observation and interactions among people. The notions of Internal Emotional State and External Emotional State form the Theory of Integral Emotional Model of the Robot.

Just as in human beings, our biological systems have internal emotional states that depend on random input, environment and persons in those environments. External Emotional States could be, but are not necessarily, the same as Internal Emotional

States. External states are fully device/robot-dependent and, on the other hand, Internal Emotional States could be device/robot-independent.

The output of the RIEEM is a Social Robot Behavior based on the relationship between IEM and EEM within an Integrated Emotional Model. These relationships are characterized by weights "w11"..."w1m" up to "wn1"..."wnm" and describe intensity and connectivity which can be a personalization factor. Thus, each social robot can be emotionally set up for the convenience of a human co-worker in the sense of better performance regarding their collaboration.

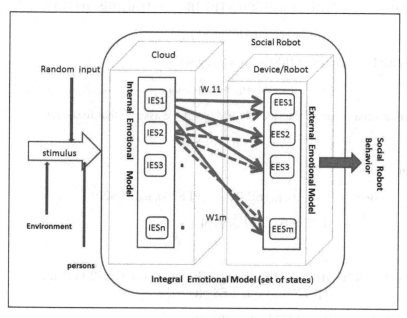

**Fig. 10.** Theoretical Model of Robot Integral Emotional Model

The above figure represents a model of RIEM that is based on fuzzy approaches to the set of Internal Emotional States and External Emotional States. Also, the time dimension of IES and EES is a very important factor of the Emotions, which means we have 2 fuzzy sets if we consider:

$$IES(t) \rightarrow IEM(t+q), \ EES(t) \rightarrow EES(t+q) \tag{3}$$

so in fact we can get a set of fuzzy sets related to time "t".
so we have set

$$IEM = ( \ IES_F(t), \ IES_F(t+1),..., \ IES_F(t+q) \ ) \tag{4}$$

the same is for

$$EEM = ( \ EES_F(t), \ EES_F(t+1),..., \ EES_F(t+q) \ ) \tag{5}$$

Following these definitions, we state the following fuzzy sets for IEM set and EEM sets as follows:

If we say that fuzzy set A(t) = IESF(t), then

$$A(t) = (\ [IES1(t), \mu_A(IES1(t))],\ [IES2(t), \mu_A(IES2(t))],\ \ldots,\ [IESn(t), \mu_A(IESn(t))]\ )$$

$$(6)$$

also

$$A(t+1) = (\ [IES1(t+1), \mu_A(IES1(t+1))],\ [IES2(t+1), \mu_A(IES2(t+1))],\ \ldots,$$

$$[IESn(t+1), \mu_A(IESn(t+1))]\ )$$

$$(7)$$

And the last member of IEM set is a fuzzy set

$$A(t+q) = IES_F(t+q)$$

$$(8)$$

Similar situation is on the side of EEM, so, if we assume that fuzzy set

$$B = EES_F(t)$$

then

$$B(t) = (\ [EES1(t), \mu_B(EES1(t))],\ [EES2(t), \mu_B(EES2(t))],\ \ldots,$$

$$[EESm(t), \mu_B(EESm(t))]\ )$$

$$(9)$$

also

$$B(t+1) = (\ [EES1(t+1), \mu_B(EES1(t+1))],\ [EES2(t+1), \mu_B(EES2(t+1))],\ \ldots,$$

$$[EESn(t+1), \mu_B(EESm(t+1))]\ )$$

$$(10)$$

And the last member of IEM set is a fuzzy set

$$B(t+q) = EES_F(t+q)$$

$$(11)$$

The relation between real sets IEM and EEM or better set of fuzzy sets

$$A = (A(t),\ \ldots A(t+q))$$

$$(12)$$

and set of fuzzy sets

$$B = (B(t),\ \ldots B(t+q))$$

$$(13)$$

as follows :

$$B\ (t)\ = function\ (\ A(t),\ W(t),\ Stimuli)$$

$$(14)$$

where

A(t) is fuzzy set IESF(t), W(t) is matrix of weight between A(t) and B(t)
and Stimuli is integration persons, environment and external random input, see Figure 9.

This general model of Theory of Integrated Emotional Model of Robot is model-free, so in fact the IEM can be represented with some well-known models and the EEM part also has the same possibility. As it is well known, the emotional states of humans were under research observation of psychologists and they have set up number of models beginning with the Ekman model [15], Izard model [16], the very popular Plutchik model [17] and many others, including PAD ( Pleasure, Arousal, and Dominance ). The emotional model proposed by Lovheim [18], the Lovheim Cube of Emotion, in fact takes into consideration the Mehrabian Worker Satisfaction Scale (WSS) to achieve a comfort level for humans working with robots (or machines in general). The concept of IEM and EEM and the fully weighted connection between these layers can present a personality model of a human model behavior. Hence, the EEM depends on the robot device and its ability to express emotions.

So, at the end of the day, we can set up a model of emotional behavioral activity of a human or, by adapting in sense of Reinforcement learning, a W(t) matrix where a criteria function is set by a human. Thus, it may enhance a control adaptability of weights for its own benefit and conformability for collaboration purposes between human and machine. Very interesting feature of the Lovheim PAD model is that it also has a response to managers and commercial community to achieve the optimum productivity of humans. In the near future we may talk about a human-machine community instead of a human community. The theoretical backgrounds for the PAD Emotional State Model have been proposed by Mehrabian [19].

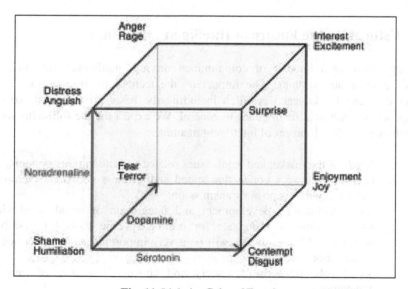

**Fig. 11.** Lieheim Cube of Emotion

'Pleasure-displeasure' defines a positively-negatively affected quality of emotional states. 'Arousal-nonarousal' defines a mental state that describes a mental involvement of a human/machine in a present situation and 'dominance-submissiveness' determines the terms of master control versus lack of control or slave-like control.

Very interesting is the distribution of emotions based on a PAD scale range from -1 to +1: e.g. angry (-.51, .59, .25), bored (-.65, -.62, -.33), curious (.22, .62, -.01), dignified (.55, .22, .61), elated (.50, .42, .23), hungry (-.44, .14, -.21), inhibited (-.54, -.04, -.41), loved (.87, .54, -.18), puzzled (-.41, .48, -.33), sleepy (.20, -.70, -.44), unconcerned (-.13, -.41, .08), violent (-.50, .62, .38). What is particularly interesting is that all those numbers can be a matter of personalization and could be adapted according to the Human Companion.

### 4.2     Consequences of RIEM to Human Robot Interaction

The application of this theoretical approach can have a number of implications. Creating a number of robots with similar IEM and different EEM (or vice versa) can create a very interesting situation where various robots may have a different IEM and similar EEM. Then, certain implications arise for other states of EEM, while maintaining the different IEM. The logic of these theoretical layouts can be intriguing and lead to a number of interesting personalization effects which are hidden in the adaptations of weights between IEM and EEM.

These adaptations can be set up for the user in a mirroring way (who likes to collaborate with machines that have similar emotions as the user) or simply setting up emotional responses of the machine for user's convenience and regarding the task in which the user collaborates with the robot.

## 5     Estimating the Future of Intelligent Machines

Generally, there is a number of communities driving intelligence and machines towards autonomous systems. The impact of the technology is enormous and it enables core and fundamental research including the basic principles of Artificial Intelligence and Ambient Intelligence in general. We expect the the following factors will influence the development of intelligent machines:

1.  Needs of the market and legal issues related to Autonomous systems;
2.  Computer network connection speed and wireless network development including wireless power transmission;
3.  Cloud technology development and integration of cloud-based virtual robotic rooms as a pre-sale for renting specific tools for case-based operations. Cloud robotics will be a very important factor in service and social robot development;
4.  Social robotic needs of society and human acceptance of robots as companions;
5.  Building of a general knowledge-base will be a matter of commercial relations based on domain-oriented pieces of knowledge for robots or groups of robots;
6.  Human-robot interactions will be multimodal, less language-dependent and introduced into everyday life. Communication with machines in every form will be natural and essential.

We are approaching an exciting era that will be influenced by technology, financial profit and human society's ability to accept negative impacts of machine-human coexistence. Will humanity embrace such increase in the quality of life in the name of harmony, prosperity and its benefits to human society?

**Acknowledgments.** The research was supported by the 'Center of Competence of knowledge technologies for product system innovation in industry and service', with ITMS project number: 26220220155 for years 2012-2015; and also supported by the Project implementation: University Science Park TECHNICOM for Innovation Applications Supported by Knowledge Technology, ITMS: 26220220182, granted by the Research & Development Operational Program funded by the ERDF, for years 2012-2015.

# References

1. Internet source, `http://www-history.mcs.st-and.ac.uk/Biographies/Archytas.html`
2. Internet source, `http://en.wikipedia.org/wiki/Jacques_de_Vaucanson`
3. Tesla, N.: My Inventions: The Autobiography of Nikola Tesla. Soho Books (2011) ISBN 10: 161293093X
4. Internet source, Gutenberg project, Translated by Paul Selver and Nigel Playfair, `http://preprints.readingroo.ms/RUR/rur.pdf`
5. Havel, I.M.: Robotika – Uvod do teórie kognitivních robotů (1980)
6. National Standard Technology and Institute, USA, Special Report 800-145, September 2011, What is Cloud Computing – definition, `http://csrc.nist.gov/publications/nistpubs/800-145/SP800-145.pdf`
7. Andersen, M.R., et al.: Kinect Depth Sensor Evaluation for Computer Vision Application, Technical report ECE-TR-6, Department of Engineering – Electrical and Computer Engineering, Aarhus University (February 2012)
8. Kirkpatrick, K.: Legal Issues and Robotics. Communications of the ACM 56(11), 17–19 (2013)
9. Arumugam, R., et al.: DAvinCi: A Cloud Computing Framework for Service Robots. In: 2010 IEEE International Conference on Robotics and Automation, Anchorage Convention District, Anchorage, Alaska, USA, May 3-8, pp. 3084–3089 (2010)
10. Tantow, M.: Cloud Computing: Current Market Trends and Future Opportunities. CloudTimes (2011), `http://cloudtimes.org/2011/06/22/cloud-computing-its-current-market-trends-and-future-opportunities/`, `http://www.cloudtimes.org`
11. Schuller, G.: Designing universal knowledge. Lars Müller Publishers (2009) ISBN 978-3-03778-149-4
12. Lorencik, D., et al.: Influence of Sci-Fi films on artificial intelligence and vice-versa. In: 2013 IEEE 11th International Symposium on Applied Machine Intelligence and Informatics (SAMI), January 31-February 2, pp. 27–31 (2013)
13. Ferrate, T.: CLOUD ROBOTICS new paradigm is near. Robotica Personal (2013), `http://www.robotica-personal.es/2013/01/cloud-robotics-new-paradigm-is-near.html`

14. iRobot's user paradigm view of autonomy levels. In: Robotics Summit, Virtual Conference & Expo (June 2011)
15. Ekman, P., et al.: Emotions Revealed. Times Books Henry Holt and Company, LLC Publishers, New York (2003) ISBN 0-8050-7275-6
16. Plutchik, R.: The Nature of Emotions. American Scientists 98, 2001,
    http://www.emotionalcompetency.com/papers/
    plutchiknatureofemotions%202001.pdf
17. Izard, C.E.: Human emotions. Plenum Press, New York (1977)
18. Lovheim, H.: A new three-dimensional Model for Emotions and monoamine neurotransmitters. Medical Hypotheses 78(2), 341–348
19. Mehrabian, A.: Framework for a comprehensive description and measurement of emotional states. Genetic, Social, and General Psychology Monographs 121(3), 339–361 (1995)

# Tacit Learning for Emergence of Task-Related Behaviour through Signal Accumulation

Vincent Berenz[1], Fady Alnajjar[1], Mitsuhiro Hayashibe[2], and Shingo Shimoda[1]

[1] BSI-Toyota Collaboration Center, RIKEN, Nagoya, Japan
{vincent,fady,shimoda}@brain.riken.jp
[2] INRIA DEMAR Project and LIRMM, UMR5506, CNRS University of Montpellier, France
mitsuhiro.hayashibe@inria.fr

**Abstract.** Control of robotic joints movements requires the generation of appropriate torque and force patterns, coordinating the kinematically and dynamically complex multijoints systems. Control theory coupled with inverse and forward internal models are commonly used to map a desired endpoint trajectory into suitable force patterns. In this paper, we propose the use of tacit learning to successfully achieve similar tasks without using any kinematic model of the robotic system to be controlled. Our objective is to design a new control strategy that can achieve levels of adaptability similar to those observed in living organisms and be plausible from a neural control viewpoint. If the neural mechanisms used for mapping goals expressed in the task-space into control-space related command without using internal models remain largely unknown, many neural systems rely on data accumulation. The presented controller does not use any internal model and incorporates knowledge expressed in the task space using only the accumulation of data. Tested on a simulated two-link robot system, the controller showed flexibility by developing and updating its parameters through learning. This controller reduces the gap between reflexive motion based on simple accumulation of data and execution of voluntarily planned actions in a simple manner that does not require complex analysis of the dynamics of the system.

## 1 Introduction

Living organisms are confronted with changes of the environment. They create new behavioural patterns using body environment interactions when carrying out tasks under unknown situations. In spite of advances in machine learning and adaptive methods, such as artificial neural networks [1][2][3], reinforcement learning [4][5][6], adaptive controls [7][8] and fuzzy control [9][10], we have so far been unsuccessful in creating artificial systems that have such adaptability. In a previous paper [11] we introduced tacit learning. Tacit learning is an unsupervised learning method in which various behavioral structures spontaneously emerge through body-environment interactions subject to certain innate rules. Computations progress by accumulating the local activities of elements. This accumulation creates behaviors adapted to the environment. We investigate the possible role of such accumulation in the

© Springer International Publishing Switzerland 2015
P. Sinčák et al. (eds.), *Emergent Trends in Robotics and Intelligent Systems*,
Advances in Intelligent Systems and Computing 316, DOI: 10.1007/978-3-319-10783-7_3

spontaneous generation of adaptive behaviors based on reflex action. The use of reflex actions is particularly appealing to the control of robotics systems; contrary to methods of traditional control theory, it does not rely on the analytical resolution of systems characterized by highly non-linear dynamics.

Previously, we used the torque as the input to be reduced, the proposed controller being used for the emergence of bipedal walking behaviors of a 36DOF humanoid robot [12]. Tacit learning allowed to reduce the complexity of designing a walking behavior by allowing most of the joints to have unspecified target angles. The gait that emerged from the learning process was highly adapted to the environment in terms of efficiency, rhythm and robustness. Our robot succeeded in learning bipedal walking in a completely model-free fashion. Balance emerged within approximately 10 minutes of tacit learning in real environments. More recently, a novel optimal control paradigm in motor learning based on tacit learning was proposed and showed that simple tacit learning can realize simultaneously environmental adaptation and optimal control [13]. In a vertical reaching task, this method systematically produced motor synergies which would induce an efficient solution in a redundant task space.

In this paper, we use tacit learning to create behaviors based on accumulation of task-space feedback information. Learning refers here to the dynamic tuning of the controller. Task-space feedback information is used in many modern robot control systems to improve robustness. While the sensory information is important to improve the endpoint accuracy in the presence of uncertainty, most sensory control schemes require the exact knowledge of the Jacobian matrix from joint-space to task-space. This approach is inconvenient if the system is too complex for finding an analytical solution online, or if some of the kinematic parameters of the system are changing or unknown, for example when the robot manipulates a tool with unknown length. Several approaches approximating Jacobian controllers for set-point control of robots with uncertainties in both kinematics and dynamics have been proposed [14][15][16].

Based on the principle of tacit learning, we propose a simple controller that avoids the usage of the Jacobian altogether in a process that does not involve any model of the system being controlled. Our objective is to design a new control strategy that can achieve levels of adaptability similar to those observed in living organisms, be plausible from a neural control view point, and does not require complex analysis of the dynamics of the system.

## 2     Tacit Learning Controllers

The general expression for a tacit controller is:

$$U = KX_c + Q \tag{1}$$

$$\dot{Q} = A \tag{2}$$

U is the control, Xc the state variable expressed in the control space, K is the proportional and derivative gain matrix, and A the effect to be minimized.

In this paper we extend this controller by expressing $\dot{Q}$, the tacit component of the controller, as:

$$\dot{Q} = W(p_1, p_2, \dots, p_m) \circ E \tag{3}$$

" $\circ$ " is the Hadamard product. E is a n × m matrix which rows vectors are all e, e being a vector of size m of errors expressed in the task space. The vector e has to construct such as e=0 for the desired state and n is the number of degrees of freedom of the system. W is also a n × m matrix. P is the set of parameters dynamically tuned by higher level algorithms and controllers such as obtaining linearization of the system described by equation (1). The parameters pj correspond for example to the norms of the row vectors of W or to the angles between them. The advantage of this approach is that values of pj can be directly related to features of the task-space, for example the speed of the end-point of a manipulator.

# 3     Two-Links Robotic System

## 3.1     Task and Control

This kind of controller can be applied to a wide range of robotic systems for which analytical solutions of a task expressed in the task-space is impractical. In this paper we provide control of a two-link robotic arm which kinematic configuration is unknown. We define the task as:

$$\begin{cases} H_{t \to \infty} = 0 \\ \dot{H}_{t \to \infty} = 0 \\ H_{|H| \to \delta} = \rho V \end{cases} \tag{4}$$

H is the vector defined by the end-point and the target. t is the time. $\delta$ is a small number and $\rho$ is a real number. As shown in Fig.1, this corresponds to reaching the target at the angle specified by the vector V. Extended to the 6D space, this definition is simple but sufficient to specify a grasping task: it requires the robot to reach the target with the end-effect at a certain position and orientation, while leaving unspecified the trajectory taken by the end-effector and the configuration of the rest of the robotic system. By specifying a succession of tasks, this approach is also suitable to define the complete trajectory. For reasons explained later on, we refer to the line defined by the target and the vector V as the stability line.

We apply the equations 1-3 to this robotic system. We define e as:

$$e = \left[ V \cdot H, rot_{\pi/2} V \cdot H \right] \tag{5}$$

H is the vector defined by the end-point and the target, '·' is the dot product and $rot_{\pi/2}$ the rotation matrix of the angle $\pi/2$. W is a 2×2 matrix and we will refer to its row vectors as w1 and w2. We define $p = [\phi, \alpha, a]$: $\phi$ is the angle between V and w1

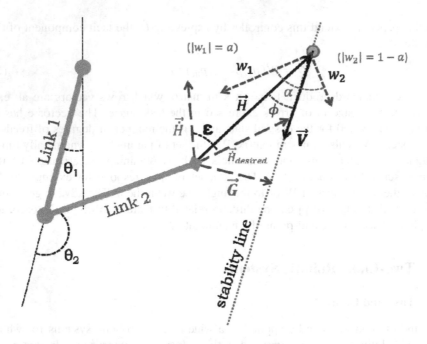

**Fig. 1.** The end-effector must reach the target following the line trajectory defined by the stability line. The controller achieves this by tuning φ (the orientation), α (the angles between w1 and w2), and a (the norms of w1 and w2) such as achieving ε=0 (see text for details).

+ w2, $\alpha$ is the angle between w1 and w2, and a is the norm of w1. W can be determined based on p:

$$w_1 = a\left(rot_{\phi+\alpha/2}V\right) \tag{6}$$

$$w_2 = (1-a)\left(rot_{\phi-\alpha/2}V\right) \tag{7}$$

For this system, equation 3 can be written:

$$\dot{Q}_c = \begin{bmatrix} a\left(rot_{\phi+\alpha/2}V\right) \\ (1-a)\left(rot_{\phi-\alpha/2}V\right) \end{bmatrix} \circ \begin{bmatrix} V \cdot H, rot_{\frac{\pi}{2}}V \cdot H \\ V \cdot H, rot_{\frac{\pi}{2}}V \cdot H \end{bmatrix} \tag{8}$$

Once the values of $\phi$, $\alpha$, and a are set, all the parameters of the controller expressed by equations 1 and 3 are determined and the system can be controlled. Based on equations 4 and 5, an informal explanation of the proposed controller can be given. The target, w1 and w2 define two lines in the task space. If $\alpha$ is zero, then these two lines coincide. If the end-point is on this line, then $\dot{Q}$ is null and the system stabilizes. Furthermore, if $\phi$ is also zero, these two lines will also coincide with the stability line on which the system will stabilize. If $\alpha$ is not zero, then $\dot{Q} = 0$ can be achieved only when the end-point reaches the target. This induced instability will result in the end-effector moving. $\phi$, $\alpha$, and a can be dynamically tuned so that this movement will follow the stability line toward the target.

We consider that the desired end-point velocity is $\dot{H}_{desired} = H + G$, G being the vector between the end-point and the stability line. a is modelled as a virtual dynamic parameter controlled such as reaching $\dot{H} \approx \dot{H}_{desired}$:

$$\ddot{a} = -k_p a - k_d \dot{a} + k_t \int \varepsilon dt \qquad (9)$$

$k_p$, $k_d$, and $k_t$ are proportional, derivative, and tacit gains. $\varepsilon$ is the signed angle between $\dot{H}$ and $\dot{H}_{desired}$.

$\alpha$ is tuned such as decreasing the speed of the endpoint when $\dot{H}$ is not converging toward $\dot{H}_{desired}$:

$$\alpha = \alpha_0 + \frac{1}{\tau_0} \times \int_{t-\tau_0}^{t} sign(\varepsilon \dot{\varepsilon}) dt \qquad (10)$$

$\alpha 0$ and $\tau_0$ are parameters of the system and sign is the function defined as:

$$\begin{cases} sign(x) = 0 & if \ x \le 0 \\ sign(x) = 1 & if \ x > 0 \end{cases} \qquad (11)$$

Finally $\phi$, tuned as the endpoint, is always positioned between the two lines defined by the target, w1 and w2 by setting it to the angle between V and H.

## 3.2    Simulation Results

The controller was tested in a simulation created using Open Dynamic Engine [17] using the configuration:

$$l1 = 0.5[m] \ l2 = 0.4[m] \qquad (12)$$

$$m1 = 0.5[kg] \ m2 = 0.4[kg] \qquad (13)$$

None of these configuration parameters are known to the controller. The gains of the controller were set to: kp=10 kd=0.6 and kt=1.0 for the controller presented in equation 1, and kp=10 kd=2.0 and kt=20 for the controller presented in equation 8. We set $\alpha 0 = \pi/12$ and $\tau 0 = 1[s]$.

The arms moved in the 2D sagittal plane subjected to gravity and were required to reach a target at [0.5,0.0] in the Cartesian reference coordinate centered on the base of the arm. The reaching vector V was the vector defined by the target and the initial positions of the end-point. The starting position of the robot is $\theta 1 = 0$ and $\theta 2 = 0$, corresponding to straight joints pointing downward (see Figure 2). Using the proposed controller, the endpoint reaches the target following the stability line. Figure 2 shows in blue circles the trajectory of the end-point. The configuration of the robotic system is also shown for the starting position $\theta 1 = 0$ and $\theta 2 = 0$), the final position when the end-point reaches the target and for an intermediate time (in lighter grey). Figure 3 shows the evolution in time of some of the parameters of the system

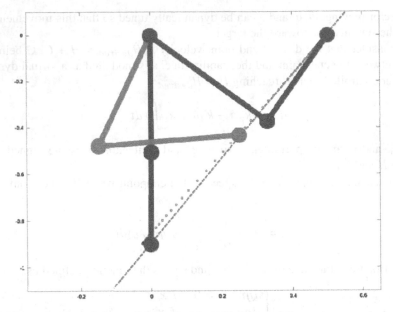

**Fig. 2.** The trajectory of the end-point (in blue). Postures of the system are shown at the beginning of the experiment, at the end of the experiment when the end-point reaches the target, and at an intermediary time. The stability line is represented by the black dashed line.

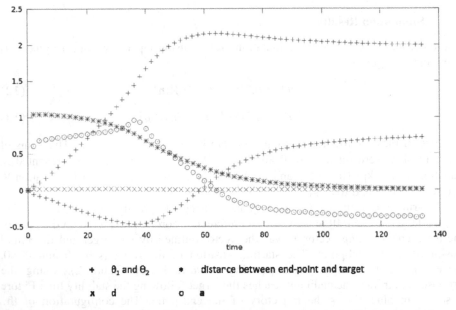

**Fig. 3.** Evolution of time in some parameters of the system. a is dynamically tuned to achieve a linear motion of the end-point, compensating the high nonlinearity displayed by the values of θ1 and θ2. It can be seen that d, the distance between the positions of the end-point and the desired line trajectory, remains very close to zero.

with time. The distance between the end-point and the target decreases while the distance between the end-point and the stability remains very close to zero. The parameter a is dynamically tuned such as finding values of $\theta 1$ and $\theta 2$ corresponding to the desired line trajectory of the end-point. The controller manages to compensate for the nonlinearities of the system, despite not using any kinematic parameter.

# 4     Discussion and Future Work

Our goal is to provide control solutions to robotic tasks for which 1) analytical solutions have high complexity, for example grasping an object while walking, or 2) the environment cannot be modelled, which is an issue for the development of rehabilitation of the robotic system that adapts to each patient. To reach this goal, we are now working towards several directions:

## 4.1     Stability

The stability of the proposed controller depends highly on the initial configuration of the system as well as on the gain values chosen. Stability analysis is required to design a systematic method for selecting the value of these gain values in a fashion that guarantees stability. First results in regards of the stability analysis of tacit controllers have already been published [18] and further characterization is on-going. Extension to higher dimensionality

We are currently working on extending the methodology presented in this paper to the six dimensional space. While in 2D the task is represented by a line expressed in the task space, in a higher dimension the task will be represented by a combination of lines and planes and virtual dynamic variables also being tuned by the tacit learning controller to control the transformation of these geometric entities.

## 4.2     Motor Control

The neural mechanisms used for mapping goals expressed in the task-space into control-space related commands without using internal models remain largely unknown. But many neural systems rely on data accumulation: the presented controller incorporates knowledge expressed in the task-space using only the accumulation of data and is plausible from a neural control viewpoint.

# 5     Conclusion

In this work, the system achieved to control the trajectory of the end-point such as following a line trajectory. The controller manages to compensate for the nonlinearities of the system, despite not using any kinematic parameter. The presented controller does not use any internal model and incorporates knowledge expressed in the task-space using only the accumulation of data. By providing control solutions that are not based on the Jacobian matrix for solving the relationships between task and control space, we target to reach levels of adaptability similar to the one observed in living organisms. Such an adaptability level will provide advantages in the control of complex over an actuated robotic system or robotic system for which the environment cannot be modelled.

# References

1. Minsky, M.L., Papert, S.A.: Perceptron. MIT Press, Cambridge (1969)
2. Rumelhart, D.E., Hinton, G.E., Williams, R.J.: Learning representations by backpropagating errors. Nature 323(6088), 533–536 (1986)
3. Kuniyoshi, Y., Yorozu, Y., Suzuki, S., Sangawa, S., Ohmura, Y., Terada, K., Nagakubo, A.: Emergence and development of embodied cognition: A constructivist approach using robots. Prog. Brain Res. 164, 425–445 (2007)
4. Barto, A.G., Sutton, R.S., Anderson, C.W.: Neuron-like adaptive elements that can solve difficult learning control problems. IEEE Trans. Syst., Man, Cybern. SMC-13(5), 834–846 (1983)
5. Doya, K.: Reinforcement learning in continuous time and space. Neural Comput. 12(1), 219–245 (2000)
6. Tedrake, R., Zhang, T.W., Seung, H.S.: Stochastic policy gradient reinforcement learning on a simple 3D biped. In: Proc. IEEE/RSJ Int. Conf. Intell. Robots Syst., pp. 2849–2854 (2004)
7. Astrom, K.J., Wittenmark, B.: Adaptive Control. Addison-Wesley, Reading (1989)
8. Slotin, J.E., Li, W.: Applied Nonlinear Control. Prentice-Hall, Englewood Cliffs (1991)
9. Zadeh, L.A.: Outline of a new approach to the analysis of complex systems and decision processes. IEEE Trans. Syst. Man, Cybern. SMC-3(1), 28–44 (1973)
10. Juang, J.G.: Fuzzy neural network control CMAC of a biped walking robot. IEEE Trans. Syst., Man, Cybern. B, Cybern. 30(4), 594–601 (2000)
11. Shimoda, S., Kimura, H.: Bio-mimetic Approach to Tacit Learning based on Compound Control. IEEE Transactions on Systems, Man, and Cybernetics- Part B 40(1), 77–90 (2010)
12. Shimoda, S., Kimura, H.: Adaptability of tacit learning in bipedal locomotion. IEEE Transactions on Autonomous Mental Development 5(2), 152–161 (2013)
13. Hayashibe, M., Shimoda, S.: Emergence of Motor Synergy in Vertical Reaching Task via Tacit Learning. In: International Conference of the IEEE Engineering in Medicine and Biology Society (EMBC), pp. 4985–4988 (2013)
14. Cheah, C.C., Hirano, M., Kawamura, S., Arimoto, S.: Approximate Jacobian control for robots with uncertain kinematics and dynamics. IEEE Transactions on Robotics and Automation 19(4), 692–702 (2003)
15. Dixon, W.E.: Adaptive regulation of amplitude limited robot manipulators with uncertain kinematics and dynamics. IEEE Transactions on Automatic Control 52(3), 488–493 (2007)
16. Ozawa, R., Oobayashi, Y.: Adaptive task space PD control via implicit use of visual information. In: Int. Sym. Robot Control, pp. 209–214 (2009)
17. Smith, R.: Open Dynamics Engine, http://www.ode.org/
18. Shimoda, S., Yoshihara, Y., Fujimoto, K., Yamamoto, T., Maeda, I., Kimura, H.: Stability analysis of tacit learning based on environmental signal accumulation. In: 2012 IEEE/RSJ International Conference on Intelligent Robots and Systems, October 7-12 (2012)

# Simulating Synthetic Emotions
# with Fuzzy Grey Cognitive Maps

Jose L. Salmeron

Computational Intelligence Lab, Km. 1 Utrera road, University Pablo de Olavide,
41013 Seville, Spain
salmeron@acm.org

**Abstract .**Autonomous robotic systems should decide autonomously without or with sparse human interference how to react to alterations in environment. Based on Thayer's emotion model and Fuzzy Grey Cognitive Maps, this work presents a proposal for simulating synthetic emotions. Thayer's proposal is based on mood analysis as a bio psychological concept. Recently, Fuzzy Grey Cognitive Maps have been proposed as a FCM extension. FGCM is mixing conventional Fuzzy Cognitive Maps and Grey Systems Theory that has become a worthy theory for solving problems with high uncertainty under discrete small and incomplete data sets. This proposal provides an innovative way for simulating synthetic emotions and designing an affective robotics system. This work includes an experiment with an artificial scenario for testing this proposal.

## 1  Introduction

Autonomous systems should decide without or with scarce human interference how to react to changes in their contexts and environments [2]. Due to the fact that self-adaptive systems are complex autonomous systems, it is hard to ensure that they behave as desired and avoid wrong behaviour [5].

For autonomous systems to make highly specialized tasks, it is sometimes needed to embed affective behavior that has not been associated traditionally with intelligence [6]. Emotions play an important role in human reasoning and its decision making.

This paper proposes Fuzzy Grey Cognitive Maps (FGCMs) as a worthy tool for forecasting artificial emotions in autonomous systems immersed in complex environments with high uncertainty. The Thayer's emotion model is used to map FGCM outputs within an emotional space. That model defines the emotion categories in a 2-dimensional Cartesian coordinate according to their valence and arousal.

The rest of the paper is structured as follows: Section 2 presents the emotional theoretical background. The next section introduces Fuzzy Grey Cognitive Maps. In Section 4 an illustrative application is given and conclusions are finally shown.

## 2  Theoretical Background

Emotions have an important impact on human decisions, actions, beliefs, motivations, and desires [4]. In this sense, if we want the robots to have real intelligence, to adapt

© Springer International Publishing Switzerland 2015
P. Sinčák et al. (eds.), *Emergent Trends in Robotics and Intelligent Systems*,
Advances in Intelligent Systems and Computing 316, DOI: 10.1007/978-3-319-10783-7_4

to the environment which humans are living in, and to communicate with human beings naturally, then robots need to detect, understand, and express emotions in a certain degree.

Affective computing assigns robots and systems in general a human-like potential of detection, understanding and generation of emotions. It is a new but promising research area dealing with the issues regarding emotions and systems. Nowadays, emotions research has become a multi-disciplinary and growing field [1]. Indeed, it could be used to make robots act according to the human emotions.

This proposal is inspired by Thayer's emotion model for defining the affective space. Next, a brief overview is shown.

## 2.1    Two-Dimensional Emotion Representation in Thayer's Model

Thayer's model [12] is based on mood analysis as a biopsychological concept. Moreover, this proposal models mood as an affective state closely related to biochemical and psycho-physiological elements.

Various emotions are divided into four quadrants of a two-dimensional Cartesian coordinate system, valence (x), arousal (y), as shown in Fig. 1. The central intersection indicates the lack of emotion.

Each model's quadrant includes three basic emotions. The first quadrant with positive valence and arousal is composed of the emotions: pleased, happy and excited. The second one with negative valence and positive arousal comprises annoying, angry and nervous. The third one with negative valence and arousal consists of sad, bored

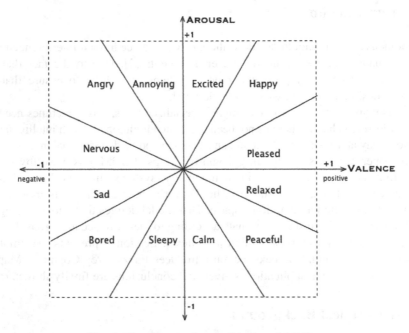

**Fig. 1.** Thayer's emotion graphical model [12]

and sleepy. Finally, the last one with positive valence and negative arousal covers calm, peaceful and relaxed. As a result, the Thayer's emotional space is composed of twelve emotions.

Regarding the intensity, the points closer to the intersection mean less intense emotions and points far away from the center model more intense emotions.

# 3    Fuzzy Grey Cognitive Maps

## 3.1    Fundamentals

Grey Systems Theory (GST) is an interesting set of problem solving tools within environments with high uncertainty, under discrete small and incomplete data sets [3]. GST has been designed to work with small data samples and poor information with successful applications in military science, business, agriculture, energy, transportation, meteorology, medicine, industry, geology and so on.

Fuzzy Grey Cognitive Map is based on FCMs and GST, and it has become a very worthy theory for solving problems within domains with high uncertainty [7]. FGCMs provide an intuitive yet precise way of modelling concepts and reasoning about them. By transforming decision models into causal graphs, decision-makers without technical background can understand all of the components in a given situation. Moreover, with a FGCM, it is possible to identify the most critical factor that impacts the expected target concept.

The FGCM nodes are modelling variables, representing concepts. The relationships between nodes are represented by directed edges. An edge linking two FGCM nodes is modelling the grey causal influence of the causal variable on the effect variable. The FGCM model is represented by an adjacency matrix ($\otimes A$).

$$\otimes A = \begin{array}{c} x_1 \\ \vdots \\ x_n \end{array} \begin{pmatrix} \otimes w_{11} & \cdots & \otimes w_{1n} \\ \vdots & \ddots & \vdots \\ \otimes w_{n1} & \cdots & \otimes w_{nn} \end{pmatrix} \begin{array}{c} \\ \end{array} \qquad (1)$$

FGCMs are dynamical systems involving feedback where the effect of change in a variable (node) may affect other variables (nodes) which, in turn, can affect the variable initiating the change. An FGCM models unstructured knowledge through causalities through grey concepts and grey relationships between them based on FCM [7][11].

Since FGCMs are hybrid methods mixing neural networks and grey systems, each cause is measured by its grey weight as

$$\otimes w_{ij} = [\underline{w}_{ij}, \overline{w}_{ij}] | \{\underline{w}_{ij}, \overline{w}_{ij}\} \in \{[-1, +1], [0, +1]\} \qquad (2)$$

where $i$ is the pre-synaptic (cause) node and j is the post-synaptic (effect) one.

FGCM dynamics begin with an initial grey vector state $\otimes C(0)$ which models a proposed initial imprecise stimuli. The initial grey vector state with n nodes is denoted as

$$\otimes C(0) = (C_1(0), C_2(0), \ldots, C_n(0)) = ([\underline{c_1}(0), \overline{c}_1(0)], [\underline{c_2}(0), \overline{c}_2(0)], \ldots, [\underline{c_n}(0), \overline{c}_n(0)]) \quad (3)$$

The updated nodes' states are computed in an iterative inference process with an activation function (usually sigmoid or hyperbolic tangent function) [7][9][10], which maps monotonically the grey node value into a normalized range $[0, +1]$ or $[-1, +1]$, depending on the selected function. Note that grey arithmetic is detailed as [7]. Each single node would be updated as follows:

$$\otimes c_j(t+1) = f\left(\sum_{i=1}^{n} \otimes w_{ij} \cdot \otimes c_i(t)\right) = [\underline{c_j}(t+1), \overline{c}_j(t+1)] \quad (4)$$

The unipolar sigmoid function is the most used one in FGCM when the nodes' value maps in the range of $[0, 1]$. If $f(\cdot)$ is a sigmoid, then the i component of the grey vector state at t+1 iteration ($\otimes C(t+1)$) after the inference would be

$$\otimes c_i(t+1) \in \left[\left(1 + e^{-\lambda \cdot \underline{c_i^*}(t)}\right)^{-1}, \left(1 + e^{-\lambda \cdot \overline{c}_i^*(t)}\right)^{-1}\right] \quad (5)$$

On the other hand, when the concepts' states map in the range $[-1, +1]$, the function used would be the hyperbolic tangent.

The nodes' states evolve along the FGCM dynamics. The FGCM inference process stops when the stability is reached. The steady grey vector state represents the final impact of the initial grey vector state on the final state of each FGCM grey node.

After its inference process, the FGCM reaches one steady state following a number of iterations. It settles down to a fixed pattern of node states, the so-called grey hidden pattern or grey fixed-point attractor.

Moreover, the state could keep cycling between several fixed states, known as a limit grey cycle. Using a continuous activation function, a third state would be a grey chaotic attractor. It happens when instead of stabilizing, the FGCM continues to produce different grey vector states for each iteration.

FGCM includes greyness as an uncertainty measurement. Higher values of greyness mean that the results have a higher uncertainty degree. It is computed as follows:

$$\phi(\otimes c_i) = \frac{|\ell(\otimes c_i)|}{\ell(\otimes \psi)} \quad (6)$$

where $|\ell(\otimes c_i)| = |\overline{c}_i - \underline{c}_i|$ is the absolute value of the length of grey node $\otimes c_i$ state value and $\ell(\otimes \psi)$ is the absolute value of the range in the information space, denoted by $\otimes \psi$. It is computed as follows:

$$\ell(\otimes \psi) = \begin{cases} 1 \; if \; \{\otimes c_i, \otimes w_i\} \subseteq [0,1] \\ 2 \; if \; \{\otimes c_i, \otimes w_i\} \subseteq [-1,+1] \end{cases} \tag{7}$$

## 3.2    FGCM Advantages over FCM

FGCMs have several advantages over conventional FCM [7][9][10]. FGCMs are able to compute the desired steady states by handling uncertainty and hesitancy present within raw data (due to noise) for causal relations among concepts as well as within the initial concepts states.

The main difference between FGCMs and FCMs is within weights design. FCM applies weights with discrete numerical values associated to edges. FGCMs uses weights with grey intensity including grey uncertainty and fuzziness to better describe the impact between the nodes.

Note that, even if the FCM dynamics would get the same steady vector state than FGCM after the whitenization process, the FGCM proposal handles the inner fuzziness and grey uncertainty of human emotions.

FGCMs are a generalization and can be applied to approximate human decision making more closely. It handles the uncertainty inherent in the complex systems by assessing greyness in the nodes and edges. The reasoning process's output would incorporate a degree of greyness expressed in grey values.

In addition, FGCMs are able to model more kinds of relationships than FCM. For instance, it is possible to run models with relations where the intensity is not known at all wi $\in [-1,+1]$ or just partially known.

## 4    Illustrative Example

With the intention of illustrating the proposal, this paper proposes an artificial experiment. The goal is the simulation of an autonomous systems emotions generated by environmental conditions. Note that the goal of the model is not to design a real-world emotional Ambient Intelligence system but to test the FGCM approach for artificial emotions forecasting for people in a queue in a hypothetic Ambient Intelligence.

The FGCM model in Fig. 3 represents an example of a FGCM-based emotional Ambient Intelligence system. Eq. 8 shows the adjacency matrix.

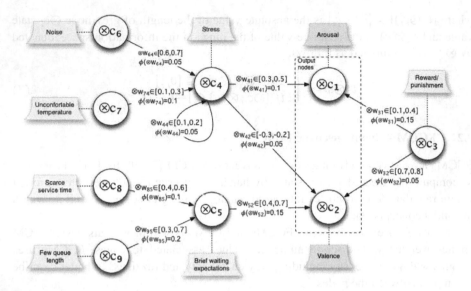

**Fig. 2.** FGCM-based model

$$\otimes A = \begin{pmatrix}
[.0,.0] & [.0,.0] & [.0,.0] & [.0,.0] & [.0,.0] & [.0,.0] & [.0,.0] & [.0,.0] & [.0,.0] \\
[.0,.0] & [.0,.0] & [.0,.0] & [.0,.0] & [.0,.0] & [.0,.0] & [.0,.0] & [.0,.0] & [.0,.0] \\
[.1,.4] & [.7,.8] & [.0,.0] & [.0,.0] & [.0,.0] & [.0,.0] & [.0,.0] & [.0,.0] & [.0,.0] \\
[.3,.5] & [-.3,-.2] & [.0,.0] & [.1,.2] & [.0,.0] & [.0,.0] & [.0,.0] & [.0,.0] & [.0,.0] \\
[.0,.0] & [.4,.7] & [.0,.0] & [.0,.0] & [.0,.0] & [.0,.0] & [.0,.0] & [.0,.0] & [.0,.0] \\
[.0,.0] & [.0,.0] & [.0,.0] & [.6,.7] & [.0,.0] & [.0,.0] & [.0,.0] & [.0,.0] & [.0,.0] \\
[.0,.0] & [.0,.0] & [.0,.0] & [.1,.3] & [.0,.0] & [.0,.0] & [.0,.0] & [.0,.0] & [.0,.0] \\
[.0,.0] & [.0,.0] & [.0,.0] & [.0,.0] & [.4,.6] & [.0,.0] & [.0,.0] & [.0,.0] & [.0,.0] \\
[.0,.0] & [.0,.0] & [.0,.0] & [.0,.0] & [.3,.7] & [.0,.0] & [.0,.0] & [.0,.0] & [.0,.0]
\end{pmatrix}$$

**Table 1.** FGCM nodes and description

| Node Label ($xi$) | Label | Description |
|---|---|---|
| $x_1$ | Arousal | State of being awake or reactive to stimuli |
| $x_2$ | Valence | The intrinsic attractiveness (positive valence) or averseness (negative valence) of an emotion |
| $x_3$ | Reward/Punishment | Reward is related to a positive queue where individuals are going to get something positive (e.g.: a lottery award). Punishment is when they are in the queue for something negative (e.g.: paying taxes) |
| $x_4$ | Stress | A person's response to a stressor, such as noise or uncomfortable temperature |
| $x_5$ | Waiting expectations | Waiting time considered the most likely to happen according to people before each one and time in service for each one |
| $x_6$ | Noise | Environmental noise |
| $x_7$ | Uncomfortable temperature | Temperature higher or lower than comfortable |
| $x_8$ | Scarce service time | Waiting time for each person |
| $x_9$ | Few queue length | People in the queue |

In the test scenario, we have an initial vector C(0) representing the initial state values of the events at a given time of the process and a final vector C(t) representing the steady state that it can be arrived at. The final vector C(t) is the last vector in the convergence region.

For the synthetic case study, the initial vector state and the steady vector state are the following:

$$A(0) = ([0,0], [0,0], [0.2,0.2], [0,0], [0,0], [0.2,0.3], [-0.2, -0.1],$$
$$[0.1,0.3], [0.3,0.4])$$
$$A(t) = ([\mathbf{0.04, 0.20}], [\mathbf{0.12, 0.42}], [0.2,0.2], [0.7,0.24], [0.13,0.43], [0.2,0.3],$$
$$[-0.2, -0.1], [0.1,0.3], [0.3,0.4])$$

According to the A(t) results, Arousal = [0.04, 0.20] and Valence = [0.12, 0.42], and the simulation emotion is mixing light/medium pleased and light happy.

## 5    Conclusions

Emotions have a critical impact on human's motivations, decisions, actions, beliefs and desires. Emotion simulation and its application in autonomous systems and robics is an emerging and promising research area. For those reasons, there is a sign of an emotion simulation module based on FGCMs and Thayer's emotion model was proposed.

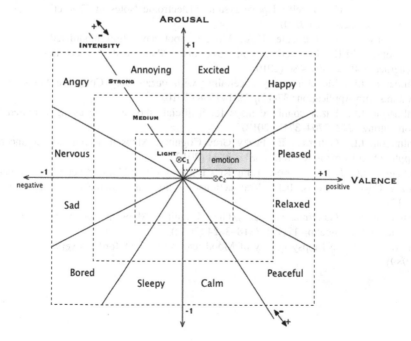

**Fig. 3.** Experiment results

This paper shows an artificial experiment of a FGCM-based emotion simulation system. FGCM is an FCM extension for representing causal reasoning within complex systems with high uncertainty. FGCM represents knowledge, uncertainty and relates states, variables, events, inputs and outputs in a similar way to that of human beings. This paper shows that it is possible to simulate the emotions generated from sensors raw data with FGCMs.

This is not an empirical research. An FGCM-based framework is based on external data. Constructs and output nodes are also presented. Indeed, the goal is not to model a real world system, but it just proposes an FGCM-based theoretical model, so that robotics practitioners or future research can use it to simulate or generate emotions.

# References

1. Albornoz, E.M., Milone, D.H., Rufiner, H.L.: Spoken emotion recognition using hierarchical classifiers. Computer Speech & Language 25(3), 556–570 (2011)
2. Brun, Y., et al.: Engineering self-adaptive systems through feedback loops. In: Cheng, B.H.C., de Lemos, R., Giese, H., Inverardi, P., Magee, J. (eds.) Self-Adaptive Systems. LNCS, vol. 5525, pp. 48–70. Springer, Heidelberg (2009)
3. Deng, J.L.: Introduction to grey system theory. The Journal of Grey System 1, 1–24 (1989)
4. Guojiang, W., Xiaoxiao, W., Kechang, F.: Behaviour decision model of intelligent agent based on artificial emotion. In: 2nd International Conference on Advanced Computer Control (ICACC), Shenyang, China, pp. 185–189 (2010)
5. Khakpour, N., Jalili, S., Talcott, C., Sirjani, M., Mousavi, M.R.: PobSAM: policy-based managing of actors in self-adaptive systems. Electronic Notes in Theoretical Computer Science 263, 129–143 (2010)
6. Lee-Johnson, C.P., Carnegie, D.A.: Mobile robot navigation modulated by artificial emotions. IEEE Transactions on Systems, Man, and Cybernetics, Part B: Cybernetics 40(2), 469–480 (2010)
7. Salmeron, J.L.: Modelling grey uncertainty with Fuzzy Grey Cognitive Maps. Expert Systems with Applications 37(12), 7581–7588 (2010)
8. Salmeron, J.L.: Fuzzy cognitive maps for artificial emotions forecasting. Applied Soft Computing 1212, 3704–3710 (2012)
9. Salmeron, J.L., Gutierrez, E.: Fuzzy Grey Cognitive Maps in Reliability Engineering. Applied Soft Computing 12(12), 3818–3824 (2012)
10. Salmeron, J.L., Lopez, C.: Forecasting Risk Impact on ERP Maintenance with Augmented Fuzzy Cognitive Maps. IEEE Transactions on Software Engineering 38(2), 439–452 (2012)
11. Salmeron, J.L., Gutierrez, E.: Fuzzy Grey Cognitive Maps in Reliability Engineering. Applied Soft Computing 12(12), 3818–3824 (2012)
12. Thayer, R.E.: The Biopsychology of Mood and Arousal. Oxford University Press, USA (1989)

# The Design and Implementation of Quadrotor UAV

Petr Gabrlik[1], Vlastimil Kriz[2], Jan Vomocil[1], and Ludek Zalud[2]

[1] Central European Institute of Technology, Brno University of Technology, Czech Republic
{petr.gabrlik,jan.vomocil}@ceitec.vutbr.cz
[2] Faculty of Electrical Engineering and Communication,
Brno University of Technology, Czech Republic
xkrizv00@stud.feec.vutbr.cz, zalud@feec.vutbr.cz

**Abstract.** The design and implementation of the four-rotor aerial mobile robot called quadrotor is described in this paper. The beginning is focused on the mechanical construction of the robot body and the hardware implementation of the main control board. The next part describes the simplified mathematical model of the quadrotor. On the base of the created model, a state space controller was designed and implemented. As a result, every quadrotor axis is controlled independently as in the case of the use of separated PI or PD controllers. The last part of the paper deals with a software solution. This can be divided into three parts: the first part describes the application for the onboard microcontroller, the second part focuses on the solution of the base station. The special part of software which solves the localization of the quadrotor using a camera is described at the end of the article.

## 1   Introduction

The quadrotors are a very popular type of unmanned aerial vehicles because of their mechanic simplicity in comparison with other flying robots. The construction shown on Fig. 1 is very simple. It is formed by four beams which are orthogonal to each other. At the end of each beam, there is a BLDC (Brushless DC) engine with a propeller. Common helicopters have very similar flight characteristics as quadrotors, but quadrotors have different methods of flight control.

The paper deals with the modelling and realization of quadrotor. This robot is used for many purposes. It can be used in civil or military applications. This type of robot allows for very fast takeoff and its operation is very low-cost. It can carry many devices, e.g. chemical sensors for inspection of air pollution during fire.

This work is divided into several parts. First of all, necessary electronics are described. This section is followed by the description of the mathematical model of the robot. This model describes a relation between forces and torques affecting the quadrotor body. Following parts deal with the design of the state space controller. The next chapter explains software solution and implementation of the microcontroller. A computer vision system for automated landing is described at the end of the paper.

P. Sinčák et al. (eds.), *Emergent Trends in Robotics and Intelligent Systems*,
Advances in Intelligent Systems and Computing 316, DOI: 10.1007/978-3-319-10783-7_5

**Fig. 1.** Quadrotor prototype

## 2    Hardware

The auxiliary stabilization must be done by electronics because this robot cannot be controlled only by man. For this reason a necessary control unit was created. The control unit was developed just for this application. It can be divided into several parts. The main part is the microcontroller and power supply modules. Another part of the control unit is the wireless communication module. On Fig. 2, there is a block diagram of the control unit and ground station with wireless modules and a laptop computer.

All electronic devices on this robot are powered by the Li-Pol accumulator. In this construction a 3-cell accumulator with a nominal voltage 11.1 V is used. All engines use this voltage, while other electronics are supplied by 3.3 V and 5 V. Voltage level reduction is done by switching and LDO (Low-dropout Regulator) stabilizers.

The base of the control unit is the microcontroller LM3S8269. It is a 32-bit ARM with Cortex M3 architecture and operating up to 50 MHz. This device provides relatively high performance in comparison with similar projects (e.g. [1]), which is useful for testing various control algorithms. The microcontroller is also equipped with required communication buses (SPI, UART, I2C) and internal A/D converters.

The main sensor is an IMU (Inertial Measurement Unit) module Vectornav VN-100. This device provides us  data about acceleration, angular speed and magnetic field in each axis, required for orientation determination [2]. Its advantage is the integrated Kalman filter. Next, an ultrasonic sensor SRF10 is used for measuring the distance to the ground in low altitudes.

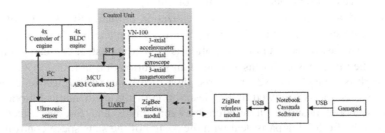

**Fig. 2.** The block diagram of control unit and ground station with user interface

# 3    Quadrotor Model

The mathematical model of the quadrotor was divided into two parts. The first part handles relations between the speed of rotors and forces and torques affecting the quadrotor rigid body. The second one handles the dynamics of the rigid body including the transformation between frames and the effect of gravity force. This separation allows to design the controller with better dynamics, because the first part can be easily linearized in the controller.

The forces and torques caused by propellers of rotors affecting the quadrotor rigid body were noted as $u_1$ to $u_4$. The meaning is following: $u_1$ - torque around x axis caused by different thrusts of rotors 2 and 4, $u_2$ - torque around y axis caused by different thrusts of rotors 1 and 3, $u_3$ - torque around z axis caused by different reaction torques from rotors rotating in the opposite direction, $u_4$ - force in z axis caused by the common thrust of all rotors.

Relation between the thrust and propeller speed of the rotor is

$$F_T = k_T \cdot n^2 \tag{1}$$

and the relation between the reaction torque and propeller speed of the rotor is

$$M_R = k_M \cdot n^2. \tag{2}$$

$k_T$ and $k_M$ are constantly determined by measurement. The first part of the model:

$$u_1 = \left(-n_2^2 + n_4^2\right)k_T l \tag{3}$$

$$u_2 = \left(n_1^2 - n_3^2\right)k_T l \tag{4}$$

$$u_3 = \left(-n_1 + n_2 - n3 + n_4\right)k_M \tag{5}$$

$$u_4 = \left(-n_1^2 + n_2^2 - n_3^2 + n_4^2\right)k_T \tag{6}$$

$l$ denotes the distance between the rotors and the center of the gravity of the quadrotor. The second part of the model can be further divided into three parts: Equations that describe rotational movement, equations of linear movement, equations describing transformation between frames and equations expressing the effect of gravity force.

Rotational movement:

$$\dot{\omega}_x = \frac{u_1 + (I_{yy} - I_{zz})\omega_y\omega_z + u_1}{I_{xx}} \tag{7}$$

$$\dot{\omega}_y = \frac{u_2 + (I_{zz} - I_{xx})\omega_x\omega_z + u_2}{I_y y} \tag{8}$$

$$\dot{\omega}_z = \frac{u_3 + (I_{xx} - I_{yy})\omega_z\omega_y + u_3}{I_z z} \tag{9}$$

$I_{xx}$, $I_{yy}$ and $I_{zz}$ represent moments of inertia around given axis.
Translation movement:

$$\dot{v}_x = \frac{m\omega_y v_z + m\omega_z v_y + G_x}{m} \tag{10}$$

$$\dot{v}_y = \frac{m\omega_x v_z + m\omega_z v_x + G_y}{m} \tag{11}$$

$$\dot{v}_z = \frac{m\omega_x v_y + m\omega_y v_x + G_z - u_4}{m} \tag{12}$$

$G_x$, $G_y$ and $G_z$ are gravity forces component in given axis:

$$G_x = -mg \sin \theta \tag{13}$$

$$G_y = mg \cos \theta \sin \phi \tag{14}$$

$$G_z = mg \cos \theta \cos \phi \tag{15}$$

Transformation between robot frame and inertial frame [3]:

$$\dot{\phi} = \omega_x + \omega_y \sin \phi \tan \theta + \omega_z \cos \theta \tan \theta \tag{16}$$

$$\dot{\theta} = \omega_y \cos \theta - \omega_z \sin \phi \tag{17}$$

$$\dot{\psi} = \omega_y \frac{\sin \phi}{\cos \theta} + \omega_z \frac{\cos \phi}{\cos \theta} \tag{18}$$

$$\dot{x} = v_z(\sin \phi \sin \psi + \cos \phi \cos \psi \sin \theta) -$$
$$-v_y(\cos \phi \sin \psi - \cos \psi \sin \phi \sin \theta) + v_x \cos \theta \cos \psi \tag{19}$$

$$\dot{y} = v_y(\cos \phi \cos \psi + \sin \phi \sin \theta \sin \psi) -$$
$$-v_z(\cos \psi \sin \phi - \cos \phi \sin \theta \sin \psi) + v_x \cos \theta \cos \psi \tag{20}$$

$$\dot{z} = v_z \cos \phi \cos \theta - v_x \sin \theta + v_y \cos \theta \sin \phi \tag{21}$$

## 4    Controller Design

To stabilize the quadrotor, the controller based on the state space representation of the quadrotor was designed.

As it was shown in Sect. 3, the model of the quadrotor was divided into 2 parts. The first part handles the non-linear relation between the rotation speed of the propellers and introduced variables $u_1$ - $u_4$ with the meaning of forces and torques affecting the rigid body. Because in this part of model there are no dynamics, this relation can be expressed by purely static Equations (3). In the controller there is an inserted block with inverse function (8).

$$n_1 = \sqrt{\frac{2k_M u_2 - k_T l u_3 + k_M l u_4}{4 k_M k_T l}} \tag{22}$$

$$n_2 = \sqrt{\frac{-2k_M u_1 + k_T l u_3 + k_M l u_4}{4 k_M k_T l}} \tag{23}$$

$$n_3 = \sqrt{\frac{-2k_M u_2 - k_T l u_3 + k_M l u_4}{4 k_M k_T l}} \tag{24}$$

$$n_4 = \sqrt{\frac{2k_M u_1 + k_T l u_3 + k_M l u_4}{4 k_M k_T l}} \tag{25}$$

This block together with the first part of the model will act as the linear section with the unitary transfer function (while operating in the working range of rotor drivers). This allows direct use of variables $u_1$ - $u_4$ as inputs to the system and treats those parts of system as linear.

The schema of the system with the controller is on Fig. 3. The classical state controller is modified with a few improvements. The first one has added bias ($u_{4-0}$) to input $u_4$ that has the meaning of thrust needed to keep the quadrotor in hover flight. The next one is adding new state $I_z$ which is integral of error in altitude z. The purpose of this is to achieve a zero steady-state error while hovering, because it is almost impossible to set up the thrust that exactly compensates gravity force.

The design of the state space controller was based on the linearized model of the rigid part of the quadrotor. Equations (6) and (7) were linearized around the working point which is hovering. Following equation applies for the equilibrium point

$$\phi = \theta = \omega_x = \omega_y = 0. \tag{26}$$

To reach this state, the thrust of rotors must compensate the gravity force in hover flight acting in z axis. Thus

$$u_{4-0} = mg. \tag{27}$$

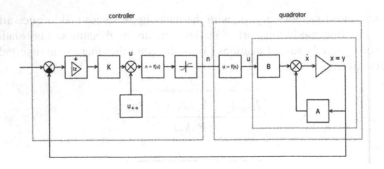

**Fig. 3.** Designed controller

Matrix **A** derived from linearized Equations (4), (5), (6) and (7):

|   | $\omega_x$ | $\omega_y$ | $\omega_z$ | $v_x$ | $v_y$ | $v_z$ | $\phi$ | $\theta$ | $\psi$ | $x$ | $y$ | $z$ |
|---|---|---|---|---|---|---|---|---|---|---|---|---|
| $\omega_x$ | 0 | 0 | 0 | 0 | 0 | 0 | 0 | 0 | 0 | 0 | 0 | 0 |
| $\omega_y$ | 0 | 0 | 0 | 0 | 0 | 0 | 0 | 0 | 0 | 0 | 0 | 0 |
| $\omega_z$ | 0 | 0 | 0 | 0 | 0 | 0 | 0 | 0 | 0 | 0 | 0 | 0 |
| $v_x$ | 0 | 0 | 0 | 0 | 0 | 0 | 0 | −9.81 | 0 | 0 | 0 | 0 |
| $v_y$ | 0 | 0 | 0 | 0 | 0 | 0 | 9.81 | 0 | 0 | 0 | 0 | 0 |
| $v_z$ | 0 | 0 | 0 | 0 | 0 | 0 | 0 | 0 | 0 | 0 | 0 | 0 |
| $\phi$ | 1 | 0 | 0 | 0 | 0 | 0 | 0 | 0 | 0 | 0 | 0 | 0 |
| $\theta$ | 0 | 1 | 0 | 0 | 0 | 0 | 0 | 0 | 0 | 0 | 0 | 0 |
| $\psi$ | 0 | 0 | 1 | 0 | 0 | 0 | 0 | 0 | 0 | 0 | 0 | 0 |
| $x$ | 0 | 0 | 0 | 1 | 0 | 0 | 0 | 0 | 0 | 0 | 0 | 0 |
| $y$ | 0 | 0 | 0 | 0 | 1 | 0 | 0 | 0 | 0 | 0 | 0 | 0 |
| $z$ | 0 | 0 | 0 | 0 | 0 | 1 | 0 | 0 | 0 | 0 | 0 | 0 |

Matrix **B**:

|   | $u_1$ | $u_2$ | $u_3$ | $u_4$ |
|---|---|---|---|---|
| $\omega_x$ | 6.67 | 0 | 0 | 0 |
| $\omega_y$ | 0 | 6.67 | 0 | 0 |
| $\omega_z$ | 0 | 0 | 4.55 | 0 |
| $v_x$ | 0 | 0 | 0 | 0 |
| $v_y$ | 0 | 0 | 0 | 0 |
| $v_z$ | 0 | 0 | 0 | −1.23 |
| $\phi$ | 0 | 0 | 0 | 0 |
| $\theta$ | 0 | 0 | 0 | 0 |
| $\psi$ | 0 | 0 | 0 | 0 |
| $x$ | 0 | 0 | 0 | 0 |
| $y$ | 0 | 0 | 0 | 0 |
| $z$ | 0 | 0 | 0 | 0 |

Matrix C is identity matrix and matrix D is zero. Matrix K that determines control law was obtained experimentally by pole placement method.

Matrix K:

$$
\begin{matrix}
1.05 & 0 & 0 & 0 & 0.29 & 0 & 2.66 & 0 & 0 & 0 & 0.12 & 0 & 0 \\
0 & 1.05 & 0 & -0.29 & 0 & 0 & 0 & 2.66 & 0 & -0.12 & 0 & 0 & 0 \\
0 & 0 & 0.66 & 0 & 0 & 0 & 0 & 0 & 0.44 & 0 & 0 & 0 & 0 \\
0 & 0 & 0 & 0 & 0 & -4.08 & 0 & 0 & 0 & 0 & 0 & -6.32 & -3.06
\end{matrix}
$$

As seen above, the linearized state space description of the system regarded the quadrotor as an independent system in every axis and the controller handles it in this manner. In fact, when looking at altitude control, the state space controller can be understood as a set of separated PI or PD controllers. The results prove that this approach using the space controller and PID controllers gives very comparable results in this mode of flight.

# 5　Software Solution

The software solution of the quadrotor can be divided into several parts (see Fig. 4). Stabilizing algorithms are computed onboard to preserve real-time control and due to operation with no radio signal. Base station, which standardly consists of a PC or laptop, is equipped with the Cassandra system. The system allows the remote control of the group of mobile robots and visualizes various information. Quadrotor can also communicate with another base station, for example for the need of image data processing and other laboratory experiments.

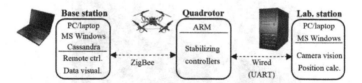

**Fig. 4.** Block diagram describing software solution

## 5.1　ARM Application

The main computing unit of the quadrotor is using the LM3S8962microcontroller which is based on the ARM Cortex-M3 architecture. The control program was written by the development environment Code Composer Studio v4 using C programming language. The microcontroller application does not use any operating system, but all tasks are handled in interrupt routines. To achieve a real-time behaviour, a system timer with the period of 1 ms was used, which is the shortest possible period to handle any application task.

　The main quadrotor stabilization (angular rotations) is computed with a period of 5 ms due to the maximum output frequency of the inertial measurement unit VN-100, which is 200 Hz. With the same frequency, actuators speeds are updated. The sample period of altitude stabilization depends on the used sensor, for example it is period of 25 ms for ultrasonic sensor.

Data from the base station are asynchronously handled by UART interrupt routine and then they are stored in a software ring buffer. For the reason of data synchronization and verification, data is transmitted in defined messages. The physical layer of communication is formed by asynchronous serial interface UART and Zig-Bee modules.

The application also checks special states, for example radio signal lost detection or battery low voltage detection and it is able to perform pre-programmed actions.

## 5.2  Base Station

The base station typically consists of a PC or laptop with a Microsoft Windows TM operating system. The computer must be equipped with a controller (gamepad or joystick), an appropriate communication interface and the Cassandra system [4].

Cassandra is a real-time robot control system for reconnaissance of previously unknown environments through a group of heterogeneous robots. It represents a relatively universal user interface program capable to control various robots in the similar way.

The user interface was designed to allow the operator to concentrate on the mission and to show the relevant information. The operator must not be flooded by the not-so-important data. The user interface, shown on Fig. 5, typically consists of the main camera image with a series of transparent overlay displays with important data.

**Fig. 5.** The screen of the user interface of the Cassandra operator system

## 5.3  Computer Vision System

This project also deals with an automated landing system. The aim is a solution that allows landing on a target. This target can be placed, e.g., on another robot on the ground. The main objective is the recognition and localization of landing mark on the image from the camera. Found coordinates of the mark are used to subsequently compute position stabilization.

In this case a camera is mounted on the flying robot. The better solution is an application with a mechanical stabilization system based on servos. The computer vision system is realized on the board. In the laboratory it is possible to use a more simple configuration. The camera is on the ground, the recognition mark is on the quadrotor and the computer vision system is realized on classic PC.

The landing mark must enable the system to find out a position, orientation and altitude.

The selected mark is comprised of a big square with three sub-squares [5]. The position is calculated by the centre of the big square. Orientation is detectable by sub-squares and the altitude is calculated by the size of the mark in the image.

Image processing includes several operations with the image. It is considered landing only during day. Pre-processing includes an elimination of a disturbance by a convolution mask. The next operation is thresholding. In this case the daylight causes the brightness value of white parts on the mark to reach up to 255. The global threshold value is sufficed. After this, the edges pixels with high gradient are founded (Figure 6). An approximation of the edges causes linking of many lines with the same direction to one solid line. The desired square is described by four orthogonal lines. Lines that are orthogonal to each other are found and corners are stored as a potential area of the landing mark. The landing mark contains three small sub-squares. The method of search is the same as in the previous case, only the area of the application is inside the big square. Around the sub-squares are circumscribed circles. The centres of these circles are also the centres of the squares. The orientation point is the corner close to these squares.

All necessary data are known and the desired information is calculated. The altitude, x, y positions and the orientation are known.

**Fig. 6.** Edges after image processing and sub-squares

# 6     Conclusion

This article described current results of the development of the quadrotor type aerial robot at the Department of Control and Instrumentation of BUT. The aim is to develop a complete robot platform which will be independent from third party products and solutions.

The applied aluminium construction is suitable for testing thanks to its strength and variability, but it will be replaced by a carbon construction in the final version because of its low weight. The created control board, equipped with the ARM Cortex-M3 microcontroller, provides sufficient computing performance for various stabilizing algorithm testing and hardware resources to work with miscellaneous peripherals.

The robot can be remotely controlled from the base station and integration to the robotic system Cassandra will allow it to participate in various robotic missions in the group of mobile robots in the future. The camera position system which is currently tested in the laboratory should facilitate autonomous landings.

The paper also briefly described the simplified mathematical model of the quadrotor which was divided into two parts. This allowed the linearization of the first part of the model and on its base the state space controller was designed. The complete form of the matrixes A, B, C and D which describe quadrotor behaviour and the form of matrix K which determines control law are shown in the paper. The resulting altitude stabilization behaves very similarly as a set of separated PI or PD controllers.

**Acknowledgments.** The work was supported by the Technology Agency of the Czech Republic under the project TE01020197 "Centre for Applied Cybernetics 3" under project CEITEC - Central European Institute of Technology (CZ.1.05/1.1.00/02.0068-) from the European Regional Development Fund.

# References

1. Gurdan, D., Stumpf, J., Achtelik, M., Doth, K.-M., Hirzinger, G., Rus, D.: Energy-efficient Autonomous Four-rotor Flying Robot Controlled at 1 kHz. In: 2007 IEEE International Conference on Robotics and Automation (2007)
2. Silhan, M.: Odhad orientace UAV v prostoru. Diploma thesis - Czech Technical University in Prague, Faculty of Electrical Engineering (2010)
3. Solc, F.: Modelling and Control of a Quadrocopter. Advances in Military Technology 5(2), 29–38 (2010)
4. Zalud, L., Florian, T., Kopecny, L., Burian, F.: Cassandra - Heterogeneous Reconnaissance Robotic System for Dangerous Environments. In: Proceedings of 2011 IEEE/ SICE International Symposium on System Integration, pp. 1–6 (2011)
5. Sharp, C.S., Shakernia, O., Sastry, S.S.: A vision system for landing an unmanned aerial vehicle. In: IEEE International Conference Proceedings ICRA 2001, vol. 2, p. 1720 (2001)

# Computers Capable of Distinguishing Emotions in Text

Martina Tarhanicova, Kristina Machova, and Peter Sinčák

Department of Cybernetics and Artificial Intelligence, Letna 9, 042 00 Kosice, Slovakia
{martina.tarhanicova,kristina.machova,peter.sincak}@tuke.sk

**Abstract.** Detecting human emotions is an important reserach task in intelligent systems. This paper in the following sections outlines the issue of sentiment analysis with emphasis on recent research direction in emotion detection in text. Firstly, we describe emotions from a psychological point of view. We depict accepted and most used emotional models (categorical, dimensional and appraisal-based). Next, we describe what sentiment analysis is and its interconnection with emotions. We take a closer look at methods used in sentiment analysis taking into consideration emotion detection. Each method will be covered by a few studies. At the end, we propose utilization of emotion detection in the text in human-machine interaction.

## 1    Introduction

Detecting emotions is an interesting and nowadays popular research topic. It connects together the field of humanities with computer science. Affective computing is an interdisciplinary field spanning over computer science, psychology and cognitive science. Psychologists and cognitive scientists supply us with theories behind emotion, such as: What is emotion? What role does it play in thinking? What is the reason behind emotion? Does everybody feel the same about a certain thing? How does emotion affect us in everyday life? On the other hand, researchers from computer science, especially in the field of artificial intelligence, are trying to take advantage of gained knowledge.

What is emotion? Even today the answer for this question is unclear. The problem is that emotion has many rather disparate and often unspecified meanings [9]. However, all modern theorists agree that emotions influence what people perceive, learn and remember and that they play an important part in personality development. We cannot be mistaken by saying that people are driven by emotions to do something or make activities to experience emotions. Such emotional behaviour is a perfect ground for research. Regarding this, a lot of work has been done on emotion analysis in speech and video ([14] (robust speech-based happiness recognition), [23] (speech emotion recognition based on multilinear principal component analysis), [19] (modular neural-SVM scheme for speech emotion recognition using the ANOVA feature selection method), [21]). Speech and video emotion detection together with text emotion detecting could be an interesting cross-connection. We see a great potential in using such contribution in human-machine interaction. Human-machine interaction has got a lot of attention in recent years. It studies a human and a machine in conjunction [17].

The paper is organized as follows. In section 2 we describe widely used emotional models. Subsequently, we describe what sentiment analysis is and its connection to

© Springer International Publishing Switzerland 2015                                                                              57
P. Sinčák et al. (eds.), *Emergent Trends in Robotics and Intelligent Systems*,
Advances in Intelligent Systems and Computing 316, DOI: 10.1007/978-3-319-10783-7_6

emotion detection. Next, we depict three basic approaches to detecting emotion and describe works done in each. In section 3 we describe our approach to the emotion detection in the text in the field of robotics and human-computer interaction. Section 4 concludes the paper.

## 2     Sentiment Analysis and Emotion

Sentiment analysis is one of the hot topics in the field of natural language processing. It incorporates emotion detection as one of its research tasks besides subjectivity detection, classification of polarity and intensity classification. To detect emotion, researchers use a generally known algorithm created for sentiment analysis. To be able to tell which emotion is in a text, we need to know emotion models according to which we can estimate emotion.

### 2.1     Emotion Models

Let's look at emotions from a psychological point of view. According to [7], three major directions to affect computing could be distinguished: categorical/discrete, dimensional and an appraisals-based approach. Despite the existence of various other models, the categorical and dimensional approaches are the most commonly used models for automatic analysis and prediction of affect in continuous input [8].

**Categorical Approach**
The categorical approach claims that there is a small number of basic emotions which are hard-wired in our brains and recognized across the world. Each affective state is classified into a single category Table 1. However, a couple of researchers proved that people show non-basic, subtle and rather complex affective states such as thinking, embarrassment or depression which could be impossible to handle [8]. Assigning text to a specific category can be done manually or by using learning-based techniques.

**Table 1.** Emotion classes according to psychologists

| | Ekman (1973) | Izard (1977) | Plutchik (1980) | Tomkins (1984) | Epstein (1984) | Shaver et al. (1987) | Frijda et al. (1995) | Oatley and Johnson - Laird (1987) |
|---|---|---|---|---|---|---|---|---|
| negative | fear | fear | fear | fear | fear | fear | fear | fear |
| | anger | anger | anger | anger | anger | anger | anger | anger |
| | sadness | distress | sadness | distress | sadness | sadness | sadness | sadness |
| | disgust | disgust | disgust | disgust | - | - | - | disgust |
| | - | contempt | - | contempt | - | - | - | - |
| | - | shame | - | shame | - | - | - | - |
| | - | guilt | - | - | - | - | - | - |
| positive or negative | surprise | surprise | surprise | surprise | - | surprise | - | - |
| positive | joy | joy | joy | joy | joy | joy | happiness/joy | happiness |
| | - | - | acceptance | - | love | love | love | - |
| | - | interest | - | interest | - | - | - | - |
| | - | - | anticipation | - | - | - | - | - |

**Dimensional Approach**

The dimensional approach is based on Wundt's proposal that feelings (which he distinguished from emotions) can be described by the dimensions of pleasantness–unpleasantness, excitement–inhibition and tension–relaxation, and on Osgood's work on the dimensions of affective meaning (arousal, valence, and potency). Most recent models have concentrated on only two dimensions - valence and arousal. Valence (pleasure-displeasure) depicts how positive or negative an emotion is. Arousal (activation-deactivation) depicts how excited or apathetic an emotion is [3].

**Appraisals-Based Approach**

Appraisals-based approach view emotion as a dynamic episode in the life of an organism that involves a process of continuous change in all of its subsystems (e.g., cognition, motivation, physiological reactions, motor expressions and feeling—the components of emotion) to adapt flexibly to events of high relevance and potentially important consequences (adopting a functional approach in the Darwinian tradition) [7].

**Sentiment Analysis**

Let's take an example review:[1]

(1) I am now trying to find words to describe this movie for an hour. (2) I couldn't. (3) You've seen it, or you haven't. (4) It's monumental and outrageously good. (5) The cast is brilliant. (6) The jokes are lovely. (7) The story and the idea behind the movie are beautiful. (8) Especially when you've worked/lived with handicapped people. (9) The music is such a perfect choice, it is unbelievable. (10) I hope this movie makes plenty of people think about how good their life is and how bad it could have been.

Looking closer at the review mentioned above, we get an idea about what sentiment analysis do. We can say sentences (4),(5),(6),(7),(9) express highly positive opinion (explicit opinion) and emotions. Emotions are usually not expressed directly but indirectly, by describing situations (1),(2),(8). At the first sight, sentences (1),(2) seem to depict neutral opinion, but in an emotional level it could be rather disturbing. Sentence (8) implies emotion depending on background knowledge about handicapped people.

There are four major approaches to detecting emotions in text: corpus-based methods, machine learning methods, knowledge-based methods and hybrid methods [3].

**Corpus-Based Methods**

This approach uses emotion lexicon. Lexicon contains weighted scores from training documents which are then used to build an emotion prediction model. Corpus-based classification uses unigrams (bag-of-words). Its key features are that it employs an emotion lexicon with weighted scores from training documents and uses unigrams (i.e. bag-of-words) [3].

In [11] authors present bootstrapping technique for identifying para phrases. They used LiveJournal blog, Text Affect, Fairy Tales and Annotated blog as datasets. To

---

[1] We marked every sentence by number for further reference.

pre-process data, they used Stanford part-of-speech tagger and chunker (identifying noun and verb phrases in the sentences). Using k-window algorithm they identify candidates which contain the target seeds (seeds are from WordNet Affect lexicon). Human judges subsequently evaluate results. This method identifies six emotions (proposed by Eckman): happiness, sadness, anger, disgust, surprise and fear.

In [1] authors used corpus of 1000 English words (English dictionary, children stories). Words are afterwards labelled with 34 different emotions (authors did not mention which one). Data was pre-processed using part-of-speech tagger. The rule-based system was used to extract emotions from input (words, sentences).

Detection emotion on newspaper and news web site headlines was a task (called Affective Text") on SemEval 2007. In [20] authors compare their research with three others participants. They used several knowledge-based 2.2.3 and corpus-based methods (variations of Latent Semantic Analysis). They follow the classification of affective words in WordNet Affect. LiveJournal and annotated blogposts were used as dataset. They conducted inter-tagger agreement studies for each of the six emotions. Blogposts were used to train a Naive Bayes classifier 2.2.2. They worked with five emotions: anger, disgust, fear, joy, sadness, surprise and neutral. At the end of the comparison with SWAT [10], UA [12] and UPAR7 [5] system is done.

In comparison with others studies, in [18] authors tried to identify happiness without previous human annotation. They used blogposts from LiveJournal as data set. Naive Bayes was used as a classifier over unigram features and evaluated the classification accuracy in a five-fold cross-validation experiment. This study revealed an interesting fact that certain hours during day and weekdays have higher happiness content than others.

In [15] authors build a system that uses a semantic role labelling tool to detect emotions within textual information. They are validating their research on English based-text with the possibility to extend their emotion detecting on French and Chinese texts. They created a table with emotional rules using publicly available tools: the semantic labelling tool developed by the Cognitive Computation Group of the University of Illinois at Urbana-Champaign and a web mining engine (as Google). Rules are defined as a combination of some selected adjectives and a verb (combination of subjects and objects with verbs). They work with seven emotions: happiness, sadness, anger, fear, disgust, surprise and neutral.

In [13] authors proposed a linguistic-driven rule-based system for emotion cause detection. They constructed a Chinese emotion cause corpus (from Sinica Corpus - corpus of Mandarin Chinese) annotated with emotions and the corresponding cause of events. They work with five emotions: happiness, sadness, fear, anger and surprise (Turners list).

## Machine Learning Methods

This approach makes use of an annotated corpus to train an emotion classifier. Its key features are that it employs an annotated corpus to train the emotion classifier. It also uses a supervised or unsupervised method to classify emotions and relies on a classifier for emotion detection, i.e. makes use of a classifier [3].

In [4] authors recognize emotions from Czech newspaper headlines. Several algorithms (SVM method with linear kernel, the SVM method with radial kernel, the SVM method with polynomial kernel with two and three degrees of freedom, k-nearest neighbour algorithm, decision trees using the J48 algorithm, Bayes networks, linear regression and linear discriminated analysis) for learning were assessed and compared according to their accuracy of emotion detection and classification of news headlines. The best results were achieved using the SVM (Support Vector Machine) method with a linear kernel. Data was pre-processed to transform texts into vector space and evaluated using 10-fold cross-validation. They worked with six emotions: fear, joy, anger, disgust, sadness and surprise.

In [6] authors detected emotion from suicide notes. Dataset consists of 900 notes (600 for training, 300 for testing). They used machine learning methodology for fine-grained emotion detection using support vector machines. To differentiate between the 15 different emotions present in the suicide notes, they experimented with lexical and semantic features, viz. bags-of-words of lemmas, part-of-speech tags and trigrams and information from external resources that encode semantic relatedness and subjectivity. They measured performance with micro-averaged F-score.

In [22] were detected emotions from suicide notes using maximum entropy classification, in [16] authors presented experiments in fine-grained emotion detection of suicide notes.

**Knowledge-Based Methods**
This approach applies linguistic rules through exploiting the knowledge of sentence structures in conjunction with sentiment resources (e.g. Word-Net Affect and SentiWordNet) for emotion classification. Its key features are that it applies linguistic rules and exploits sentence structures in conjunction with sentiment resources like Word-Net Affect and SentiwordNet. Additionally, it employs keyword spotting [3].

In [2] authors proposed a new approach to emotion detection. Their approach defines a new knowledge base called EmotiNet. They extracted descriptions of situations between family members from ISEAR databank for seven emotions: joy, fear, anger, sadness, disgust, shame and guilt. Subsequently, the examples were POS-Tagged using Treeagger. Within each category, they then computed the similarity of the examples with one another using Ted Pedersen Similarity Package. This score is used to split the examples in each emotion class into six clusters using the Simple K-Means implementation in Weka. The next step was to extract action chains from these examples and assign an emotion to them. They use Robert Plutchik's wheel of emotion and Parrot's tree-structured list of emotions.

# 3    Proposal of System for Detecting Emotions

We propose a system that can detect emotions from text. Data acquisition from social networks, news and blogs is crucial in order to increase our knowledge about what is going on in the world. We propose to employ such a system on social networks and inhuman-machine interaction (Figure 1). As it is depicted on the picture, humans (users) communicate and get information from social networks, rss readers. We propose to program an application which would be able to gather data. Subsequently, the data will be analysed and processed on the cloud. Cloud would be offering the detecting emotion service which would identify following emotions: happiness, sadness, anger, disgust, surprise and fear. After that, users would be able to decide which news they want to know according to the offered emotion. The same principle would work with robots (nao module). Humans would communicate with robots and ask them to tell them news according to the emotion they want. Robots (speech to text) will eventually communicate with an application which acquires information from the web and communicate with the cloud. Finally, the application will send the robot the news depending on the preferences of users. The robot would tell (text to speech) the news and fit motion of its body (motion related to emotion) to the expressed emotion. The last block which we did not describe is learning algorithm. It will be learning habits of each user and, according to that, users would not need to manually do what they used to do. They will become observers on their "personal application".

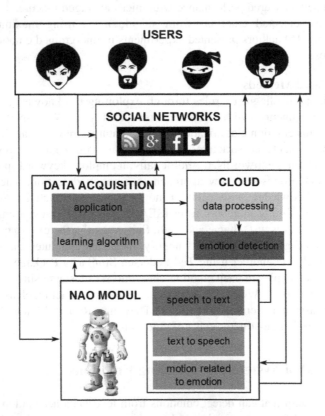

**Fig. 1.** Proposal of System for Detecting Emotions Using Humanoid Robot

# 4    Conclusion

We made an overview about emotions in general. We described three approaches to emotion classes: categorical, dimensional and appraisal-based, from which the first and the second are mostly used, respectively. Exploration of the field of emotion detection has become very popular recently. Regarding the emotion detection, a lot of research has been done in video and speech processing and there is some progress in the field of text processing. We introduced four approaches regarding emotion detection in text: corpus-based, machine learning, knowledge approach and hybrid methods. We described the most relevant works done up until today. At the end we propose utilization of detecting emotions in useful application for human-machine interaction.

**Acknowledgments.** Research supported by the "Center of Competence of knowledge technologies for product system innovation in industry and service", with ITMS project number: 26220220155 for years 2012-2015.

# References

1. Chandak, M.B., Bhutekar, S.: Corpus based Emotion Extraction to implement prosody feature in Speech Synthesis Systems (August 2012)
2. Balahur, A., Hermida, J.M., Montoyo, A.: Building and Exploiting EmotiNet, a Knowledge Base for Emotion Detection Based on the Appraisal Theory Model. IEEE Transactions on Affective Computing 3(1), 88–101 (2012)
3. Binali, H., Potdar, V.: Emotion detection state of the art. In: Proceedings of the CUBE International Information Technology Conference on CUBE 2012, p. 501. ACM Press, New York (2012)
4. Burget, R., Karasek, J., Smekal, Z.: Recognition of Emotions in Czech Newspaper Headlines. Radioengineering 20(1), 1–39 (2011)
5. Chaumartin, F.R.: UPAR7: a knowledge-based system for headline sentiment tagging, pp. 422–425 (June 2007)
6. Desmet, B., Hoste, V.: Emotion Detection in Suicide Notes. Expert Systems with Applications 40(16), 6351–6358 (2013)
7. Grandjean, D., Sander, D., Scherer, K.R.: Conscious emotional experience emerges as a function of multilevel, appraisal-driven response synchronization. Consciousness and Cognition 17(2), 484–495 (2008)
8. Gunes, H., Schuller, B.: Categorical and dimensional affect analysis in continuous input: Current trends and future directions. Image and Vision Computing 31(2), 120–136 (2013)
9. Izard, C.E.: More Meanings and More Questions for the term "Emotion". Emotion Review 2(4), 383–385 (2010)
10. Katz, P., Singleton, M., Wicentowski, R.: SWAT-MP: the SemEval-2007 systems for task 5 and task 14, pp. 308–313 (June 2007)
11. Keshtkar, F., Inkpen, D.: A corpus-based method for extracting paraphrases of emotion terms, pp. 35–44 (June 2010)
12. Kozareva, Z., Navarro, B., Vázquez, S., Montoyo, A.: UA-ZBSA: a headline emotion classification through web information, pp. 334–337 (June 2007)

13. Lee, S.Y.M., Chen, Y., Huang, C.-R.: A text-driven rule-based system for emotion cause detection, pp. 45–53 (June 2010)
14. Lin, C.-H., Siahaan, E., Chin, Y.-H., Chen, B.-W., Wang, J.-C., Wang, J.-F.: Robust speech-based happiness recognition. In: 2013 1st International Conference on Orange Technologies (ICOT), pp. 227–230. IEEE (March 2013)
15. Lu, C.-Y., Hong, J.-S., Cruz-Lara, S.: Emotion Detection in Textual Information by Semantic Role Labeling and Web Mining Techniques. In: Third Taiwanese-French Conference on Information Technology, TFIT 2006 (2006)
16. Luyckx, K., Vaassen, F., Peersman, C., Daelemans, W.: Fine-grained emotion detection in suicide notes: a thresholding approach to multi-label classification. Biomedical informatics Insights 5(suppl. 1), 61–69 (2012)
17. Microsoft Corporation. Being Human: Human-Computer Interaction in the Year 2020. Technical report (2008)
18. Mihalcea, R., Liu, H.: A Corpus-based Approach to Finding Happiness. In: AAAI Spring Symposium: Computational Approaches to Analyzing Weblogs 2006, pp. 139–144 (2006)
19. Sheikhan, M., Bejani, M., Gharavian, D.: Modular neural-SVM scheme for speech emotion recognition using ANOVA feature selection method. Neural Computing and Applications 23(1), 215–227 (2012)
20. Strapparava, C., Mihalcea, R.: Learning to identify emotions in text. In: Proceedings of the 2008 ACM Symposium on Applied Computing, SAC 2008, p. 1556. ACM Press, New York (2008)
21. Waibel, A.H., Polzin, T.S.: Detecting Emotions in Speech
22. Wicentowski, R., Sydes, M.R.: Emotion Detection in Suicide Notes using Maximum Entropy Classification. Biomedical informatics insights 5(suppl. 1), 51–60 (2012)
23. Xin, M.-H., Gu, W.B.: Speech emotion recognition based on multilinear principal component analysis. International Journal of Advancements in Computing Technology 5(8), 452–459 (2013)

# Grammar Representation Forms in Natural Language Interface for Robot Controlling

László Kovács and Péter Barabás

University of Miskolc, Department of Information Technology, Miskolc, Hungary
{kovacs,barabas}@iit.uni-miskolc.hu

**Abstract.** Grammar parsing and transformation of the incoming sentence into a corresponding semantic model are the kernel modules in natural language interface engines. This paper presents an analysis of the candidate grammar representation forms. The proposed grammar formalism is a combination of regular grammar and dependency grammar extended with predicate oriented annotations. The paper presents an implemented natural language interface for controlling humanoid robots in the environment of Hungarian language and similar languages.

**Keywords:** grammar models, human machine interface, natural language processing.

## 1 Introduction

The most important and flexible interface channel between humans and computers is the natural language. The role of the natural language interface (NLI) is steadily increasing in intelligent interface systems. The input source can be either speech or textual data that are converted into corresponding machine commands. The task of the NLI engine is to accept user's command formulated in natural language sentences and convert these commands into low level program function calls. The first proposals [3] on implementation of NLI systems were published in the late 1970s. Nowadays, the web-based and mobile-based client applications [1] are the main markets for NLI engines. The NLI engine integrates different operational units to perform the text to command conversion. The engine contains the following main modules [2]:

— Text parser API unit;
— Morphologic analyzer unit;
— Semantic analyzer unit;
— Function mapping unit.

One of the most important related application areas is the implementation of natural language interfaces for robot controlling. This application area has a significant history and its importance is still increasing. Regarding the pioneer works, Sato and Hirai [4] can be mentioned who implemented a language-aided instruction interface for teleoperational control. The applied language model was based on the keyword

© Springer International Publishing Switzerland 2015
P. Sinčák et al. (eds.), *Emergent Trends in Robotics and Intelligent Systems,*
Advances in Intelligent Systems and Computing 316, DOI: 10.1007/978-3-319-10783-7_7

identification mechanism. Later, Torrance[5] developed a natural language interface for an indoor office-based mobile robot in navigation area. RHINO [6] is a robotic guide in a museum which can move and describe particular exhibits. It can only recognize simple phrases without supporting true dialogues. RHINO's ancestor was Polly, a vision-based robot which interaction mechanisms were more primitive. Jiro-2 [7] is a mobile office assistant which can convey information and guide people through an office environment. A frame-based speech dialog system is used by Jiro-2 and it communicates in Japanese. AESOP 3000 [8] is a surgical robot which is controlled by voice in heart surgery and does not provide a full dialog system. The implemented prototype systems mainly support only the dominating languages. The goal of our project was to implement an NLI module for the Hungarian language and to develop a flexible and efficient language processing model.

## 2      Grammar Representations

The meaning of sentences is partially encoded with syntax of a language. The rules of sentence generation are defined by grammar. The theoretical background of grammar analysis is the theory of formal grammar where a language is defined as a set of sentences:

$$L = \{s^*\}, s \in A \tag{1}$$

where A denotes an alphabet of a language. The elements of the alphabet are called terminal symbols. The grammar describes how to construct valid sentences from the alphabet. The main operators are the

> concatenation (ab) ;
> Kleene star operator (a*) .

The set of related complex structures generated by concatenation operations are usually encoded with non-terminal symbols. The symbol for valid sentence is denoted by S. The grammar consists of production rules in the form

$$\alpha \to \beta$$

where $\alpha, \beta$ are sequences of terminal and non-terminal symbols. The complexity of valid sentences depends on the complexity of the grammar rules. According to the Chomsky classification, the following main grammar classes are defined in the literature:

— regular grammar,
— context-free grammar,
— context sensitive grammar,
— unrestricted grammar.

The main shortcoming of formal grammar formalism is that it ignores the semantic roles of the terminal and non-terminal symbols. As in many applications, as in the case of NLI engines, the management of the semantic context is a key element of the analysis, there are some extensions of the standard formal grammars to include the semantic aspects, too. The Word Grammar proposed by Hudson [9] belongs to this category. Word grammar represents a grammar with a knowledge graph including all four levels of human language, namely the semantic level, the syntax level, the morphology and the phonology.

A different approach is implemented in the Dependency Grammar proposed by Tesniere [10]. Dependency grammar uses dependency description between the words of a sentence. The dependency has a head (verbs) and some dependents. The dependency relation corresponds to grammatical functions. The dependency relationship is described with a dependency tree. A similar formalism is used in the Link Grammar. The link grammar uses only binary relationships, i.e. complex relationships within a sentence are represented with relationships between two words of the sentence.

Considering the standard grammar representations, another shortcoming is the dominance of the sequence-oriented structure. In many dominating languages, the ordering of the different parts of speech units has an important role in encoding the semantics. On the other hand, in many other languages, there is no dominating word order, many different orders are valid. To cope with the free-order approach, some extensions of the standard grammars are proposed.

One of the approaches of this domain is the FAwtl-grammar (Finite Acceptor with translucent letter)[11], where the base regular automaton is extended with some new features. A node in the automaton is assigned to a set of translucent letters. These letters are skipped during the parse operation. The automaton processes the first symbol which is not a member of the current translucency set. The processed symbol is then removed from the string and the head of the automaton jumps back to the beginning of the sentence. Using this technique, it is possible to model some order-free structures.

The conversion of the input word sequence into semantic structure is called function mapping. The goal of this step is to transform sentence analysis into a function description in order to be able to execute the real action related to the description. Two main tasks can be defined to satisfy the function mapping:

1. Find the proper function description to sentence analysis.
2. Map nodes (concepts) of sentence analysis to function parameters.

To be able to map the nodes of the sentence analysis tree into parameters, description should also be completed with some constituent information, as

$$p = (c, \{\kappa\}, f) \tag{2}$$

where c is the concept belonging to the parameter, $\{\kappa\}$ is the set of constituents belonging to the parameter, f is the compulsory function which assigns mandatory value to the parameter. However, there is a problem which was yet not handled properly with the previous approaches: when more instances of the same concepts can be found in the analysis tree, their mapping is ambiguous.

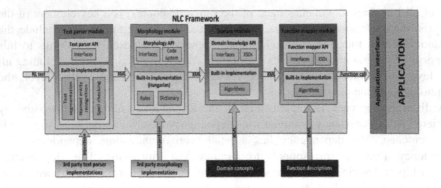

**Fig. 1.** Operational framework of the model

# 3    Efficiency of the Representation Models

The simplest solution is to use a keyword language where sentences consist of only one word. In this case, a simple regular grammar can be used. The cost of the parsing operation depends on the applied search mode, where exact or fuzzy search can be used. In the case of fuzzy matching, some differences are tolerated. Using an exact keyword matching and a corresponding index structure, the parsing requires only O (log(N)) cost, where N denotes the number of keywords in the lexicon. In the case of fuzzy matching, a more complex index structure is used, like the vantage-point structures [12]. In this case, a distance function should be defined to measure the similarity of the words. Usually, the edit distance is selected for dissimilarity measure. The calculation of the edit distance belongs to the O(m2) family, where m denotes the average length of the words.

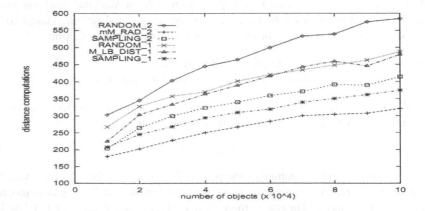

**Fig. 2.** Cost function for NN search (source: [13])

The cost of matching in the index tree is higher than in the case of exact matching; as several paths should be parsed to locate the nearest neighbor element, it can be estimated with $O(\log k(N))$. A good cost analysis for the M-tree variant is given in [13].

The functionality of this approach is very limited, there is no way to submit additional parameters to the operation.

The next complexity level is the application of restricted natural language structure, i.e. only flat structures are allowed with simple POS components. This means that the sentence can be modeled with a star-like dependency graph. This model enables the passing of the required execution parameters and it is simple enough to meet the efficiency requirements. If a fixed order exists in the applied language, this language can be modeled with a regular grammar. Using an exact matching, the cost of parsing is only $O(n\log(N))$, where n denotes the average length. In the proposed prototype system, a free-order language variant was tested, thus a star-graph model was selected as a grammar representation. In this model, the central node is the predicate word which has some dependent POS components. It is assumed that every POS component has a simple linear structure and has a unique semantic role. Based on this simplification assumption, the parsing cost could be reduced to $O(n2\log(N))$. As n is a small number, this representation model provides an efficient execution of the parsing process. The parsing process returns a distance value for the matching of an incoming sentence and a graph model. This distance value describes the degree of similarity. Enabling a fuzzy search for the predicate part, an NN-search operation is required, where the cost value will be equal to $O(n2\log k(N))$.

## 4    Robot Control Application

In the case of Hungarian language, the speech recognition and the speaker-independent ASR are in their childhood and there is no exact, fast and accurate solution which could meet the requirements of real-time usage. There are some by-pass solutions with restricted functionality. Nuance has some promising products like Nuance Recognizer or Dragon which increasingly support Hungarian language. In sample application, the Nuance Recognizer is used to convert speech into written text. Well-defined grammar XML files should be built which describe the word sequences that the robot is able to recognize. The benefit of this representation is that the grammar can be arbitrarily recursive so the number of combinations can be increased. It comes for a certain price: the process of recognition will be slower. The speech recognizer or the keyboard generates the input of the NLC framework which is well-adapted with domain knowledge and function descriptions. The structure of application can be seen in Figure 1.

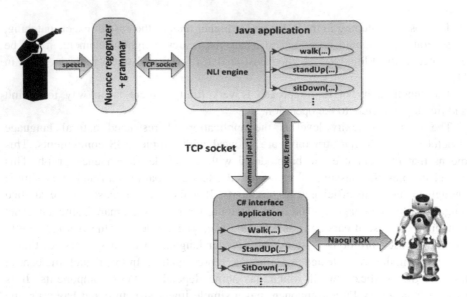

**Fig. 3.** The structure of Robot Controlling Application

The domain knowledge is encoded in a set of available functions. The NLI engine supports 18 standard functionalities of the robot, which includes, among others, the following elements:

— Robot can „close its eyes" which means hiding camera window in application;
— Robot can lay down from any position;
— Robot can walk specified length in any direction.

The applied Aldebaran's Nao[14] humanoid robot has a software development kit which provides an application programming interface in several languages like C++, Python, C#. At the time of development, there was no Java support yet. A by-pass solution had to be found to communicate with Nao from the natural language interface framework written in Java. The solution was to develop an interface application in C# which connects to the robot and calls its API methods. The C# application gets commands via socket communication where the interface application acts as a socket server and the NLC framework has a socket client part.

The socket messages have simple structure which consists of a command name and parameters. A sample socket message for the walk function is as follows:

$$Walk|0.4|forward\#$$

This instructs the robot to walk 40 centimeters forward. The delimiter is 'I' between function names and parameters and the '#' sign is the terminate symbol of the message. The interface application calls the corresponding NaoQi API function and Nao makes the movement.

Based on the performed tests, the implemented architecture provides a suitable NLI framework with basic functionality. Regarding the execution-cost factor, the STT (speech to text) module is the dominating cost component. The generation of the command text required usually 1-2 seconds, while the location of the corresponding semantic graph took only 0.2 seconds. The knowledge base of the test system contains the following concepts:

stand, sit, lay, sorry, wave, nod, exclaim, make excercises, see, close eyes, say hello, walk, turn, introduce, welcome, raise, lower, twist, look, body_part, limb, head, leg, arm, hand, elbow, direction, unit, person, side, forward, backward, left, right, back, up, down, here, around, meter, centimeter, degree, step, Ági, Peti, Szabi, Attila, left , one, right one.

The main drawback of the limited word-set was the lower accuracy factor. The recognized language contains only a small subset of real-life domain, thus, for human commands containing untrained words, the user should reformulate the query command. For the query refinement step, the architecture contains a dialog module. If a command has missing parameters or the application does not understand a command, a question entry will be shown in the conversation panel.

# References

1. Lee, G., Lee, J.H., Rho, H., Park, Y.T., Choi, J., Seo, J.: Interactive NLI agent for multiagent Web search model. In: 4th World Congress on Expert Systems (1998)
2. Barabás, P.: Domain- and Language-adaptable Natural Language Controlling Framework, PhD dissertation, University of Miskolc (2013)
3. Sondheimer, N.K.: Spatial Reference and Natural Language Machine Control. Int. Journal of Man-Machine Studies, 329–336 (1976)
4. Sato, T., Hirai, S.: Language-Aided Robotic Teleoperation System (LARTS) for Advanced Teleoperation. IEEE Journal on Robotics and Automation, 476–480 (1987)
5. Torrance, M.C.: Natural Communication with Robots. Master's thesis. MIT, Department of Electrical Engineering and Computer Science, Cambridge, MA (1994)
6. Burgard, W., Cremers, A.B., Fox, D., Hahnel, D., Lakemeyer, G., Schulz, D., Steiner, W., Thrun, S.: The interactive museum tour-guide robot. In: Proceedings of the Fifteenth National Conference on Artificial Intelligence, Madison, Wi (1998)
7. Asoh, H., Matsui, T., Fry, J., Asano, F., Hayamizu, S.: A spoken dialog system for a mobile office robot. In: Proc. of Eurospeech 1999, pp. 1139–1142 (1999)
8. Versweyveld, L.: March: Voice-controlled surgical robot ready to assist in minimally invasive heart surgery. Virtual Medical Worlds Monthly (1998)
9. Hudson, R.: Language Networks: The new Word Grammar. Oxford University Press (2007)
10. Tesniere, L.: Elements de syntaxe structurale. Paris, Klincksieck (1959)
11. Nagy, B., Kovács, L.: Linguistic Applications of Finite Automata with Translucent Letters. In: ICAART 2013: 5th International Conference on Agents and Artificial Intelligence, Barcelona, vol. 1, pp. 461–469 (2013)

12. Kovács, L.: Reduction of Distance Computations in Selection of Pivot Elements for Balanced GHT Structure. Transaction on Machine Learning and Data Mining 6, 14–37
13. Ciaccia, P., Patella, M., Rabitti, F., Zezula, P.: Indexing Metric Spaces with M-tree. In: Sistemi Evoluti per Basi di Dati SEBD, pp. 67–86 (1997)
14. Aldebaran: NAO. Aldebaran, http://www.aldebaran-robotics.com (retrieved June 6, 2013)

# Basic Motion Control of Differential-Wheeled Mobile Robot ALFRED

Ján Jadlovský and Michal Kopčík

Department of Cybernetics and Artificial Intelligence, Technical University of Košice
jan.jadlovsky@tuke.sk, michal.kopcik@student.tuke.sk

**Abstract.** This article deals with a motion control system of a mobile robot with differential drive. The system for motion control was designed to allow the mobile robot to be driven in several ways, namely by setting speed of the wheels, by setting speed and angular velocity of the robot and drive from origin point to destination along arc of a circle. The essential part of the motion control is the speed regulation of the wheels and the quality of the regulation dependant accuracy of motion control of the mobile robot. Current position of the mobile robot is calculated by using odometry based on the kinematic model of the robot. The motion control system described in this article was applied and tested on a laboratory model of a mobile robot ALFRED (Autonomous Line Following Robot for EDucation).

## 1    Introduction

Mobile robots with differential drive are widespread due to their simple design, motion control and calculation of the position. In this article we will discuss the motion control of mobile robots with differential drive and two wheels. These robots are mostly used for educational purposes and for robotic competitions as line follower competition, micromouse competition, robot soccer, etc. The robots with this design are also used as household appliances (robotic vacuum cleaners).

For educational purposes, there are several models of mobile robots with differential drive on the market, like Khepera I, II, III, Hemisson, E-puck, etc. The motion control that is described in this article is applicable to most of them.

The motion is a fundamental activity of a mobile robot. The method of its control depends, for example, on construction of the robot, on environment in which it will move, or on task that it will perform. In our case, the robot ALFRED's motion control can be logical, but, in order to improve speed and accuracy of motion along the line, we decided to use the motion control using regulation of the speed of wheels.

## 2    Mobile Robot ALFRED

ALFRED is a mobile robot (MR) with differential drive designed especially for the purpose of line following. This MR features two DC motors with built-in gearboxes, five infrared reflective sensors to detect line, an ultrasound distance sensor to detect obstacles and a Bluetooth module to communicate with computer, smartphone or other robot. A picture of MR ALFRED is shown in Fig. 1.

© Springer International Publishing Switzerland 2015
P. Sinčák et al. (eds.), *Emergent Trends in Robotics and Intelligent Systems,*
Advances in Intelligent Systems and Computing 316, DOI: 10.1007/978-3-319-10783-7_8

## 2.1 Mechanics

The basis of mechanics of MR is a chassis made up of two layers of cuprextite between which are two motors, a board with sensors and a battery. The front part of MR is made of plexiglass to see LEDs placed on the sensor board. Wheels are made of two layers of cuprextite with holes used for speed sensing. A rubber O-ring serves as a tire.

## 2.2 Electronics

Electronics of ALFRED consists of two main parts, namely a control board and a sensor board. On the control board, there is a 8-bit microcontroller PIC 16F1827 with 32MHz clock frequency, a voltage regulator, a motor driver, an LED, two buttons, two infrared gap incremental sensors and connectors for ultrasound distance sensor and a Bluetooth communication module.

The sensor board features five infrared reflective sensors for line detecting and six LEDs. All sensors and LEDs are connected to I/O I2C expander PCA9555. A block diagram of MR is shown in Fig. 2.

**Fig. 1.** Picture of mobile robot ALFRED

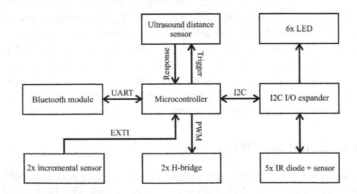

**Fig. 2.** Block diagram of ALFRED's electronics

# 3     Wheel Speed Sensing

To regulate the circumferential speed (CS) of the wheel, we need to know its current speed. This is achieved by two infrared incremental gap sensors placed on control board which give us information about the wheel rotation. Since there is only one sensor per wheel, changes of rotation of the wheel need to be treated by a program. Speed sensing can be achieved in two ways, namely by counting impulses per time constant or by measuring time between two impulses. In the first case, the period of speed reading is constant, but the number of quantization levels is limited by the number of impulses per revolution ($n = 28 imp/rev$), maximal speed of the wheel ($v_{max} = 0.82 m/s$), diameter of the wheel ($d = 0.0243 m$) and by time constant ($T_{vz} = 0.01 s$). Number of quantization levels can be easily calculated using equation (1).

$$n_{kv} = \frac{v_{max} \times n \times T_{vz}}{\pi \times d} \qquad (1)$$

In our case, the number of quantization levels from equation (1) is 3, what is absolutely insufficient for regulation, so we were forced to use a second method. Unlike the first method, the period of sampling is not constant, so the wheel must have a certain speed to achieve the desired sampling rate. This minimal speed ($v_{min}$) can by calculated using equation (2).

$$v_{min} = \frac{\pi \times d}{n \times T_{vz}} \qquad (2)$$

Regulator is an operation even under $v_{min}$, but the quality of regulation decreases proportionally with desired speed. Current speed of the wheel can be calculated using equation (3), where $t_v(k)$ is the measured time between the two latest impulses.

$$v(k) = \frac{\pi \times d}{n \times t_v(k)} \qquad (3)$$

$$v(k) = \frac{\pi \times d \times i(k)}{n \times T_{vz}} \qquad (4)$$

# 4     Wheel Speed Regulation

Circumferential speed (CS) regulation of the wheels is the most fundamental and the most important part of the motion control of MR. The accuracy of movement depends on the quality of CS regulation. This type of a system can be regulated using several types of regulators. In our case, we used discrete PI regulator enhanced by feed forward regulator. The control action is applied to the system by PWM signal, where we control the duty cycle in the range of 0-100%.

The output from the system is the CS of the wheel and the method of its acquiring is described in Sect. 3. A block diagram of the regulator with regulated system is shown in Fig. 3.

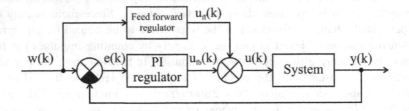

**Fig. 3.** Block diagram of circumferential speed regulator

Control action (CA) consists of two elements - feedforward CA and feedback CA.

$$u(k) = u_{ff}(k) + ufb(k) \tag{5}$$

Feedforward CA is determined by the steady state function $(F_{ff}(x))$ of the system described in Sect. 4.2.

## 4.1 Feedback Regulator

According to the step responses shown in Fig. 4, we can assume that the system we want to regulate is stable and can be approximated in any operating point by a linear differential system with one or two real negative poles.

Such a system can be easily regulated using PI, PS, PID or PSD regulator. In our case, we decided to use a discrete PI regulator because of easy manipulation of its individual components. The calculation of the feedback regulator control action is shown in equations (6) [1].

$$u_{fb}(k) = u_i(k) + u_p(k) \tag{6}$$

$$u_i(k) = u_i(k-1) + e(k) \times I \tag{7}$$

$$u_p(k) = e(k) \times P \tag{8}$$

Parameters of the regulator were determined experimentally using a PI controller tuning map [3].

## 4.2 Feedforward Regulator

Feedforward regulator helps to compensate nonlinearity in the regulated system. This regulator is based on the steady state function that we obtained by averaging the measured data from step responses in steady states. Measured step response characteristics are shown in Fig. 4 [4].

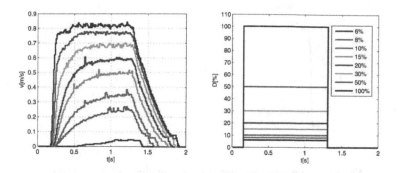

**Fig. 4.** Step response characteristics of the regulated system

After processing the data from step responses, we obtained steady state function shown in Fig. 5. Feedforward control action is described by equation (7)

$$u_{ff}(k) = F_{ss}(w(k)) \tag{9}$$

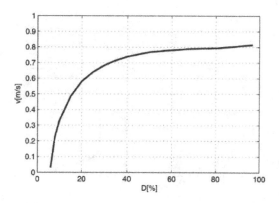

**Fig. 5.** Steady state function of the system $F_{SS}(x)$

## 5    Acceleration and Deceleration (A&D) of MR

The change of reference trajectory of the wheels speed can be step or gradual. With Step change of the desired speed, skidding between the wheel and base may occur, which reduces the quality of motion control, therefore, we decided to use constant A&D which leads to linear speed response. Since we don't want to break motor neither we want to drive it in the opposite direction, we can obtain the value of maximal A&D from step response shown in Fig. 6 [2].

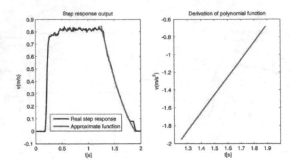

**Fig. 6.** Step response with approximate function and its derivation

From this step response we can see that the falling edge is much milder, so we can obtain the value of maximal A&D by differentiating this edge. In order to get smoother course, we approximated the falling edge with second order polynomial function within its range. The maximum value of A&D is equal to the maximum value of the derivative approximated polynomial function.

In the Fig. 7 we can see a comparison of two responses to linear A&D and step reference trajectory.

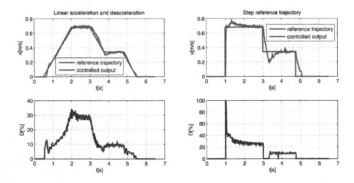

**Fig. 7.** Comparison of two responses of the controlled system to different reference trajectories

# 6    Calculation of the Position of the MR

For the purposes of the point to point motion control, MR must have information about its actual position. The mobile robot ALFRED was designed for motion on a flat surface, therefore we can use odometry to calculate ALFRED's position. This method uses information from incremental sensors to calculate current relative position and angle of MR. Unlike the case of the speed regulation of wheels, we use information about distance travelled by the wheels over time $(\Delta s_r(k), \Delta s_l(k))$. In the first step we need to determine robot's current angle $\varphi(k)[rad]$ using equation ( 8), where $d = 0.063m$ is distance between wheels [2].

$$\varphi(k) = \varphi(k-1) + \frac{\Delta s_r(k) - \Delta s_l(k)}{d} \tag{10}$$

In the second step we calculate current position ($x_R(k)$ and $y_R(k)$) using equations (9).

$$x_R(k) = x_R(k-1) + \cos(\varphi(k)) \times \frac{\Delta s_r(k) + \Delta s_l(k)}{2} \tag{11}$$

$$y_R(k) = y_R(k-1) + \sin(\varphi(k)) \times \frac{\Delta s_r(k) + \Delta s_l(k)}{2} \tag{12}$$

## 7    Point to Point Motion Control along the Arc of a Circle

Mobile robot can get from a current point to a destination point along various trajectories. One of them is, for example, arc of the circle. This method of the motion control is easily enforceable and have many uses. The calculation of reference trajectories for wheel speed regulators is implemented as follows. In the first step we need to calculate relative coordinates ($x'_A(k)$ and $y'_A(k)$) of destination point according to robot's current position and angle ($x_R(k)$, $x_R(k)$ and $\varphi(k)$) using equations (10) [4].

$$x'_A(k) = \sin(\varphi(k)) \times x_A - \cos(\varphi(k)) \times y_A - x_R(k) \tag{13}$$

$$y'_A(k) = \cos(\varphi(k)) \times x_A + \sin(\varphi(k)) \times y_A - y_R(k) \tag{14}$$

In the second step we need to determine the radius ($r(k)$) of circle along which the robot will move

$$r(k) = \frac{x'_A(k)^2 + y'_A(k)^2}{2 \times x'_A(k)} \tag{15}$$

In the next step there is angular velocity of the robot ($\omega(k)$) calculated according to its actual speed ($v(k)$).

$$\omega(k) = \frac{v(k)}{r(k)} \tag{16}$$

In the last step the desired speed of the wheels ($v_r(k)$, $v_l(k)$) are calculated from current speed and angular velocity of the MR using equations (13).

$$v_r(k) = v(k) + \frac{\omega(k) \times l}{2} \tag{17}$$

$$v_l(k) = v(k) - \frac{\omega(k) \times l}{2} \tag{18}$$

# 8    Conclusion

In this paper we have discussed basic motion control of mobile robots with differential drive which we implemented to the laboratory model ALFRED. The advantage of the described control is that it is applicable to most mobile robots with similar constructions and it has relatively low computational demands on the control element, which, in our case, was an 8-bit microcontroller.

We can assume from the output characteristics of regulated speed that used regulator is capable to regulate the given type of system, although in further work we want to test different types of regulators and compare the quality of their regulation.

The improvement of the current motion control of MR might be the implementation of active motor braking which would ensure higher deceleration and could improve the quality of regulation.

# References

1. Krokavec, D., Filasova, A.: Diskrétne systémy. Elfa, Kosice (2006) ISBN 80-8086-028-9
2. Novak, P.: Mobilní roboty - pohony, senzory, řízení. BEN - technická literatura, 1st edn. Prague (2005) ISBN 80-7300-141-1
3. Cooper D. J.: The Challenge of Interacting Tuning Parameters (2006), Available via DIALOG http://www.controlguru.com/wp/p72.html (cited June 15, 2013), Lucas, G. W.: A Path Following a Circular Arc To a Point at a Specified Range and Bearing (2006), Available via DIALOG http://rossum.sourceforge.net/papers/CalculationsForRobotics/CirclePath.htm (cited June 15, 2013)
4. Dolinsky, K., Jadlovska, A.: Implementácia výsledkov experimentálnej identifikácie v riadení výukového modelu helikoptéry. In: Electrical Engineering and Informatics: Proceedings of the Faculty of Electrical Engineering and Informatics of the Technical University of Kosice, Kosice, pp. 546–550 (2010) ISBN 978-80-553-0460-1
5. Dolinsky, K., Jadlovska, A.: Application of Results of experimental Identification in Control of Laboratory Helicopter Model. Electrical and Electronic Engineering, Scientific Reviewed Journal Published in Czech Republic 9(4), 157–166 (2011) ISSN 1804 3119

# Teach Your Robot How You Want It to Express Emotions

## On the Personalized Affective Human-Humanoid Interaction

Mária Virčíková and Peter Sinčák

Department of Cybernetics and Artificial Intelligence, Technical University of Košice
{Maria.Vircikova,Peter.Sincak}@tuke.sk

**Abstract.** We believe that in order for robots to interact naturally with humans, they should be able to express affective behavior. This paper deals with the development of an affective model for social robotics in which the resulting robotic expressions adapt according to the human subjective preferences. We have developed a method which can be used by non-technical individuals to design the affective models of humanoid robots. Our vision of the future research is that the proposed personalization will be treated, from user's perspective, as an empathic response of the machine. We see the major contribution of this unique approach especially in long-term human-robot relationships and it could ultimately lead to robots being accepted in a wider domain.

## 1    Introduction

Machines become social entities and they start to engage in social interactions where the humans and robots interact as companions. We agree with Schaal 1, that the New Robotics (as he names this field) is going to be human-centered, thus it needs to adhere to certain social behaviors and standards that we, as humans, find acceptable. The domain of human-centered robotics prefers an approach which says that a person who is in contact with a robot should not be required to learn a new form of interaction and should enjoy the long-term collaboration with the machine.

This paper proposes a system for socially engaging robots and interactive technologies that provide people with long-term social and emotional support. The goal is to create a personalized cognitive model with emotional aspects which allows the robot to identify a human's cognitive state and then adjust its expressions for enjoyable communication and, in general, to adapt its actions to the human partner. We believe that this personalized approach leads to the simulation of empathy, which may be facilitated through a process of automatic mapping between self and other – in our case between a human and a robot.

Our model will represent empathy as a process composed of perception and, consequently, of reaction of the system. The perception of user's behavior is the assessment of affective state of others - as the system has to recognize user's state.

© Springer International Publishing Switzerland 2015
P. Sinčák et al. (eds.), *Emergent Trends in Robotics and Intelligent Systems*,
Advances in Intelligent Systems and Computing 316, DOI: 10.1007/978-3-319-10783-7_9

System's reactions present robotic reactions which take into account user's actual state. The reactions have the form of affective expressions displayed by humanoid's body as a non-verbal communication.

## 2    Similar Approaches in the World

The field of social robotics can be viewed as an application domain of human-robot interaction which brings a new trend - to program robots that respond and adapt to people's needs accordingly. In the past, most machine earning applications operated 'off–line', where a set of training data would be collected and used to fit a statistical model. Robots or, in general, artificial agents should possess the ability to interpret their interactions with humans, to remember these interactions and to recall them appropriately as a guide for future behavior.

The concept of social robots is not new - the boom started more or less in the nineties. In the last decade, researchers have developed physically embodied mobile robots, such as robotic tour guides, that are meant to interact socially with people. Minerva was an autonomous robot developed at CMU moving daily through crowds in a museum. It used reinforcement learning to adapt appropriately. Kismet was a robot made in the late 1990s at MIT which purpose was to interact socially with people. It used feedback look of affective perception and affective behavior in motion or speech to explore social exchanges with humans. Vikia and Valerie were mobile robots designed for social interaction using social cues in speech and movement to create social responses among people which were made possible through speech and visual recognition, a realistic, developing personality and a rich, friendly dialogue and appearance. Paro, a seal robot designed by Shibata (AIST, Japan), has clearly shown that robot companions can bring some moral and psychological comfort to fragile people.

The common goal of these examples of social robots is to overcome the human-robot social barrier. Researchers construct robots that behave more like people so that people do not have to behave like robots when they interact with them.

We present a software module tested on the Nao platform, but it can be used for any humanoid-type hardware. It has a stage of affect recognition in human, internal state management – emotional dynamics creation in the robot – and the affect expression of the robot. The additional part is a simple system to measure user's interest in interaction with the robot.

The affect recognition systems have a long history. Suwa et al.[24] presented an early attempt in 1978 to automatically analyze facial expressions. Vocal emotion analysis has even longer history starting with study of Williams and Stevens in 1972[23]. Nowadays, multimodal affect recognition systems play an important role. For example, the system of Lisetti and Nasoz[25] combines facial expressions and physiological signals to recognize user's emotions like fear and anger, and then adapts an animated interface agent to mirror user's emotions. The multimodal system of Duric et al.[26] applies a model of embodied cognition that can be seen as a detailed mapping between user's affective states and types of interface adaptations.

Picard[27] arouses considerable awareness of the role of emotions in human-computer interaction due to its numerous potential applications, such as criminal investigation, sickness diagnosis, e-learning homecare alerting system, emotion accoutrement, stress detection system or entertainment. Research of systems with internal architecture based on emotions has experienced many different approaches, e.g.[18 -22]. Their common goal is to integrate emotions into machine control processes and the evolution of artificial, synthetic emotion. The incorporation of an emotional model should improve machine performance – its decision making, action selection, and management of behavior, autonomy and interaction with people.

To evaluate the human-robot interaction, we need appropriate techniques that give the robot (and us as designers of the system) information about people's interest in establishing interaction. Some approaches try to measure mechanical forces and displacements during a physical interaction with a robot, e.g.[33-36]. The second type of systems monitor communication signals from a human. Visual monitoring systems capture video data of the human involved in robot-machine interaction and use this data to guide machine's response to the interaction. This can include visual tracking of user's eye-gaze direction and head position or classifying facial expressions and hand and body gestures, e.g. [37-39].

We were inspired by the work of Salinas et al.[40] and their fuzzy system that establishes a level of possibility of the degree of interest that people around a robot have in interacting with it. The method uses a height map of the environment and a face detector besides Kalman filter to detect and track persons in the surroundings of a robot. Then, the interest of each person is computed using fuzzy logic by analyzing its position and its level of attention to the robot. The level of attention is estimated by analyzing whether the person is looking at the robot or not.

## 3    Design of the Personalized Affective Model

The motivation to construct the model falls under our belief that if a robot can express emotions understandable for its human partner and, more importantly, learns to recognize people's intentions, the robot will act as a better team partner in collaboration tasks. The incorporation of the emotion technology in HRI systems helps to determine goals and communicate internal states of humans and robots. Emotions comprise subjective experience and expressive behavior and different users of the system also need the robot to adapt to their preferences. With this goal, we constructed a model that consists of following parts, as illustrated in Fig. 1.:

- Web interface for the creation of a personalized dictionary where users assign emotional affect to the words.
- Word recognition with assigned emotional meaning for the most common words using modified speech recognition by Microsoft; if a human uses a word that is situated in the dictionary, the robot expresses assigned emotion.
- Giving emotional meaning to a gesture, posture or a motion pattern of people.
- Gesture (or body-based posture/motion) recognition using ARTMAP-like neural networks combined with fuzzy approach; the system recognizes gestures performed by the specific human and reacts according to the stimuli.

- Learning framework for body-based expressions from a human to a robot (user can add body-based emotional expressions to the system without any programming knowledge – a robot imitates users' expressions) – as the source of personalization of body-based expressions of emotions.
- Emotional model where basic emotions are combined into a complex emotional spectrum using Plutchik's psychoevolutionary theory of emotions which combines eight basic emotions to the entire emotional spectrum.
- Visual system for monitoring the interest of people based on fuzzy logic in the interaction with a robot to test the system.

Our Personalized Affective Model (PAM) has two phases. During the first phase, the robot learns its partner's emotional expressions. The second phase consists of the robot adapting the learned emotions to express its state to the interacting human.

### 3.1    The Phase of Emotional Mirroring

During this phase, the robot reflects the emotional expressions of the human partner. The human can see whether the robot performs the expressions correctly. We designed a learning framework for body-based expressions from human to robot, in which a user can add body-based emotional expressions to the system without any programming knowledge.

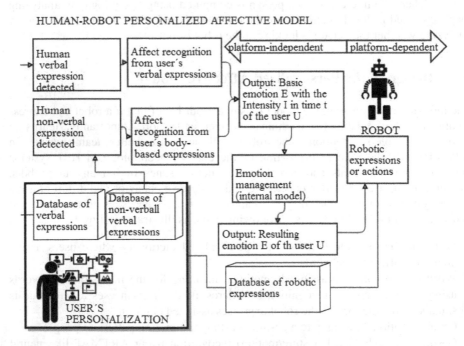

**Fig. 1.** The schematic block of the Personalized Affective Model

The implementation is based on body movement expressions, as the humanoid robot Nao does not have a face capable of mimicking facial expressions. The advantage is that it can be replicated by any other humanoid robot. Humanoid robotics is a challenging area, both in terms of engineering human-like movements and expressions, and in terms of the challenges that arise when a robot takes a human form. With such a form, social and emotional aspects of interaction become paramount as a human expects more from a human-like robot than from other kinds of devices. According to the specific positioning that the body takes during a time frame, studies report that the body-based expressions represent an effective medium for expressing emotion [16],[17]. Many emotions are differentiated by characteristic body movements and these are effective cues for judging emotional states of other people even in the absence of facial and vocal cues.

Each input pattern is a vector where each component represents the position of one point of a human body in time. We use Microsoft Kinect sensor device which features not only an RGB camera but also a depth sensor which provides full-body 3D motion capture capability. To describe an emotional expression, we use a depth sensor which gives us a 3D description of motion in time. To handle the time factor of the expressions, we use a Dynamic Time Warping algorithm (DTW), in general an algorithm for measuring similarity between two sequences which may vary in time or speed. It is based on dynamic programming and is extremely powerful in comparing different signals with minimizing their shift or distortion over time.

**Fig. 2.** Mirroring user's expression of joy

Users can also personalize the verbal part of the communication with the robot. We designed an interface for creation of a personalized dictionary of words where a user assigns an emotional affect based on his/her preferences.

The personalized database of emotional affects for the most common words used in conversation with a humanoid robot is used during the proximate interaction – the audible stimuli. The system identifies a speech command from a user, which is followed by expressed emotion of the robot (for example, if the human assigns the emotional affect of "joy" to the word "mother", the robot performs the expression of joy every time it recognizes the word "mother" during the communication with the human).

**Fig. 3.** A specified user assigns the emotion of "anger" to the word "war." His/her body-based emotional expression will be used to express the emotion of anger. During the "Affective Acting Phase", the robot will perform this emotion every time it recognizes the word "war" during the HRI.

## 3.2   Internal Model

During this phase, we handle user's different affects recognized by the robot in the same time period. The robot has to cope with multimodal information and the perceived emotions can be often contradictory. The work is based on Plutchik's theories and fuzzy logic approach is used for blending of basic emotions to primary mixed emotions. Every input (recognized emotional state of a user) is characterized by three variables: the name of the emotion (e.g. joy), the intensity of the emotion that can be of values between 0 and 1 and the duration of the emotional state.

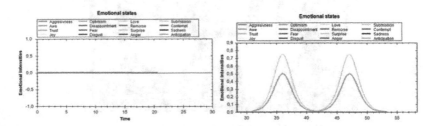

**Fig. 4.** Graph representing the internal model. Left – no emotional state detected. All emotional states are of value 0. Right – emotion of joy of 0.75 intensity and of duration of 5 units of time detected two times. If the primary emotion of joy is detected, the emotion of optimism is also elevated.

The inference system is composed of rules represented by the following code:

```
IS = new InferenceSystem(fuzzyDB, new CentroidDefuzzifier(800));
IS.NewRule("Rule 1", "IF Anger IS Medium AND Anticipation IS Medium THEN
Aggressiveness IS Low");
...
Anger IS High AND Anticipation IS High -> Aggressiveness IS Medium;
Joy IS Medium AND Anticipation IS Medium -> Optimism IS Low";
Joy IS High AND Anticipation IS High -> Optimism IS Medium";
Joy IS Medium AND Trust IS Medium -> Love IS Low";
Joy IS High AND Trust IS High -> Love IS Medium";
Trust IS Medium AND Fear IS Medium -> Submission IS Low";
Trust IS High AND Fear IS High -> Submission IS Medium";
Fear IS Medium AND Surprise IS Medium -> Awe IS Low";
Fear IS High AND Surprise IS High -> Awe IS Medium";
Surprise IS Medium AND Sadness IS Medium -> Disappointment IS Low";
Surprise IS High AND Sadness IS High -> Disappointment IS Medium";
Sadness IS Medium AND Disgust IS Medium -> Remorse IS Low";

Sadness IS High AND Disgust IS High -> Remorse IS Medium";
Disgust IS Medium AND Anger IS Medium -> Contempt IS Low";
Disgust IS High AND Anger IS High -> Contempt IS Medium";
```

The internal model of the system represents the "life and the evolution of the robotic personality" during the time. Every stimulus can increase or lower the corresponding emotion and this forms the memory of the system which develops.

## 3.3    The Phase of Affective Acting

During this phase, the robot uses personalized behaviors to communicate its internal state to the human partner.

The perception of user's state activates the robot's corresponding representations, which, in turn, activate responses. To perceive user's state, the system has to be able to recognize his/her expressions.

To recognize some parts of verbal communication, we use word recognition with assigned emotional meaning using a modified speech recognition algorithm developed by Microsoft. The Kinect sensor captures the word, compares it to the words saved in the database and if a human uses a word that is situated in the dictionary, the robot expresses assigned emotion.

To recognize the non-verbal body-based expressions, we use the gesture (or body-based posture/motion) recognition using ARTMAP-like neural networks combined with fuzzy approach. The recognition of personalized emotion expression is performed using an ART-like neural network, described in detail in our previous paper 41. In general, the network compares a current input (a vector of points of the skeleton parts in time) with a trained class representation (emotion) using a training set (performed representations of emotions of each of the users). When the network receives a feature vector, it deduces the best-matching category by evaluating a distance measured against all memory category nodes. Each new pattern is compared to all of the nodes (categories). Before the learning starts, the memory is empty, which means no categories have been yet recorded. With the first pattern to arrive, the first category is created. The created category (which is in fact the pattern itself) is then saved in the memory. When the second pattern reaches the network, a comparison between the arrived pattern and the saved category(s) is established. The output of the recognition layer of the neural network is of following form, e. g.: Anger of User A with the degree of membership 0.8, where the degree of membership is between the values 0 (does not belong to the category) and 1(certainly is of that category).

To evaluate the proposed method, we designed a set of events which represent the stimuli for the robot to increase or lower the corresponding emotion.

The examples for the model are as following:

- Seeing a familiar face - Joy increases AND Sadness decreases
- Seeing an object the robot knows - Interest increases AND Surprise decreases
- Not seeing anything, in darkness - Fear increases
- Not seeing a human AND feeling a touch - Surprise AND Fear increases
- Seeing a human AND feeling a touch - Trust AND Joy increases

If some of the emotion exceeds a threshold, the emotional expression, learned previously from observing human, is performed.

### 3.4    Monitoring the Interest in the Interaction: Evaluation of the System

We designed a visual system for monitoring the interest of people in the interaction with a robot to test the system. It should help to conduct various experiments to prove whether the mirroring of emotional affects has a positive influence on the human-robot interaction.

Using the camera with the same sensor used for gesture and word recognition, we measure user's attention, which can be, depending on the number of elements found on the face, Low, High or None. The second input variable is the angle of the face, which can be Center, Sideways or N/A. The third input variable is the distance of the user from the robot, simply calculated depending on the size of the face on the screen and can be of values High, Medium and Low.

We calculate the interest of a user in the robot which can be of values Low, Medium, High and Very High. The system can serve as a feedback for the robot to bring information about the characteristics of the interest of human partner in collaboration.

In current research stage, we ask people to teach robots to express their emotions. The preliminary results of the experiment show that even non-technical individuals are able of developing affective behaviors for the robot based on their preferences. We believe that by personalization the service and interactive robots are easier for people to interact with and people report that they enjoy these personalized behaviors.

# 4    Conclusion: Towards Emphatic Robotics

This paper introduces our approach to construct an interactive human – robot system with elements of emotions. The idea is that communication and interaction should be easy and enjoyable, both for unfamiliar users and trained professionals.

There are countless possibilities where a robot which can express empathy could help. Future work will explore the contribution of the simulation of empathy in real environments, measuring and comparing the interest of human subjects in the interaction with and without the elements of artificial empathy. Machines have begun to adapt to their user rather than the other way round, and the interface is becoming a responsive entity rather than a passive portal.

Brothers[2] views empathy as a means of understanding and relating to others by willfully changing one's own emotional and psychological state to mirror the states of others. It is a fundamental mechanism for establishing emotional communication with others.

Inspired by human society, by empathic computing we mean an emergent paradigm that enables a system to understand human states and feelings. We agree with Francis[3] that robots or, in general, artificial systems should possess the ability to interpret their interactions with humans, to remember these interactions and to recall them appropriately as a guide for future behavior. Current research beyond the hypothesis that giving machines the capability of expressing empathic emotions towards users demonstrates its great potential on improving the overall interaction, e.g. in [4],[5].

The main application domain of such systems is in the area of social robotics. This term includes service robots – e.g. tour guide, assistive – e.g. medical care, elderly care-taking, physically impaired and personal robots – e.g. household, companions and entertainment. In our opinion, in the near future the fields like robotics for elder people, rehabilitation robots used in various processes of therapy, robots used in education and entertainment and artistic applications will play a role.

In general, the domain of human-robot interaction (HRI) explores the potential for partner robots to interact with people in daily life. It raised a significant number of studies which report observations of non-technical individuals interacting with robots with the aim to assist people in the daily real-world, e.g. [6-15]. Such projects believe that, during the process of machine migration to the human society, machines will be considered as beneficial and intuitive partners. The future cooperation between machines and us, how we believe it will be, can be summed up with these words: machines fully adapting to man – that man no longer has to adapt his behavior to machines.

**Acknowledgments.** Research supported by the "Center of Competence of knowledge technologies for product system innovation in industry and service", with ITMS project number: 26220220155 for years 2012-2015.

# References

1. Schaal, S.: The New Robotics-towards human-centered machines. HFSP Journal 1(2), 115–126 (2007)
2. Brothers, L.A.: Biological perspective on empathy. Am. J. Psychiatry 6(1), 10–19 (1989)
3. Francis, A.G., Ram, A.: Emotional Memory and Adaptive Personalities
4. Ochs, M., Paris, U., Pelachaud, C., Sadek, D.: An Empathic Virtual Dialog Agent to Improve Human-Machine Interaction, pp. 89–96
5. Prendinger, H., Ishizuka, M.: The Empathic Companion: a Character-Based Interface That Addresses Users Affective States. Applied Artificial Intelligence 19(3–4), 267–285 (2005)
6. Tanaka, F., Movellan, J.R.: Behavior Analysis of Children's Touch on a Small Humanoid Robot: Long-term Observation at a Daily Classroom over Three Months. In: The 15th IEEE International Symposium on Robot and Human Interactive Communication, RO-MAN 2006 (2006)
7. Kozima, H., Nakagawa, C., Kosugi, D., Yano, Y.: A Humanoid in Company with Children. In: 4th IEEE/RAS International Conference on Humanoid Robots (2004)
8. Kanda, T., Ishiguro, H.: Reading human relationships from their interaction with an interactive humanoid robot. In: Orchard, B., Yang, C., Ali, M. (eds.) IEA/AIE 2004. LNCS (LNAI), vol. 3029, pp. 402–412. Springer, Heidelberg (2004)
9. Ros, R., et al.: Child-Robot Interaction in The Wild: Advice to the Aspiring Experimenter. In: Proceedings of the ACM International Conference on Multi-modal Interaction (2011)
10. Tapus, A., Mataric, M., Scassellati, B.: Socially Assistive Robots, The Grand Challenges in Helping Humans Through Social Interaction. IEEE Robotic and Automation Magazine, 35–42 (2007)
11. Pierce-Jordan, S., Lifter, K.: Interaction of social and play behaviors in preschoolers with and without pervasive developmental disorder. Topics in Early Childhood Special Education 25(1), 34 (2005)
12. Hobson, R., Lee, A., Hobson, J.: Qualities of symbolic play among children with autism: A social-developmental perspective. Journal of Autism and Developmental Disorders 39, 12–22 (2009)
13. Robins, B., Dautenhahn, K., Dickerson, P.: From isolation to communication: A case study evaluation of robot assisted play for children with autism with a minimally expressive humanoid robot. In: Int. Conf. Adv. in Computer-Human Interactions, pp. 205–211 (2009), Kozima, H., Nakagawa, C.,Yasuda, Y.: Interactive robots for communicationcare: A case-study in autism therapy. In: IEEE Int. Workshop Rob. and Human Interactive Com., pp. 6–341 (2005)
14. Rasconi, R.: RoboCare: An integrated robotic system for the domestic care of the elderly. In: Proceedings of Workshop on Ambient Intelligence AI*IA-2003, Pisa, Italy (2003)
15. Broadbent, E., Stafford, R., MacDonald Acceptance, B.: of healthcare robots for the older population: Review and future directions. Int. J. Soc. Robot. 1, 319–330 (2009)
16. Kipp, M., Martin, J.C.: Gesture and Emotion: Can basic gestural form features discriminate emotions? In: 3rd International Conference on Affective Computing and Intelligent Interaction and Workshops, ACII 2009 (2009)ISBN 978-1-4244-4800-5
17. Gunes, H., Schuller, B., Pantic, M., Cowie, R.: Emotion representation, analysis and synthesis in continuous space: A survey. In: IEEE Int. Conf. on Automatic Face & Gesture Recognition and Workshops (2011)
18. Bailenson, J.N., Yee, N.: Brave: Virtual Interpersonal Touch: Expressing and Recognizing Emotions Through Haptic Devices. Human–Computer Interaction 22, 325–353 (2007)

19. Gadanho, S., Hallam, J.: Emotion-triggered learning in autonomous robot control. Cybernetics and Systems: An International Journal 32, 531–555 (2001)
20. Davis, D.N.: Modelling emotion in computational agents (2000), http://www2.dcs.hull.ac.uk/NEAT/dnd/papers/ecai2m.pdf (revised: April 15, 2012)
21. Scheutz, M.: Useful roles of emotions in artificial agents: A case study from artificial life. In: Proceedings of AAAI 2004, pp. 42–48. AAAI press (2004)
22. Velásquez, J.: Modeling emotion-based decision-making. In: Proceedings of the 1998 AAAI Fall Symposium, Emotional and Intelligent: The Tangled Knot of Cognition, pp. 164–169 (1998)
23. Williams, C., Stevens, K.: Emotions and Speech: Some Acoustic Correlates. J. Acoustic Soc. of Am. 52(4), 1238–1250 (1972)
24. Suwa, M., Sugie, N., Fujimora, K.: A Preliminary Note on Pattern Recognition of Human Emotional Expression. In: Proc. Int'l Joint Conf. Pattern Recognition, pp. 408–410 (1978)
25. Lisetti, C.L., Nasoz, F.: MAUI: A Multimodal Affective User Interface. In: Proc. 10th ACM Int'l Conf. Multimedia (Multimedia 2002), pp. 161–170 (2002)
26. Duric, Z., Gray, W.D., Heishman, R., Li, F., Rosenfeld, A., Nschoelles, M.J., Schunn, C., Wechsler, H.: Integrating Perceptualnand Cognitive Modeling for Adaptive and Intelligent Human-nComputer Interaction. Proc. IEEE 90(7), 1272–1289 (2002)
27. Picard, R., Vyzas, E., Healy, J.: Toward machine emotional intelligence: Analysis of affective physiological state. IEEE Trans. Pattern Anal. and Machine Intell. 23, 1175–1191 (2001)
28. Davis, D.N., Lewis, S.C.: Computational models of emotion for autonomy and reasoning. Informatica, Special Edition on Perception and Emotion based Reasoning (2003)
29. Picard, R., Vyzas, E., Healy, J.: Toward machine emotional intelligence: Analysis of affective physiological state. IEEE Trans. Pattern Anal. and Machine Intell. 23, 1175–1191 (2001)
30. Siow, S., Loo, C., Tan, A., Liew, W.: Adaptive Resonance Associative Memory for Multichannel Emotion Recognition. In: IEEE EMBS Conf. on Biomedical Engineering & Sciences (2010)
31. Breazeal, C.: Function Meets Style: Insights From Emotion, Theory Applied to HRI. IEEE Trans. on Systems, Man and Cybernetics 34(2) (2004)
32. Kipp, M., Martin, J.C.: Gesture and Emotion: Can basic gestural form features discriminate emotions? In: 3rd International Conference on Affective Computing and Intelligent Interaction and Workshops, ACII 2009 (2009) ISBN 978-1-4244-4800-5
33. Gunes, H., Schuller, B., Pantic, M., Cowie, R.: Emotion representation, analysis and synthesis in continuous space: A survey. In: IEEE Int. Conf. on Automatic Face & Gesture Recognition and Workshops (2011)
34. Maeda, Y., et al.: Human-Robot Cooperation with Mechanical Interaction Based on Rhythm Entrainment. In: ICRA 2001, pp. 3477–3482 (2001)
35. Fernandez, V., et al.: Active Human-Mobile Manipulator Cooperation Through Intention Recognition. In: ICAR 2001, pp. 2668–2673 (2001)
36. Yamada, Y., et al.: Construction of a Human/Robot Coexistence System Based on A Model of Human Will Intention and Desire. In: ICRA 1999, pp. 2861–2867 (1999)
37. Traver, V.J., et al.: Making Service Robots Human-Safe. In: IROS 2000, pp. 696–701 (2000)
38. Matsumoto, Y., et al.: The Essential Components of Human – Friendly Robot Systems. In: Int. Conf. on Field and Service Robotics, pp. 43–51 (1999)

39. Bien, Z.Z., Kim, J.-B., Kim, D.-J., Han, J.-S., Do, J.-H.: Soft Computing Based Emotion / Intention Reading for Service Robot. In: Pal, N.R., Sugeno, M. (eds.) AFSS 2002. LNCS (LNAI), vol. 2275, pp. 121–128. Springer, Heidelberg (2002)
40. Munoz Salinas, R., Aguirre, E., Garcia-Silvente, M., González, A.: Fuzzy System for visual detection of interest in human–robot interaction. In: ICMI 2005 (2005)
41. Vircikova, M., Pala, M., Smolar, P., Sincak, P.: Neural approach for personalized emotional model in HRI. Neural Networks, IJCNN (2012)

# A Talking Robot and Its Autonomous Learning of Speech Articulation for Producing Expressive Speech

Hideyuki Sawada

Kagawa University, Takamatsu-city, Kagawa, Japan
sawada@eng.kagawa-u.ac.jp

**Abstract.** The author is developing a talking robot by reconstructing a human vocal system mechanically based on the physical model of human vocal organs. The robotic system consists of motor-controlled vocal organs such as vocal cords, a vocal tract and a nasal cavity to generate a natural voice imitating a human vocalization. By applying the technique of the mechanical construction and its adaptive control, the robot is able to autonomously reproduce a human-like vocal articulation using its vocal organs. In vocalization, the vibration of vocal cords generates a source sound, and then the sound wave is led to a vocal tract, which works as a resonance filter to determine the spectrum envelope. For the autonomous acquisition of the robot's vocalization skills, an adaptive learning using an auditory feed-back control is introduced. In this manuscript, a human-like expressive speech production by the talking robot is presented. The construction of the talking robot and the autonomous acquisition of the vocal articulation are firstly introduced, and then the acquired control methods for producing human-like speech with various expressions will be described.

## 1 Introduction

Auditory and speech functions play an important role in the human communication, since vocal sounds instantly and directly transmit information using our vocal and auditory organs. In vocal communication, we present verbal information using words and phrases, and also emotional expressions in voices are simultaneously transmitted to listeners, so that smooth and flexible communication would be established among speakers and listeners. Various vocal sounds are generated by the complex movements of speaker's vocal organs, and this mechanism contributes to generate different voices that include speech expressions and speaker's individuality. By realizing the mechanisms of human speech and auditory systems using mechanical systems and computers, a new innovative robotic system will be introduced, which are utilized in the flexible speech communication with humans.

Various techniques have been reported in the research of human speech productions. For example, in the production of human voices, algorithmic syntheses historically have taken the place of analogue circuit syntheses, and became widely used techniques [1]-[4]. Sampling methods and physical model based syntheses are typical techniques these days, which are expected to provide realistic vocal sounds. In addition to these algorithmic synthesis techniques, a mechanical approach using a phonetic or vocal model imitating the human vocal system would be valuable and notable solutions.

Vocal sounds are generated by the relevant operations of the vocal organs such as a lung, trachea, vocal cords, vocal tract, tongue and muscles. The airflow from the lung causes a vocal cord vibration to generate a source sound, and the glottal wave is led to the vocal tract, which works as a sound filter as to form the spectrum envelope of a particular voice. The voice is at the same time transmitted to the auditory system so that the vocal system is controlled for the stable vocalization. Different vocal sounds are generated by the complex movements of vocal organs under the feedback control mechanisms using an auditory system.

The articulation of vocal organs for the appropriate vocalization is acquired by repeating trials and errors of hearing and vocalizing vocal sounds in the age of infants. If any part of the vocal organs or the auditory system is injured or disabled in the age of infants, we might be involved in the impediment in the vocalization. Even in an adult age, we can learn new vocalization skills such as the mimicry of speech of other people and the utterance of foreign words that include different sounds from the mother language.

Mechanical constructions of a human vocal system to realize human-like speech have been reported so far [5]-[7]. In most of the researches, however, the mechanical reproductions of the human vocal system were mainly directed by referring to X-ray images and FEM analysis, and the adaptive acquisition of control methods for natural vocalization have not been considered so far. In fact, since the behaviors of vocal organs have not been sufficiently investigated due to the nonlinear factors of fluid dynamics yet to be overcome, the control of mechanical system has often the difficulties to be established.

The authors are developing a talking robot by reproducing a human vocal system mechanically [8],[9]. An adaptive learning using an auditory feedback control for the acquisition of vocalizing skill is introduced. The fundamental frequency and the spectrum envelope determine the principal characteristics of a sound. The former is the characteristics of a source sound generated by a vibrating object, and the latter is operated by the work of the resonance effects. In vocalization, the vibration of vocal cords generates a source sound, and then the sound wave is led to a vocal tract, which works as a resonance filter to determine the spectrum envelope.

The robot consists of motor-controlled vocal organs such as vocal cords, a vocal tract and a nasal cavity to generate a natural voice imitating a human vocalization. By introducing the auditory feedback learning with an adaptive control algorithm of pitch and phoneme, the robot is able to autonomously acquire the control skill of its mechanical system to vocalize stable vocal sounds imitating human speech. After the learning, the relations between articulatory motions and their produced sounds are established in the robot brain, which means that by listening to a certain vocal sound, the robot estimates its articulatory motion to autonomously generate vocal sounds.

In this study, we try to realize a talking robot that mimics the human-like expressive speech to establish a speech communication with a human. The speech expression is important for the smooth speech communication to transmit emotions to human listeners, and the robotic speech with human-like expressions would realize flexible vocal communication. In the first part of the paper, the structure and the adaptive control method of mechanical vocal cords and vocal tract are briefly described, and then the control method to reproduce human-like speech with various expressions is presented. In this study, the robot recognizes speech expressions given by a human speaker, and reproduces the expression in the robotic speech by articulating its mechanical vocal systems.

# 2    Construction of a Talking Robot

The talking robot mainly consists of an air pump, artificial vocal cords, a resonance tube, a nasal cavity, and a microphone connected to a sound analyzer, which respectively correspond to a lung, vocal cords, a vocal tract, a nasal cavity and an auditory system of a human. The construction and the overview of the talking robot are shown in Figure 1.

An air from the pump is led to the vocal cords via an airflow control valve, which works for the control of the voice volume. The resonance tube as a vocal tract is attached to the vocal cords for the manipulation of resonance characteristics. The nasal cavity is connected to the resonance tube with a rotary valve settled between them for the control of nasal sounds. The sound analyzer plays a role of the auditory system. It realizes the pitch extraction and the analysis of resonance characteristics of generated sounds in real time, which are necessary for the auditory feedback learning and control. The system controller manages the whole system by listening to the vocalized sounds and calculating motor control commands, based on the auditory feedback control mechanism employing neural network learning. The relation between the voice characteristics and mo-tor control commands are stored in the system controller, which are referred to in the generation of speech articulatory motion.

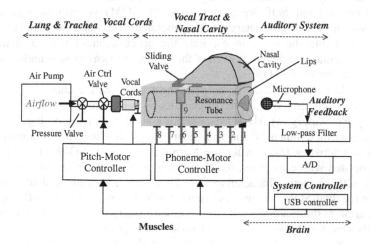

**Fig. 1.** System configuration of the talking robot

The characteristics of a glottal wave which determines the pitch and the volume of human voice are governed by the complex behavior of the vocal cords. It is due to the oscillatory mechanism of human organs consisting of the mucous membrane and muscles excited by the airflow from the lung. We employed an artificial vocal cord used by people who had to remove their vocal cords because of a glottal disease. The vibration of a rubber with the width of 5 mm attached over a plastic body makes vocal sound source. We measured the relationship between the tensile force and the fundamental frequency of a vocal sound gener-ated by the artificial vocal cord, and found that the fundamental frequency varied from 110 Hz to 350 Hz by the manipulations of a force applying to the rubber. The relation between the produced frequency and the applied force was not stable but tended to change with the repetition of experiments due to the fluid dynamics. The artificial vocal cord is, however, considered to be suitable for our system not only because of its simple structure, but also its frequency characteristics to be easily controlled by the tension of the rubber and the amount of airflow. The instability of the vibration would be compensated by employing an auditory-feedback control as humans do in daily speech and singing songs.

## 3    Learning of Vocal Articulations

In this study, we introduce an adaptive learning algorithm of the robotic voice articulations for achieving a talking and singing performance of the robot. The algorithm consists of two phases. First in the learning phase, the system acquires two maps in which the relations between the motor control values and the char-acteristics of generated voices are described. One is a motor-pitch map, which associates motor control values with fundamental frequencies. It is acquired by comparing the pitches of generated sounds with the desired pitches included in speaking phrases. The other is a motor-phoneme map, which associates the con-trol values of vocal tract motors with the phonetic characteristics of words to be generated. Then in the performance phase, the robot gives a speaking perfor-mance by referring to the obtained maps while pitches and phonemes of produced voices are adaptively maintained by hearing its own output voices.

A three-dimensional Self-Organizing Map (3D-SOM) is employed to autono-mously associate vocal tract shapes with generated vocal sounds. The associated relations will enable the robot to estimate the articulation of the vocal tract for generating particular vocal sounds even if they are unknown vocal sounds, owing to the inference ability of the SOM.

The 3D-SOM has three-dimensional mapping space, and the characteristics could be located three-dimensionally, so the probability of miss location could be decreased. In this study, we employ two 3D-SOMs, one for constructing the topo-logical relation among the control commands and the other for establishing the relations of the phonetic characteristics [8-9]. After the learning of two 3D-SOMs, two 3D-SOMs will be associated with each other, based on the topological relations of motor control commands with phonetic characteristics. We call this algorithm a dual-SOM. The structure of the dual-SOM is shown in Figure 2, which consists of two self-organizing maps that arrange the mapping cells three dimensionally. One is a 3D-Motor_SOM that describes the topological relations of various shapes of the vocal tracts, in which close shapes are arranged in close locations with each other, and the other is

3D-Phonetic_SOM, which learns the relations among phonetic characteristics of generated voices. The talking robot generates various voices by changing its own vocal tract shapes. Generated voices and vocal tract shapes have the physical correspondence, since different voices are produced by the resonance phenomenon of the articulated vocal tract. This means that similar phonetic characteristics are generated by similar vocal tract shapes.

By adaptively associating the 3D-Phonetic_SOM with the corresponding 3D-Motor_SOM, we could expect that the talking robot autonomously learns the vocalization by articulating its vocal tract. After the learning of the relationship between the 3D-Phonetic_SOM and the 3D-Motor_SOM, we inputted human voices from microphone to confirm whether the robot could speak by mimicking human vocalization. Figure 3 shows the spectra of vocalized Japanese vowels /a/ and /i/. The first and second formants, which presented the characteristics of different vowel sounds, were formed as to approximate the human vowels, and the sounds were well distinguishable by human listeners. The robot also successfully acquired the other Japanese vowels /u/, /e/ and /o/ and nasal sounds, and the first and second formants were formed as to appear in vowels vocalized by a human. Figure 4, on the other hand, presents the obtained vocal tract shapes for /a/ and /i/ vocalization given by the robot. We verified that all the shapes given by the robot reproduced the actual human vocal tract shapes by the comparison with MR images.

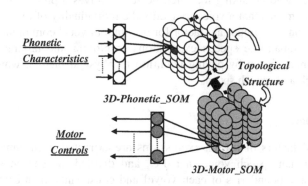

**Fig. 2.** Structure of the dual-SOM

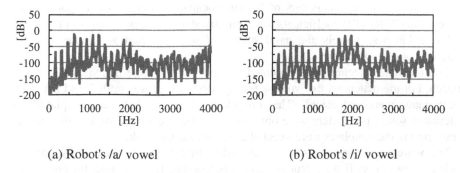

(a) Robot's /a/ vowel                    (b) Robot's /i/ vowel

**Fig. 3.** Spectra of generated vowels

(a) /a/ vocalization          (b) /i/ vocalization

**Fig. 4.** Two different vocal tract shapes given by the talking robot

# 4    Mimicry of Human Expressive Speech

A human generates speech by controlling their own vocal organs not only for just uttering words, but for changing the speech expressions such as the volume, speech speed and the intonation. The speech expression is important in the speech communication to transmit emotions to listeners. These expressions generated by the articulation of vocal organs also present the individuality of a speaker, and a listener instantly recognizes who is talking to, when he just receives a phone call and hears the first voice. This means that a voice transmits the individuality of the speaker, and the speech expression is an important factor for the smooth vocal communication.

To realize the human-like speech expressions employing the talking robot, we firstly studied human speech by paying attention to physical factors, which are the voice volume, pitch and speech speed.

## 4.1    Human Speech

The three physical factors of speech expressions are extracted from human speech. We recorded a human speech voice in a PC, and divided the speech signal into phonemes to find the boundaries of each vowel and consonants, then calculated the volume, pitch and speech speed for each phoneme. A template matching method was employed to find the boundaries of each phoneme. The templates were obtained by calculating the sound parameters of human vocaliza-tion. Tenth-order partial auto-correlation (PARCOR) coefficients were employed as sound parameters in this study, and five Japanese vowels and nasal sound /n/ were selected as templates for the matching of each phoneme.

The template matching was executed to find the phoneme boundaries in human speech. For the matching, the window frame of 64 [msec] was settled, and PARCOR coefficients were calculated. The Euclidean distances between templates and calculated sound parameters were obtained, and the one which marked the smallest value among the templates were selected as the matching result.

The vowels and consonants are distinguished by referring to the average power of each window frame. If the calculated power is less than the threshold, the phoneme is defined as a consonant. On the other hand, if the power is greater than the threshold, the phoneme is defined as a vowel or a nasal sound, and the template matching was executed to recognize the vowel name. Figure 5 shows the analysis results of a speech /kon-nichi-wa/, a Japanese greeting, given by a human. The sampling frequency was set to 8 [kHz] in this study. The abscissa shows the sampling data points, and the ordinate shows the amplitude. As a result, the speech sequence was well partitioned into phonemes by the introduced method using the template matching. The speech speed and the change of the volume are corresponded with the time duration of each phoneme and the amplitude, respectively. In addition, we found that when the power becomes greater, the pitch also becomes higher. This result means that there is a relation between the pitch and the power of a vocal signal.

We examined the vocalizations of a human and the talking robot to validate the relation between the power and the pitch. The randomly uttered voices with changing the volume and pitch given by one subject, together with vocal sounds given by the talking robot with the random operations of vocal articulations were recorded. From the results, both human and the talking robot indicated the similar characteristics, in which the power increases its value in accordance with the higher pitch. With this result, we understood that as the opening and closing values of the airflow control motor were associated with human speech expressions, the talking robot could repro-duce the human-like speech expressions.

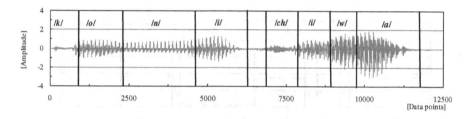

*a)* Sound wave and the result of phoneme separation

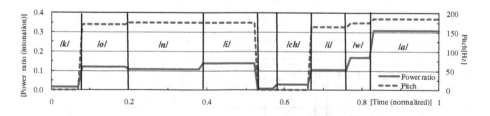

*b)* Temporal change of power and pitch

**Fig. 5.** Analysis results of human speech /kon-nichi-wa/

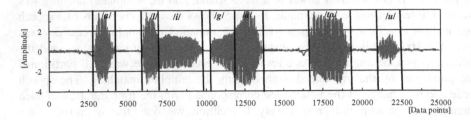

a-1) Without speech expressions /arigatou/

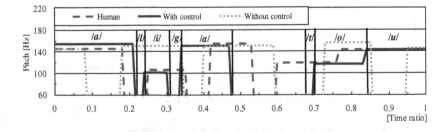

a-2) With speech expressions /arigatou/

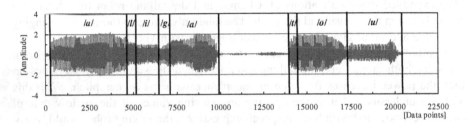

b-1) Change of power in speech /arigatou/

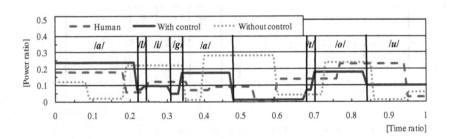

b-2) Change of pitch in speech /arigatou/

**Fig. 6.** Comparison of speech expressions

## 4.2    Reproduction of Human-Like Expressive Speech

The time duration, pitch and power would be controlled to reproduce the human-like speech expressions. The time duration can be managed by the temporal planning of opening and closing the motor for the airflow control, and the power and pitch will be manipulated by the opening and closing values of the airflow motor. The extraction of stable pitch from consonants is not easy due to their non- periodicity, so we pay attention to the signal power. If the appropriate relation between the motor angle and the power is established, the estimation of the motor angle from the human speech would be possible. An experiment for the estimation was conducted, and we calculated the powers of phonemes obtained by the talking robot. A regression formula was obtained by the method of least squares as presented below;

$$P = 0.02 \, \theta + 1.10, \quad 0 < \theta < 110. \tag{1}$$

In the formula, $\theta$ indicates the motor angle and P indicates the power. If the angle becomes 110 degrees or greater, the generated sound becomes unnatural, and we set the maximum angle with 110 degrees.

The motor angle is calculated by referring to the power of phonemes, and the time durations of each phoneme are obtained by measuring the speech speed of a human speech. For human-like speech expressions, the calculated power and time durations are reflected to the talking robot. Figure 6 shows the sound signal of a speech /arigatou/ reproduced by the talking robot. An initial robotic speech without speech expressions is presented in the figure a-1), and we find that all the vowel and consonant vocalizations are made by the equal time durations. On the other hand, the figure a-2) shows the expressive robotic speech, in which the extracted volume and the time scale are reflected. Comparing with a human speech, the sound wave gets similar with each other by the proposed control method. The comparison of the temporal intonation and pitch changes among the three vocalizations are shown in the figures b) and c) respectively, and we find that both characteristics of robotic speech with expressive control get similar to human speech. Furthermore, by adequately controlling the timing of opening and closing the motors, the talking robot has successfully generated the human-like speech expressions.

For the assessment of the expressive speech vocalized by the robot, listening experiments were conducted and the speech was evaluated by questionnaires. Ten able-bodied subjects listened to 3 Japanese speeches with and without expressions given by the robot, which were /kon-nichi-wa/, /arigatou/ and /sayo-nara/, and evaluated them from the viewpoint of the naturalness and their preference. All the subjects preferred the expressive speech, and reported that the clarity of expressive speech is much higher than the one without expressions.

## 5    Conclusions

This paper introduced a talking robot constructed mechanically with human-like vocal chords and a vocal tract. By employing the adaptive learning and control-ling of the mechanical model based on the auditory feedback, the robot success-fully acquired the vocal articulation as a human baby did when he grew up, and autonomously generated vocal sounds whose pitches and phonemes were uniquely specified.

The human-like expressive speech production was also realized and evaluated in the study. The human speech was firstly analyzed, and physical factors for the human-like speech were extracted. The control method of the airflow motor which determines the pitch and volume of a speech was acquired to be given to the robotic speech, and the talking robot successfully reproduced the human-like expressive speech. The mimicry of human speech was evaluated by the listening experiment. All the subjects preferred the expressive speech in comparison with speech without expressions, and reported that the clarity of expressive speech is much higher than the one without expressions.

For the future work, we will study emotional factors from various human speeches, and try to let the talking robot to understand and mimic the expressive speech for realizing smooth speech communication.

**Acknowledgements.** This work was partially conducted by the doctoral study project of the late Dr. Mitsuki Kitani. He passed away one year later from his receipt of the doctoral degree. The author would like to dedicate this paper to Dr. Kitani for his great contribution to the research outcome.

# References

1. Flanagan, J.L.: Speech Analysis Synthesis and Perception. Springer (1972)
2. Rodet, X., Benett, G.: Synthesis of the Singing Voice. In: Current Directions in Computer Music Research. PIT Press (1989)
3. Hirose, K.: Current Trends and Future Prospects of Speech Synthesis. Journal of the Acoustical Society of Japan, 39–45 (1992)
4. Smith III, J.O.: Viewpoints on the History of Digital Synthesis. In: International Computer Music Conference, pp. 1–10 (1991)
5. Fukui, K., Shintaku, E., Honda, M., Takanishi, A.: Mechanical Vocal Cord for Anthropomorphic Talking Robot Based on Human Biomechanical Structure. The Japan Society of Mechanical Engineers 73(734), 112–118 (2007)
6. Sawada, K., Osuka, K., Ono, T.: For the Realization of Mechanical Speech Synthesizer - Proposal of a model of tongue for articulation. Robotic Society of Japan 17(7), 1001–1008 (1999)
7. Miura, K., Asada, M., Yoshikawa, Y.: Unconscious Anchoring in Maternal Imitation that Helps Finding the Correspondence of Caregiver's Vowel Categories. Advanced Robotics 21(13), 1583–1600 (2007)
8. Kitani, M., Hara, T., Hanada, H., Sawada, H.: A Talking Robot and Its Singing Performance by the Mimicry of Human Vocalization. In: Human-Computer Systems Interaction: Backgrounds and Applications Part 2. Advances in Intelligent and Soft Computing, vol. 99, pp. 57–73 (2012) ISSN 1867-5662
9. Kitani, M., Hara, T., Sawada, H.: Voice articulatory training with a talking robot for the auditory impaired. International Journal on Disability and Human Development 10(1), 63–67 (2011)

# Declarative Language for Behaviour Description

Imre Piller, Dávid Vincze, and Szilveszter Kovács

Department of Information Technology, University of Miskolc,
Miskolc-Egyetemváros, H-3515, Miskolc, Hungary
{piller,david.vincze,szkovacs}@iit.uni-miskolc.hu

**Abstract.** The Fuzzy Rule Interpolation (FRI)-based Fuzzy Automaton is an efficient structure for describing complex behaviour models in a relatively simple manner. The goal of this paper is to introduce a novel declarative behaviour description language which is created for supporting special needs of ethologically inspired behaviour model definition. For the sake of simplicity, the grammar is created with as few keywords as possible, keeping the ability to describe complex behavioural patterns as well. The language is a declarative language mainly supporting the behaviour models built upon structures of interpolative fuzzy automata. The paper firstly presents the formal structure of the behaviour description language itself, then gives an overview of the interpreting and processing engine designed for the language. Finally, an application example, a definition of a set of behaviours and a simulated environment is also presented.

## 1 Introduction

The idea of ethologically inspired behaviour models is originated from the existence of descriptive verbal models based on numerous observations of living creatures' reactions in different situations. Many of them are built around a predefined sequence of environmental situations and events where the reactions are noted and evaluated in detail. The knowledge representation, in this case, is a series of observations of various facts and action-reaction rules. For mathematical modelling of such a system, the rule-based knowledge representation is straightforward. Moreover, the need of following the observed sequences requires a model structure of a state machine or an automaton. Adding the fact that the observable or hidden states are continuous measures, the situation is quite complicated. Summarizing the above requirements, the model has to have a rule-based knowledge representation and the ability of describing event sequences of continuous values and continuous states. The suggested modelling method is the adaptation of the fuzzy automaton where the state is a vector of membership values, the state-transitions are controlled by a fuzzy rule-base and the observations and conclusions are continuous values.

The goal of this paper is to introduce a declarative behaviour description language supporting the simple definition of ethologically inspired behaviour models, i.e. various structures of fuzzy automata.

## 2    FRI-Based Fuzzy Automaton

There are numerous versions and understandings of the fuzzy automaton which can be found in literature (a good overview can be found in [1]). In our case, we started from the most common definition of Fuzzy Finite-state Automaton (FFA, summarized in [1]), where the FFA is defined by a tuple (according to [1], [2] and [3]):

$$\tilde{F} = (S, X, \delta, P, O, Y, \sigma, \omega),$$

where

- $S$ is a finite set of fuzzy states, $S = \{\mu_{s1}, \mu_{s2}, \dots, \mu_{sn}\}$.
- $X$ is a finite dimensional input vector, $X = \{x_1, x_2, \dots, x_m\}$.
- $P \in S$ is a fuzzy start state of $\tilde{F}$.
- $O$ is a finite dimensional observation vector, $O = \{o_1, o_2, \dots, o_p\}$.
- $Y$ is a finite dimensional output vector, $Y = \{y_1, y_2, \dots, y_l\}$.
- $\delta: S \times X \to S$ is a fuzzy state-transition function which is used to map current fuzzy state into the next fuzzy state upon an input value.
- $\sigma: O \to X$ is an input function which is used to map the observation to the input value. See, e.g., Figure 1.
- $\omega: S \times X \to Y$ is an output function which is used to map the fuzzy state and input to the output value. See, e.g., Figure 1.

In case of fuzzy rule-based representation of the state-transition function $\delta: S \times X \to X$, the rules have $n + m$ dimensional antecedent space and $n$ dimensional consequent space. Applying the classical fuzzy reasoning methods, the complete state-transition rule-base size can be approximated by the following formula:

$$|R| = n \cdot i^n \cdot j^m, \tag{1}$$

where $n$ is the length of fuzzy state vector $S$, $m$ is the input dimension, $i$ is the number of the term sets in each dimension of the state vector and $j$ is the number of the term sets in each dimensions of the input vector.

According to (2), the state-transition rule-base size is exponential with the length of the fuzzy state vector and the number of the input dimensions. Applying FRI methods to the state-transition function, the fuzzy model can dramatically reduce the rule-base size (see, e.g., [4]).

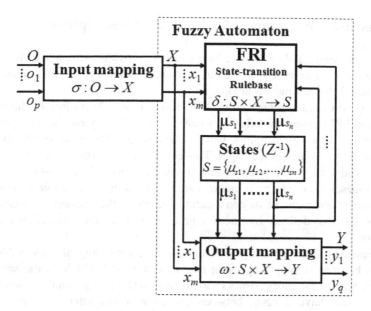

**Fig. 1.** FRI-based Fuzzy Automaton

# 3    Behaviour Modelling

In this paper, we consider behaviour modelling as a task of setting up a behaviour-based control system (see, e.g., [5] for an overview). In behaviour-based control the actual behaviour of the system is formed as a fusion of component behaviours appeared to be the most appropriate in the actual situation. According to this concept, the behaviour model is built upon the component behaviours, the behaviour coordination and the behaviour fusion. The behaviour coordination decides which component behaviour is needed, i.e. it calculates the weights for each component behaviours required in solving the actual situation. Then, the behaviour fusion combines the component behaviours to form the actual behaviour according to their corresponding weights. Turning the suggested FRI-based Fuzzy Automaton to be a behaviour model, the output function $\omega : S \times X \to Y$ has to be decomposed to parallel component behaviours and an independent behaviour fusion. In this case, we can get a very similar structure as it is expected in a behaviour-based control. (More detailed description of the suggested Fuzzy Automaton-based behaviour model structure can be found in [6]).

For setting up the structure of the behaviour model by a declarative language, the set of component behaviours and the state-transition rule-base are needed to be defined. In this case, the component behaviours act as symbols and the relation of the state-transition rules forms the model structure.

## 4     Declarative Language for Behaviour Description

The task of the suggested declarative language is to provide a simple way for defining the state-transition rule-base size in a human readable form close to the original verbal form, as the ethological models are given.

The proposed model is built from various rule-bases. According to the common decomposed FRI models, the rule-bases can have an arbitrary number of input values (antecedents) and a single output value (consequent). The state of the system is determined by the values of the edges. This enables the description of the structure of the system without restrictions to the inner implementation of the rule-bases.

All rule-bases must possess a unique name; also the name of a rule-base needs to be the same as the name of its consequent. Therefore, the connections between the rule-bases can be defined by these unique names of the antecedents and the consequent.

The presented language is a structured language consisting of various blocks. A block can be opened simply by using any valid keyword which, at the same time, defines the type of the block. Blocks should be closed with the 'end' keyword. Some types of blocks have a name between quotation marks after the type of block keyword. All types of blocks can optionally contain a 'description' keyword which is followed by a documentation comment. Depending on the type, the contents of the blocks are slightly different. Considering the block definitions presented hereinafter, our goal was to make them verbose and readable by humans.

In order to provide a formal description of the proposed behaviour description language, we present syntax diagrams. The text element marks a quoted string which can contain arbitrary characters except the quote character itself. Hereinafter, the language will be introduced in a top-down manner.

The most important type of block is the 'rule-base' type (Figure 4).

At the start of the rule-base definition a 'method' block (Figure 4) defines which of the supported consequent calculation methods should be used with its corresponding parameters. Currently, two methods are supported – the FRI method called FIVE (Fuzzy rule Interpolation based on Vague Environment, introduced in [7], [8] and [9]) and direct Shepard interpolation [10]. Of course, other methods can be applied in the future.

The 'term' element (Figure 4) is similar to the function call mechanism found in computer programming languages. After defining the name and the type of the term (term_name, term_type), one can set its parameters with an argument list in parenthesis. The 'term' also defines the name for used measures (see Figure 4), which helps to keep the rules simple but verbose. The form of 'term' is a triplet of term name, term type and value list. For example, the "large" triangle (15, 10, 20) denotes a triangle-shaped fuzzy set definition for the large linguistic term. The available parameter types are integer, logical value and string. The possibilities of the parametrization are determined by the exact method selected previously.

The syntax diagram of the antecedent definition can be seen in Figure 4 and the diagram of the consequent definition in Figure 4. The cardinal difference between the two block types is that all antecedents have the 'name' part. The accessible term types depend on the previously selected method ('method' block described above).

The antecedent and consequent definitions attach symbols, which are real values, to the inputs and outputs, the connection between them is defined with the help of the 'rule' keyword. The general form of 'rule' is shown in Figure 4.

The first 'text' defines the name of the output term in the consequent block. The predicates (see Figure 4 for the form of predicates) are given in a form of a list which contains the antecedent name and the term name pairs. A rule explains that the consequent value is the value after the 'rule' keyword when the predicates are true.

# 5     The Interpreter and the Behaviour Engine

The processing of the rule-base files follows the classical method with two stages. The first stage is the tokenization which has the task of extracting the tokens from the raw text input. In this implementation five different token types are considered.

The tokenizer distinguishes two types of string tokens: the keyword and the previously introduced text type. The keywords of the proposed language and also the term types are read as keyword tokens. Other strings are quoted and handled as text types. For parametrization the tokenizer provides numerical and parenthesis token types. There is also an empty type which marks the end of the file.

The language employs a line comment feature which is denoted by a hashmark symbol. In the first step, the comments are eliminated. The tokenizer ignores special or incorrect characters, in fact, those are treated as simple whitespace characters which are semantically important only for the purpose of separating words.

An unterminated text token causes the whole definition file to be invalidated. Error handling is an important part in the processing of the text input.

The second stage of the processing is the parsing or grammar analysis. Parsing is based on the output received from the tokenizer. The language interpreter checks the grammar of the rule-base definition with a recursive parser. All significant structural blocks have a corresponding method for parsing, for example, the whole definition file is read with the read RuleBase method of the Parser class. For the inner elements, these can be called recursively.

The grammar analysis part is more prone to errors than the tokenization steps. In the case when an error occurs, an information from the tokenizer is required for the parser for the location of the error. This is necessary because the parser itself does not have any knowledge about the position of the error in the definition file.

The rule-base definition file is represented by the rule-base class. This class stores the method name and its parameters, the antecedents, the consequent and the rules themselves. It also includes methods for creating the described rule-base object. After the syntax analysis was successfully completed, it is necessary to check the integrity of the rule-base object.

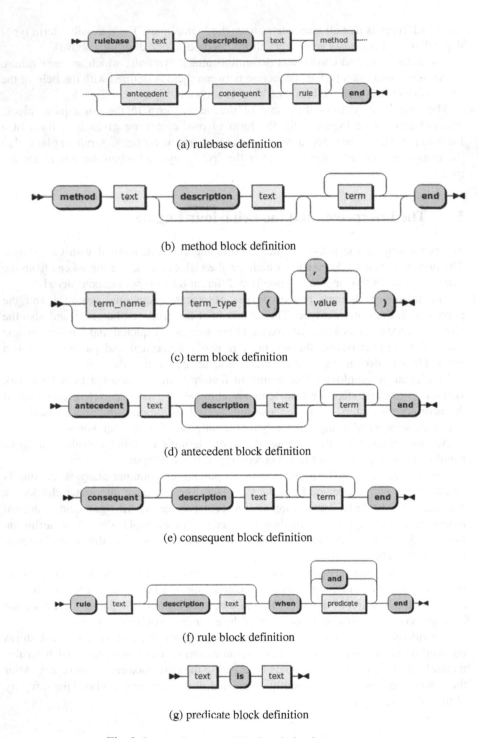

(a) rulebase definition

(b) method block definition

(c) term block definition

(d) antecedent block definition

(e) consequent block definition

(f) rule block definition

(g) predicate block definition

**Fig. 2.** Syntax diagrams of the description language

The following checks are made:

- All the terms in the antecedent are unique.
- All the antecedent names are unique.
- In consequent part, all names are unique.
- The output term in the rule is defined.
- When the predicate is valid, the antecedent and the term are existing.

The rule-base class contains only information which is available from the description. For calculations, the node object should be instantiated according to the properties of the method. It also requires parsing because the meaning and the usage of the method and term function parameters depend on the node type.

When all the rule-base definition files have been processed, the connections between rule-bases are automatically discovered based on the unique names of rule-bases and inputs. The antecedent names which are not used as rule-base names will be the inputs of the system, and the outputs will be the consequences of the rule-bases.

For the calculations to succeed, it is necessary to set the initial values. Omitting initialization can cause problems because inappropriate values possibly render the system unstable.

# 6    Application Example

As a demonstration, a simple behaviour set consisting of three rule-bases is presented. The inputs and outputs of the system are simulated by a small application developed for this purpose. The simulation environment employs an agent which is able to explore its environment (move around). Also, there are rewards distributed in the environment which can be collected by the agent. Furthermore, there are objects in the environment which are considered to be a source of danger for the agent, therefore, these should be avoided by the agent. Based on the described behaviours (rule-bases), the agent is controlled by the behaviour engine. After a certain time of operation, it is necessary for the agent to have a rest for a short while because the tiredness inner state of the agent is increasing while it is moving. The main action type of the agent is the collection of reward objects (the number of collected rewards is counted).

The inputs supplied for the behaviour engine are the distances measured from the closest reward object, distances from the place where the agent can have a rest, distances from the dangerous object, measure of tiredness and the number of collected rewards. These values are registered and provided by the simulation environment.

The above described behaviour is defined by the following rule-bases. The interest rule-base shows the measure of interest for the exploration action. Its value depends on the number of rewards, tiredness of the agent and the distance from the nearest reward object. The second rule-base is the fear, which depends on the distance from the dangerous object and on the value of tiredness. The third rule-base is the activity. Its output determines where the agent should head towards: go to the reward object, go to the resting place, or avoid the dangerous object. The heading direction is determined based on the derived interest and fear values.

According to the dependencies, the interest rule-base has three antecedent definitions. For example, the distance of the reward object looks as follows:

```
antecedent "reward distance"
description "The distance from the closest reward."
"small"    stem(0)
"average"    stem(50)
"large"    stem(100)
end
```

The *stem* in the definition is a control point on a $(n + 1)$-dimensional surface of the rule-base. The output has three possible values, these are defined in the consequent block.

```
consequent description "The measure of the interest."
"low"    stem(0)
"average"    stem(10)
"high"  stem(20)
end
```

The connection between the input and output values can be defined by the following rule blocks:

```
rule "low" when
"reward distance" is "large" and "danger di tance" is
"low"
end
rule "high" when "tiredness" is "low" end
```

For the simulation of the above presented behaviour, an application with a graphical user interface was created (a screenshot of the simulated environment is shown in Figure 3). The demonstration application is available in [11].

In the first frame of Figure 3, the agent is in the centre of the operating area. The small yellow circles denote the rewards to be collected and the larger dark yellow circle near the agent is the place for resting. The darker purple points represent the dangerous objects. During the first four steps, the agent collects the rewards in its vicinity. When it arrives at the position in the fifth frame, it gets tired, therefore it goes to its resting place. In the last three steps, the agent collects four rewards at the bottom of the area, then goes back to rest again. This will be the final position because the dangerous objects are too close, hence the remaining rewards won't be collected.

The antecedent definitions for distance-like values are similar to the already presented values in the case of 'reward distance', also the other definitions are easy to read. The output of the activity rule-base determines the current behaviour and the rules are based on the previously calculated internal properties, namely on the interest and the fear. This complex behaviour can be defined with simple and verbose rules. Examples of some rules from the activity rule-base follow:

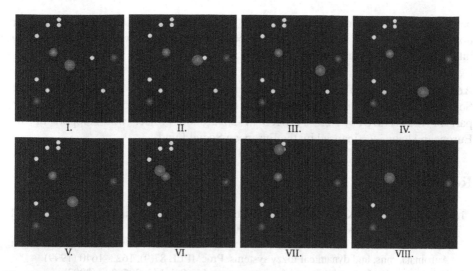

**Fig. 3.** The simulation environment for testing the behaviour descriptions. The path of the agent can be followed on the above stated frame sequence. The agent takes a rest, which can be seen on Frame V. The last frame (VIII) shows the stabilized state.

```
rule "collect reward" when

     "interest" is "high" and "fear" is "low"
end

rule "collect reward" when
     "interest" is "normal" and "count of collected
rewards" is "few"
end

rule "go to rest" when
     "count of collected rewards" is "a lot of"
end
```

## 7    Conclusion

The presented declarative behaviour description language is mainly intended to be used by non-technical users, especially ethologists. An interpreter for the proposed behaviour description language was implemented as a complete behaviour engine with the ability of handling inputs, outputs and states. Currently, for evaluating the behaviour descriptions, a fuzzy rule interpolation-based fuzzy automaton calculation method and direct Shepard interpolation are implemented, but there are no restrictions to the usable calculation methods within the behaviour engine, other methods can be implemented, too. Furthermore, the engine enables the construction of hybrid systems with different types of rule-bases, which allows for the construction of complex behaviour-based systems. Overall, the main goal of the paper is to introduce a tool for non-technical users to describe observed behaviour sets in a way which is directly interpretable by using automated processing.

**Acknowledgments.** This research was supported by the Hungarian National Scientific Research Fund grant no: OTKA K77809. This research was carried out as part of the TAMOP-4.2.2.C-11/1/KONV-2012-0002 project with support of the European Union, co-financed by the European Social Fund.

# References

1. Doostfatemeh, M., Kremer, S.C.: New directions in fuzzy automata. International Journal of Approximate Reasoning 38, 175–214 (2005)
2. Omlin, W., Giles, C.L., Thornber, K.K.: Equivalence in knowledge representation: Automata, rnns, and dynamical fuzzy systems. Proc. IEEE 87(9), 1623–1640 (1999)
3. Belohlavek, R.: Determinism and fuzzy automata. Inf. Sci. 143, 205–209 (2002)
4. Kovács, S.Z.: Interpolative Fuzzy Reasoning in Behaviour-based Control. In: Reusch, B. (ed.) Computational Intelligence, Theory and Applications. Advances in Soft Computing, vol. 2, pp. 159–170. Springer, Germany (2005) ISBN 3-540-22807-1
5. Pirjanian, P.: Behaviour Coordination Mechanisms - State-of-the-art, Tech-report IRIS-99-375, Institute for Robotics and Intelligent Systems, School of Engineering, University of Southern California (October 1999)
6. Kovács, S.: Fuzzy Rule Interpolation in Embedded Behaviour-based Control. In: 2011 IEEE International Conference On Fuzzy Systems (FUZZ-IEEE 2011), Taipei, Taiwan, June 27-30, pp. 436–441 (2011)
7. Kovács, S.: New Aspects of Interpolative Reasoning. In: Proceedings of the 6th International Conference on Information Processing and Management of Uncertainty in Knowledge-Based Systems, Granada, Spain, pp. 477–482 (1996)
8. Kovács, S., Kóczy, L.T.: Approximate Fuzzy Reasoning Based on Interpolation in the Vague Environment of the Fuzzy Rule base as a Practical Alternative of the Classical CRI. In: Proceedings of the 7th International Fuzzy Systems Association World Congress, Prague, Czech Republic, pp. 144–149 (1997)
9. Kovács, S., Kóczy, L.T.: The use of the concept of vague environment in approximate fuzzy reasoning. In: Fuzzy Set Theory and Applications, vol. 12, pp. 169–181. Tatra Mountains Mathematical Publications, Mathematical Institute Slovak Academy of Sciences, Bratislava, Slovak Republic (1997)
10. Shepard, D.: A two dimensional interpolation function for irregularly spaced data. In: Proc. 23rd ACM Internat. Conf., pp. 517–524 (1968)
11. The simulation application is, http://users.iit.uni-miskolc.hu/~piller/

# Usage of ZCS Evolutionary Classifier System as a Rule Maker for Cleaning Robot Task

Tomáš Cádrik and Marian Mach

Department of Cybernetics and Artificial Intelligence, Technical University of Košice
tomas.cadrik@gmail.com, Marian.Mach@tuke.sk

**Abstract.** This paper introduces the Cleaning robot task which is a simulation of the cleaning of a room by a robot. The robot must collect all the junk in the room and put it into a container. It must take out the junk sequentially, because the amount of carried trash is limited. The actions of this robot are selected by using the Michigan style classifier system ZCS. This paper shows the capability of this system to select good rules for the robot to perform the cleaning task.

## 1 Introduction

Learning classifier systems were introduced by John Holland in 1986 [1]. It is a technique which creates classifiers using evolutionary algorithms [2]. Later, Wilson created two successful classifier systems. They were named ZCS [3] and XCS [4] [5]. In these systems, each individual (also called classifier) represents only one rule. This style is called the Michigan style. Another style is called the Pittsburgh style, where each individual contains all rules. The learning classifier systems also have close links to reinforcement learning because they use fitness calculating algorithms which are similar to reinforcement learning techniques.

Classifier systems were tested on many problems. Some of them were single-step problems like the k-multiplexor problem [4]. But these systems show good results especially on multi-step problems like Animat [4] or Corridor problem [6]. This paper focuses on ZCS (Zeroth level classifier system) and on a simple multi-step task which was mentioned in the cleaning robot task (CR) introduced in this paper. The CR task consists of a room which must be cleaned by the robot and all of the junk scattered in the room must be put into a container. The usage of ZCS to solve the CR task can show the ability of this method on a multi-step problem which is more difficult than the Animat problem, because the robot must pick up junk and simultaneously take out this junk periodically and put it into the container, because the robot can carry only a limited amount of it.

This paper is organized as follows: First, the ZCS classifier system is described. In section III the CR task is introduced. Section IV contains experiments which show the capability of ZCS to solve the CR task.

© Springer International Publishing Switzerland 2015
P. Sinčák et al. (eds.), *Emergent Trends in Robotics and Intelligent Systems*,
Advances in Intelligent Systems and Computing 316, DOI: 10.1007/978-3-319-10783-7_12

## 2    Description of ZCS

ZCS was introduced in [3]. Each ZCS individual contains a rule consisting of conditions and action parts. The condition part reflects a particular state(s) of the environment. It is coded with symbols {0, 1, #}, where # means 'does not care', it is equal to 0 and 1 at the same time. The action part can be coded with any symbols.

ZCS also contains a genetic algorithm that is used to discover new individuals and a covering operator that is employed when no condition part matches the input. The following figure shows the schematic illustration of ZCS.

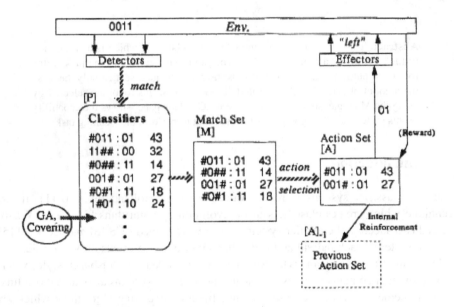

**Fig. 1.** Schematic illustration of ZCS by Wilson [2]

The ZCS starts with a randomly generated population [P]. Each individual starts with fitness equal to S0. First, the environment returns string corresponding to the current state. Then, ZCS makes a match set [M] to the input string. If [M] does not contain any classifier, the covering operator is activated. The covering operator creates a new classifier which matches the input and has a random action part. Then, each element in the condition part is changed with probability P # to # symbol. The initialization value of fitness is set to the average of fitness in the population. This classifier is inserted into the population and into [M]. Then, a classifier is deleted from the population according to the conversion of the fitness (1/fitness).

According to the fitness, a roulette wheel is used to choose an action from [M]. Then, each classifier from [M], which action is equal to the selected action, is copied to action set [A]. Each classifier in [M] that is not in [A] will update its fitness according to formula 1.

$$fitness_j = fitness_j - fitness_j * \tau \tag{1}$$

$\tau$ is a parameter that contains values from interval (0, 1>. Then, for each classifier in [A], a value will be calculated according to formula 2.

$$value_j = fitness_j * \beta \tag{2}$$

$\beta$ can contain values from the same domain as $\tau$. Each classifier in [A] will decrease its fitness with the calculated value. Then, the bucket B is calculated using the following formula.

$$B = \sum_{j=1}^{n} value_j \tag{3}$$

When the environment returns a reward, each classifier in [A] will update his fitness according to formula 4.

$$fitness_j = fitness_j + \beta * \frac{reward}{|A|} \tag{4}$$

Where |A| is the number of classifiers in [A]. Each classifier in the previous action set updates its fitness according to formula 5.

$$fitness_j = fitness_j + \gamma * \frac{B}{|Aprev|} \tag{5}$$

B is the value in the bucket, |Aprev| is the cardinality of the previous action set and $\gamma$ is a parameter and has value from interval (0, 1>.

The last part of ZCS is the evolutionary algorithm. It is applied within each cycle with probability $\psi$. When it appears, two individuals are selected from the population by a roulette wheel according to their fitness. They are used as parents. Two new individuals are created using one point crossover (crossover starts with probability $\chi$) and mutation (probability for each position in the string to mutate is $\mu$). New individuals (children) start with fitness equal to the average fitness of their parents. Then, two individuals are deleted and these two new individuals are added to the population.

## 3    The Cleaning Robot Problem

The environment in the CR problem consists of a robot, junk and a container. The goal is to collect the junk and put it into the container. The robot can carry only a limited amount of junk. So when the robot carries the maximal amount of junk, it must go to the container and then it can continue collecting more junk on the map.

The robot can only see the surroundings and it knows how far the container is from it. The input from the environment is coded using 25 bits. The surrounding of the robot is coded in 16 bits. The first two bits represent the cell to the north from the robot. Each other pair of bits is the clockwise surrounding cells. The pair of bits "00" means that there is a wall. "01" means that there is an empty cell. "10" means that there is some junk in this cell. "11" means that there is a container in this cell. For example, the 16 bits of a robot in Figure 2 will be 0000011010000000.

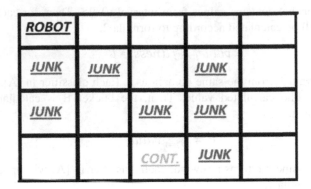

**Fig. 2.** Environment of a Cleaning robot problem

The distance of the robot from the container is coded in 8 bits. The first four bits are for the x distance and the second group represents the y distance. The first bit in these four bits means that the container is left or right from the robot (up or down if it is the y distance). The other three bits are the absolute value of the distance (if the maximal absolute distance is more than 7, we need more bits). In Figure 2, the distance of the robot from the container is 10101011 in bit representation. The last bit of the input string shows whether the robot can carry more junk.

When the robot steps on a cell with junk in it and it can take more junk, it will take it. Then the environment will return payoff 1000. When the robot steps on a cell with a container, it will put all junk it carries into it and the environment will return payoff 1000. If not, the returned payoff is 0.

The robot starts on a start cell (On Figure 2, the start cell is x=0, y=0). It can move in four directions (left, right, up, down). When the robot moves in a direction where there is a wall, it will stay on his position but he will lose one turn. When it puts all junk into the container, the environment will be restarted and the robot will start on its start cell. The goal of this problem is to take all the junk from the environment and put it into the container with usage of a minimum number of steps. The amount of junk doesn't always need to be the same in the environment after each cleaning. Figures show where the junk can appear but whether it appears on this position is random. So after restarting the environment, the amount of trash can be different. ZCS needs to learn how the environment looks and which action is the best for the input returned from the environment even if the environment is not the same for every cleaning cycle.

## 4    Experiments

The experiments were made on the environment displayed in Figure 2. Parameters of ZCS were set to values found in [7]. These values are shown in Tab 1.

**Table 1.** Parameter values of ZCS found in [7]

| Parameter | $S_0$ | $\tau$ | $\psi$ | $\mu$ | $\chi$ | $P_\#$ | $\beta$ | $\gamma$ |
|-----------|-------|--------|--------|-------|--------|--------|---------|----------|
| Value | 20 | 0.1 | 0.25 | 0.02 | 0.5 | 0.33 | 0.2 | 0.4 |

In the first experiment, the maximum of carried junk was set to 1 (optimum is 31). Population was set to 10000. The number of cycles was 3000. The result while using the maximum amount of carried trash set to one is shown in Figure 3.

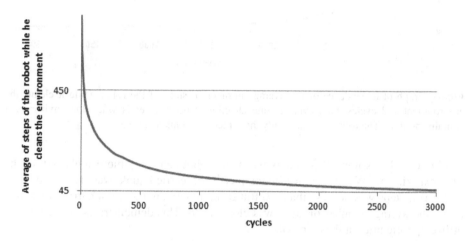

**Fig. 3.** Graph of dependence of the average number of steps of the robot while it cleans the environment and cycles. One cycle is one cleaning process, it ends when the environment contains no junk. The robot can carry only one piece of junk at a time.

In the first cycles the robot moves randomly. It can be seen in the graph, because these cycles have a big amount of steps while it cleans the environment. When the ZCS starts to learn how the environment looks like, the average of steps begins to decrease and the decreasing does not stop until the end of the last cycle.

The experiment shows the capability of ZCS to create rules in a more complicated problem than the animat problem is. It can gradually adapt to a changing environment (when the robot takes junk from a cell, the next time it visits that cell, it is clean) and choose good actions while the environment isn't clean (does not contain any junk). The number of steps is not optimal but the system shows progress in each cycle. Even if the cycles have different numbers of junk, it wasn't any problem for the ZCS.

The robot in the next experiment can carry three pieces of junk at once. The result while using the maximum amount of carried trash set to three (optimum is 15) is shown in Figure 4.

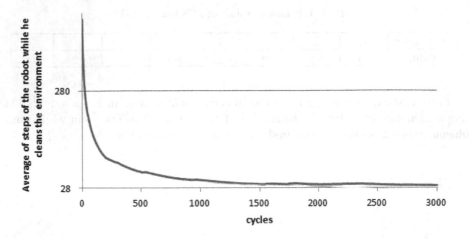

**Fig. 4.** Graph of dependence of the average number of steps of the robot while it cleans the environment and cycles. One cycle is one cleaning process, it ends when the environment contains no junk. The robot can carry only three pieces of junk at a time.

Figure 4 shows how ZCS adapts when the robot can carry more junk than in the first experiment. While the maximum pieces of carried junk was 1, the average number of steps was not less than 47. But if the maximum pieces of carried junk was three, the average number of steps was close to 28. This demonstrates how ZCS can utilize opportunities at the disposal.

## 5    Conclusion

This paper showed how ZCS can handle a multi-step problem named the Cleaning robot problem. Although this paper tested ZCS only on a simulation of a robot in a discrete world, it has a practical use. If we teach a cleaning robot where the most frequent places are, where junk or dirt can show up, we can use this robot to clean these places and it will clean only those where dirt is. Meanwhile, the robot will learn better rules. Some modifications can help ZCS to approve learning in larger and more complicated environments.

## References

1. Holland, J.H.: Escaping brittleness: The possibilities of general-purpose learning algorithms applied to parallel rule-based systems. In: Michalski, R.S., Carbonell, J.G. (eds.) Machine Learning, an Artifiial Intelligence Approach, vol. II, Morgan Kaufmann, Los Altos (1986)
2. Mach, M.: Evolutionary algorithms: Elements and principles. Elfa, Kosice (2009)
3. Wilson, S.W.: ZCS: A Zeroth Level Classifier System. Evolutionary Computation 2(1), 1–18 (1994)

4. Wilson, S.W.: Classifier Fitness Based on Accuracy. Evolutionary Computation 3(2), 149–175 (1995)

5. Butz, M.V., Wilson, S.W.: An Algorithmic Description of XCS. In: Lanzi, P.L., Stolzmann, W., Wilson, S.W. (eds.) IWLCS 2000. LNCS (LNAI), vol. 1996, pp. 253–272. Springer, Heidelberg (2001)

6. Tang, K.W., Jarvis, R.A.: Is XCS Suitable For Problems with Temporal Rewards? In: Proceedings of the International Conference on Computational Intelligence for Modelling, Control and Automation and International Conference on Intelligent Agents, Web Technologies and Internet Commerce (CIMCA-IAWTIC 2006), vol. 2. IEEE (2006)

7. Computer Society, Washington, DC, pp. 258–264 (2005)

8. Cádrik, T.: Evolučné klasifikačné systémy, Diplomová práca. Technická univerzita, Košice, 74s (2013)

# Comparison Study of Robotic Middleware
# for Robotic Applications

Gergely Magyar[1], Peter Sinčák[1], and Zoltán Krizsán[2]

[1] Center for Intelligent Technologies, Department of Cybernetics and Artificial Intelligence,
Technical University of Košice
gergely.magyar6@gmail.com, peter.sincak@tuke.sk
[2] Department of Information Technology, University of Miskolc
krizsan@iit.uni-miskolc.hu

**Abstract.** Developing a robot system is a hard task and it requires a special programming knowledge. Moreover, the development and the operation of every robot system needs common tasks such as state management, communication among parts, synchronization, etc. Several software platforms were introduced for supporting these, hence the researcher and developer can concentrate the novel ideas. This document gives an overview of various robotic middleware. Strictly defined, robotic middleware serves for realizing the communication between various software components. In a wider sense, they are helping the process of development of robotic applications. The document contains descriptions of systems such as Robot Operating System (ROS), RT-Middleware, OPRoS and Orocos. The report also contains a comparison of the above-listed.

## 1 Introduction

There are a lot of manufacturers offering various robotic hardware in the robotic market in recent years. These firms offer different software for their products and make it difficult for developers to create their own applications because they are supposed to use different tools, programming languages, etc. The idea of robotic middleware was introduced to solve this global problem. The robot middleware is a software framework containing tools and libraries that alleviate the development process of robot systems. Generally, the robot middleware has some tools for generating the source code of components and it is glue code that establishes the connection among components. Many projects came up with different architectures, but the component based model was common. It means that robotic applications would look like a complex system built from blocks called components. This architecture has many benefits [1]:

- Reusability – the same component can be used in various robotic systems. When developers need a component, they do not need to implement it from scratch, they can use one which already exists.

© Springer International Publishing Switzerland 2015
P. Sinčák et al. (eds.), *Emergent Trends in Robotics and Intelligent Systems*,
Advances in Intelligent Systems and Computing 316, DOI: 10.1007/978-3-319-10783-7_13

- Editability – the components can be easily edited and then used for different tasks.
- Facilitation of development – when testing algorithms, developers use multiple components. If they want to change a parameter or a part of a system, it is enough to change the parameters of some components. This way they can save time and be more efficient.

The above-listed robotic middleware can be used for creating components using various algorithms for artificial intelligence. These components can be used not only in research areas but also in non-robotic fields, for example, home appliances (such a system was created using RT-Middleware). The application of this idea can create a new section of the robotic market. Besides manufacturers of hardware and software, creators of components can play a significant role.

## 2    RT-Middleware Concept for Robotics

One of the most popular robotic middleware is RT-Middleware, which was created in Japan in 2002 as a project for NEDO (New Energy and Industrial Technology Development Organization). The project's original name was Robot Challenge Program. Its main purpose was to establish essential technologies for easy integration of robot systems having advanced functions of modularized software components [2].

As it was mentioned before, systems based on RT-Middleware consist of components – RT-components or simply RTCs. These components can be written in different programming languages (C++, JAVA, Python, .NET), run on different computers and communicate with each other via CORBA (Common Object Request Broker Architecture).

As shown on Fig. 1, an RT-component has a key part (Activity) which performs different tasks over data. The data processing is realized by input and output ports [3]. The main activity of the component is implemented via state    machine, so the developer has to map the steps of the algorithm into these states: BORN, INITIALIZE, READY, STARTING, ACTIVE, STOPPING, ERROR, ABORTING, FATAL ERROR, EXITING, UNKNOWN [4].

The leading developer tool for creating applications in RT-Middleware is OpenRTM-aist, which is freely available. It has two main tools [5]:

- RTC Builder – serves for creating new RT-components through generating the source code
- RT System Editor – serves for connecting RT-components into an RT-System and controlling the runtime.

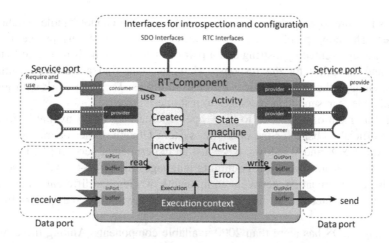

**Fig. 1.** Architecture of an RT-component [5]

OpenRTM-aist supports three programming languages: Python, C++ and Java. Besides OpenRTM-aist, some other implementations exist for the RT-Middleware technology, such as OpenRTM.NET, which uses Microsoft's .NET framework and has support for Visual Basic, C#, F# and C++.

Nowadays, there are over 300 available RT-components. Among these components, we can find RTCs for humanoids (Choromet, HRP) and various tasks using artificial intelligence, such as navigation, face recognition, speech processing.

## 3    ROS Concept for Robotics

Maybe the most popular robotic middleware is ROS, the Robot Operating System. This project is the result of collaboration between Stanford University and Willow Garage, and it was part of the STAIR project. The goals of this system are the following [6]:

- Peer-to-peer
- Tools-based
- Multi-lingual
- Thin
- Free and open-source

The main parts of systems based on ROS are nodes, messages, topics and services. Nodes are processes which are performing various tasks. These nodes are forming complex systems. In this context, we can describe these nodes as components, like RT-components. Nodes are communicating with each other via messages. Messages are data structures which consist of primitive data types, like integer, float, Boolean and others. Messages can be represented as arrays of primitive types or sets of other messages. Nodes send their messages by assigning them to topics, represented as strings. Topics can receive messages from more nodes, which can also send their messages to more topics. This architecture is efficient but cannot handle synchronous processes. The concept of services was introduced to solve this problem. These services are represented by a string and a pair of messages. One is responsible for the request and the other for the response. Here, we can find a similarity with web services. On the other hand, services differ from topics. They can work with just one node and nodes can send their messages to just one service [6].

As it was mentioned above, ROS is open-source and freely available from the project's official website for the operating system Ubuntu. Other operating systems, such as Mac OS X and Windows, have just an experimental version [7]. Developers can also choose from several programming languages like C/C++, Python, Octave or Lisp. The communication between components written in different languages is ensured via IDL (interface definition language), which was created as part of the ROS project.

Nowadays, ROS has more than 2000 available components. Among these, we can find many libraries and ready-to-use components. Robot Operating System has also many components for robots, like Aldebaran's NAO, Willow Garage PR2, Lego NXT, iRobot Roomba, Qbo and others [7].

## 4     OPRoS Concept for Robotics

OPRoS (Open Software Platform for Robotic Services) serves for creating software modules from low level controlling to controlling complex systems. This middleware has its own interface which allows cooperation between components.

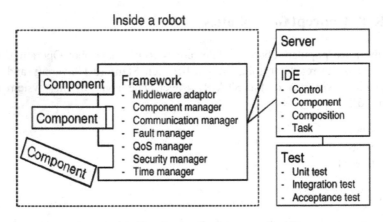

**Fig. 2.** Architecture of an OPRoS system [8]

As shown in Fig. 2, the robot contains a framework which serves for executing and managing components. This framework executes the components periodically or aperiodically, processes activities and errors of the components. It also contains various managers which ensure various functions of the components [8].

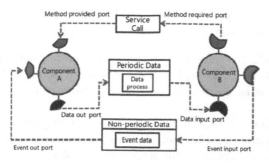

**Fig. 3.** A component in OPRoS [8]

As we can see in Fig. 3, a component in OPRoS, unlike components in RT-Middleware, has three types of ports: data ports, service ports and event ports. Periodic and aperiodic method calls are processed by data and service ports. Components can have more ports of the same type which are communicating with other ports via connectors [8].

The activities of the components are, like in RT-Middleware, realized by a state machine. However, in OPRoS it is simpler than in RTM. This structure is useful for monitoring and debugging the components. It has six states: CREATED, READY, ACTIVE, INACTIVE, ERROR, DESTROYED [8].

Users of the OPRoS platform have access to a server which contains already existing components or users can add their own components. In some cases, this server can be used for direct cooperation with robots. For example, we can execute the OPRoS components on the server and the robots just communicate with them. In this case, the robots are acting as clients. The significant benefit of this feature is that robots with low hardware specifications can also use the OPRoS platform.

Since OPRoS is a relatively young project, it does not have as many available components as RT-Middleware or ROS. Among supported robots, we can mention iRobiQ or FURo [8].

## 5    Orocos Concept for Robotics

Another interesting project for making the development of robotic applications easier is Orocos (Open Robot Control Software). Its primary purpose is to create controlling modules for robots with the following attributes [9]:

- Open-source – source codes are freely available for use and editing
- Modular and flexible – chance to change just parts of the whole system
- High quality – modules have high quality from the technical view, have a comprehensive documentation and as software are robust
- Available for all computer platforms (Linux, Mac OS, Windows)
- Localized for all languages

Orocos uses the divide and conquer method for the development of robotic systems. These are containing libraries and modules. Modules can be [9]:

- Supporting modules – software without robotic elements but needed for a working robotic system. As an example, we can provide simulators, 3D visualization or tools for generating documentation. These modules can be parts of other open-source projects.
- Robotic modules – are specific robotic algorithms implemented as modules, for example, motion planners for mobile robots, kinematics and dynamics of systems, etc. These modules cooperate with supporting modules.
- Components – are CORBA objects needed for creating robotic applications (Fig. 4).

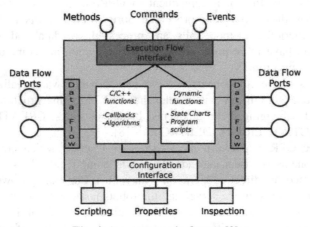

**Fig. 4.** A component in Orocos [9]

For creating components based on the Orocos platform, we need Orocos Toolchain, which is freely available on the project's official site. Toolchain is available on various distributions of Linux (Fedora, Debian, Ubuntu), Mac OS X and Windows (cooperation with Visual Studio). It contains extensions for systems such as ROS, Rock and Yarp. The Toolchain consists of the following parts [10]:

- AutoProj – tool for downloading and compilation of libraries
- Real-Time Toolkit – framework needed for creating components in C++ programming language
- Orocos Component Library – set of components needed for working with the application
- OroGen and TypeGen – tools for generating source codes of components

Among the applications of this framework, we can mention motion specification for robots, software components for an autonomous car, 3D motion tracking or dynamic parameter identification of a robot and others.

# 6    Comparison of Mentioned Robotic Middleware

On the previous pages, we described four robotic middleware, namely RT-Middleware, ROS, OPRoS and Orocos. These platforms can be compared in many ways which are important both from the view of the user and the developer. As the main criteria of the user, we considered the price of the systems, whether they are freely available and what operating systems are supported. From the point of view of the developer, a graphical IDE for managing systems and a simulator for experimenting with the developed systems are important. Table 1 also contains other criteria such as the composite component, which means that we can create one big component containing smaller components. Another important attribute of these systems is independence from the robot software architecture. It means that we can apply the components for robots and it doesn't matter what kind of software is added to them. The real-time criteria describe if we can change the parameters of the components during the run

**Table 1.** Comparison of robotic middleware

|  | RT-Middleware | ROS | OPRoS | Orocos |
|---|---|---|---|---|
| Open-source | yes | yes | yes | yes |
| Windows | yes | no | yes | yes |
| Linux | yes | yes | yes | yes |
| Composite component | yes | no | yes | no |
| Robot software architecture independent | yes | yes | yes | yes |
| Graphical IDE | yes | no | yes | no |
| Simulation environment | yes – OpenHRP, Choreonoid | yes – Stage, RVIZ | yes – OPRoS Simulator | no |
| Real-time | yes | no | planned | yes |

As we can learn from Table 1, all robotic middleware have their advantages and disadvantages. Based on this table, RT-Middleware can be considered as the perfect middleware, but it has fewer components than ROS and does not support as many robots as ROS. A big weakness of ROS can be the fact that Windows is not among the supported operating systems and it does not have a graphical IDE. The developers usually like to have a simulation environment for experimenting with the components. From this point of view Orocos, is not a smart choice, on the other hand, it has a lot of useful libraries.

# 7    Conclusion

In this paper, we focused on describing four robotic middleware which can be used for creating robotic applications. These were RT-Middleware, ROS, OPRoS and Orocos. In Chapter 6, using a table, we compared them through various attributes. Based on this table, we can choose a platform which is the best for various purposes. Besides the above-mentioned platforms, there are many middleware which can be used for the same tasks, such as MIRO, Player/Stage, URBI, ORiN, Orca and others.

**Acknowledgments.** This research was supported by the Hungarian National Scientific Research Fund grant no: OTKA K77809.

# References

1. Ando, N., et al.: RT(Robot Technology)-Component and its Standardization. In: International Joint Conference (2006)
2. Ando, N., et al.: RT-Component Object Model in RT-Middleware – Distributed Component Middleware for RT (Robot Technology). In: International Symposium on Computational Intelligence in Robotics and Automation (2005)
3. Ando, N., et al.: RT-Middleware: Distributed Component Middleware for RT (Robot Technology). In: International Conference on Intelligent Robots and Systems, pp. 3933–3938 (2005)
4. Ando, N., et al.: Composite Component Framework for RT-Middleware (Robot Technology Middleware). In: International Conference on Advanced Intelligent Mechatronics, pp. 1330–1335 (2005)
5. OpenRTM-aist, http://www.openrtm.org
6. Quigley, M., et al.: ROS: An open-source Robot Operating System. In: ICRA Workshop on Open Source Software (2009)
7. ROS Wiki, http://www.ros.org/wiki.
8. Soohee, H.A.N., et al.: Open Software Platform for Robotic Services. In: Transactions on Automation Science and Engineering (2012)
9. Bruyninckx, H.: Open Robot Control Software: The OROCOS Project. In: International Conference on Robotics and Automation, pp. 2523–2528 (2001)
10. The Orocos Project, http://www.orocos.org/

# Communication Engine in Human-Machine Alarm Interface System

Tomáš Lojka[1], Milan Zolota[2], Roman Mihaľ[1], and Iveta Zolotová[1]

[1] Technical University of Košice /Department of Cybernetics and Artificial Intelligence,
Košice, Slovakia
{Tomas.Lojka,Roman.Mihal,Iveta.Zolotova}@tuke.sk
[2] University of Bristol, Faculty of Engineering,
Department of Computer Science, United Kingdom
Milan.Zolota.2012@my.bristol.ac.uk

**Abstract.** Humans need to communicate with the society. If the society is a group of machines, there should be a language that creates a human-machine interface. This interface uses technological sources to communicate with humans. One way to realize an independent human-machine interface is service-oriented architecture (SOA). Services are realized on server and create an independent interface for clients. A client may then be a smart phone, tablet or PC. How the human-machine interface is presented only depends on the technical realization of the client and the functionality of the device. One of the most important things in a human-machine interface is alarms. Alarms directly involve a human user in processes which cannot be understood easily.

## 1 Introduction

People have the need to talk not only with one machine but also with a large complex of machines and robots. A human-machine interface (HMI) is often a subject to variable amounts of error and uncertainty.[1] Supervision Control and Data Acquisition/Human Machine Interface (SCADA/HMI) is used to process communication by visualization of abstract or real objects from the technological layer which consists of robots and machines. A SCADA/HMI system supports visualization and decision-making in real time. Creating an adequate interference needs to have a good base for supplying clients with data. This supplier should not be dependent on one type of technical equipment or operating system of its clients. Different technical equipment has a different ability to realize human-machine interface. The purpose of good human-machine interface is to create an open supplier of data which will cooperate with different and new implantations of human-machine language devices. Human-machine interface should become less a matter of pushing buttons or pulling levers with some physical result, and more a matter of specifying operations and accessing their effects through the use of a common language. [2,3,4]

## 2    Communication Language for Humans and Machine in Interactive Computing

The technical definition of "interactive computing" is simply that the real-time control over computing is placed in the hands of the user through immediate and available interrupt facilities whereby the user can override and modify the operations in the process. [2]

Interactive computing needs a language for human-machine communication which will be adequate for both sides. The best language for humans is a language which is primitive enough to understand. Conveyed information should be compact and easily understandable to support better control of the technological processes and robots. Human cognitive senses can quickly understand graphical objects and text elements with colour decorations. Therefore, making a decision for humans is simpler when all information is readably grouped in front of them. Also the language between a human and a machine consists of graphical objects which are interacting with data.

The data for a human-machine interface is processed by an engine. This engine is supported by a "brain" which gives this engine information in a suitable form to be processed in a human-machine interface. Example of human-machine interface is in Fig. 1.

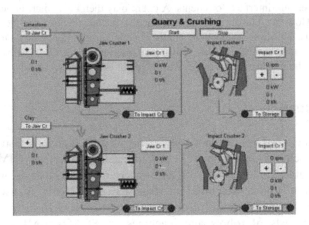

**Fig. 1.** Example of a human-machine interface with sample graphic and text visualization

SCADA/HMI systems find ways to use cognitive perception of human senses. The principle of cognitive action and human reaction is described in Fig. 2. On the left side of the diagram, there are cognitive actions which act on human receptors of senses. The human processes information from the receptors, make a decision and make an action through the visualizing system.[5] For example, a person who did not work with the system must spontaneously find the right button to stop the machine. The engine of human-machine interface accepts commands and sends them to the system with alarms which process all of the alarms and distribute alarms to clients with human-machine interface.

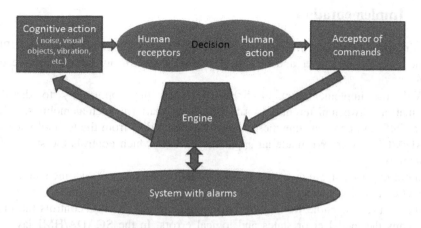

**Fig. 2.** Human-machine interaction scheme

The implementation of a human-machine interface has good application in processes with risk where alarms are important. Alarms request a human to make a decision and make them aware of possible and real problems [6].

## 3    Alarms in Human-Machine Interactions

We made an experiment where we use alarms to analyze alarms in human-machine interfaces and create communication support for alarms. They involve a human into dynamic processes with robots and machines. It is the way how the machine warns and gives them appropriate information to make decisions and solve a problem.

The alarms are used to warn or inform about processes which are problematic. [6]

The active alarms are usually acknowledgements (ACK) by the operator and are actual even if the alarm's value has normalized. Alarms are described by these three states:

- ACK - if the alarm was acknowledged by user.
- ALM - if the alarm has occurred for the first time.
- RTN - if the alarm occurred, but automatically returns back to normal state.[6]

The alarms can have different meanings and importance. The importance of alarms is defined by priority. Then the alarms can be grouped or filtered according to their priority.

Alarms can be also classified differently:

- Tripped alarms – defines the category of alarms where alarms indicate an event of actions, for example when voltage supply for robots goes down accidentally. Tripped alarms have a higher priority than non-tripped alarms. [7]
- Non-tripped alarms – defines the category of alarms that warn the operator that there is an imminent failure.[7]

## 4     Implementation

We simulated alarms in the physical model, Programmable Logic Controller (PLC) in the technological layer and the SCADA/HMI layer. In our experiment we used SOA.

With the implementation of SOA, we used the opportunity to distribute information between different platforms [8]. Every platform, such as mobiles, PC and other platforms, can consume these alarms and other data from the technological and SCADA/HMI layer. We made an application server which controls the states of all errors in our system.

In our experiment, we created a simple model of process control which simulates a magnetic arm that collects metal components from a rotating plate. This model simulates a real machine for classifying things in medicine. PLC controls the model and scans the model error states and logical errors. In the SCADA/HMI layer, we simulated alarms in communication and software base.

The errors are visualized to the client to inform the user about the states in our system. The main error with the highest priority is the error of communication with the application server.

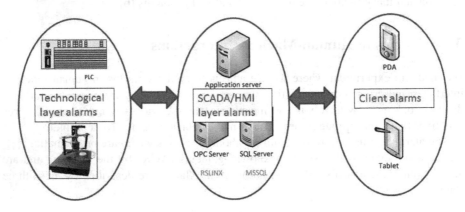

**Fig. 3.** Diagram of an implemented system of alarms

From the technological layer are states of errors written in PLC. For real time data acquisition OPC (OLE for Process Control) standard can be used, more precisely OPC DA (OPC Data Access). [9]

Error states in the OPC, database and application server are processed in SCADA/HMI layer.

In this experiment the application server process all alarms and create communication and data base for the client human-machine interface.

Example of errors is in the Table 1.

**Table 1.** Example of alarms in realized architecture

| Alarm | Layer | Priority |
|---|---|---|
| Component has been lost | Technological | 1 |
| Component has not been caught by arm | Technological | 1 |
| Initialization process has not been run | Technological | 3 |
| Communication error with OPC | SCADA/HMI | 1 |

# 5    WCF Server Architecture

This part of the solution triggers the alarm and is indicated on the user interface. The form of data was designed to achieve the best presentation of alarms in human-machine interface. We developed the application server on SOA using Windows Communication Foundation (WCF). WCF is a unified programming model for building SOA applications [10].

The client connects to the server through an endpoint and sends requests to the server service. The server has defined methods and uses them to process the requested action from clients. After processing the method, the server sends an answer to the client. SOA is used for accessing a Web Service via HTTP and, hence, passes through firewalls. In addition, it allows the communication between applications running on different operating systems with different programming languages. [11]

The server cooperates with the database and the OPC server and does the application logic for remote control. In this application, the server was inserted an additional alarm functionality. Inside the usual communication of commands and answers, the diagnostic communication for error statuses of the whole system has been added. This diagnostics on the server has 3 parts:

- Scanning – an independent process on the server that controls the functionality of the whole system in the technological, logical and communication areas. Each alarm has its own priority. This priority shows the client the importance of the operator's reaction.
- Writing – error states are saved for later diagnostics.
- Signalizing – the main part of alarming is signalizing. In this action, a table of errors if created.

The whole process from getting the errors after their sending starts with invoking an event action of error. In this process, the error is written inside an inner error table. After that, the error and its attributes are written to the database. The error table is sent to the client during the following communication. The client diagnoses this table and reacts with a graphical object in the client's application.

In our solution, we found three ways of communication and processing alarms on the side of the server. The first is to save every error in the inner error table and then send the table to the client. The second solution was to integrate alarms inside every XML message that is sent to the client. The third solution was to create a data contract with its own structure. The first and the third solution can be combined.

We chose the first solution to make processing on the client side easier. The second solution would be slower due to serialization and deserialization in adding the additional xml code. In the third solution it would be confusing to scans and processing the whole alarm structure of defined data contract. It was not necessary to combine the first and the third solution due the fact that control commands would be slowed down during additional alarm messages.

The alarm was necessary to be distributed to clients because every client might have his/her own functionality with signalizing, grouping, ACK the alarm. The alarms are distributed between clients and application server. But the alarms are logically handled centrally, i.e., the information only exists in one place and all users see the same status (e.g. the acknowledgement) [12].

Statuses of alarms are changing between ALM, ACK and RTN. The server controls the ACK status from user replies. The ALM status is set to occur when the alarm has been triggered. Sometimes an alarm occurs but immediately disappears. This status is set as RTN.

# 6      Client Side of Alarming System

The client side is the consumer of alarms. It depends on the implantation of the client application and what the alarms system will look like. We used a web application. We divided alarms into processes for communication, processing delivered data and visualizing the alarms.

When the client gets the table of errors, it is processed by the defined criteria to reach the best expression of status inside machines and robots.

The client was developed for mobile devices such as tablets, PDAs or smart phones.

In Fig. 4. alarms of the realized system are displayed. Each alarm is coloured by its priority and the operators can ACK the alarm.  Alarms in the client web application are divided into SCADA/HMI, PLC, model group errors. Another division into groups is by a priority criterion. There are three groups made for this criterion:

- Group A – highest priority. In the grid, they are red. A new alarm is signalized in the menu panel.
- Group B – middle priority. It is not necessary to notify the user immediately, so a new alarm is only displayed in the panel. In the grid, they are yellow.
- Group C – lowest priority. In the grid, they are blue.

The errors can be ACK. After they are ACK, the status of the alarm is sent to the server.

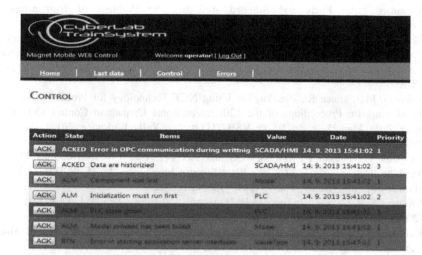

**Fig. 4.** Web application in ASP.NET with simulated errors

This client side of alarm system is focused on representation alarm data to human at the right moment. The human has the opportunity to see inside the control process of machines and robots.

# 7     Conclusion

A human-machine interface allows mutual interaction between both sides. Like every language makes some evolutionary steps, technological innovations make the communication between humans and machines easier and more convenient. A human-machine interface includes a "translating" environment which can be applied to communication with robots. This interface is based on service and can improve the control and involve operators inside the machine process by creating integration of different clients' devices which perform human-machine interface and help to analyze the human and machine interactions.

**Acknowledgments.** This article was supported KEGA no. 021TUKE -4/2012  (50%) and in the framework of the operational program Research and Development project: University Science Park Technicom for innovative applications to support knowledge technologies, code ITMS: 26220220182, co-funded by European regional Development Fund (50%).

# References

1. Perez-Cortes, J., Llobet, R., Navarro-Cerdan, J.R., Arlandis, J.: Improvement of Embedded Human-Machine Interfaces Combining Language, Hypothesis and Error Models. In: 22nd International Workshop on Database and Expert Systems Applications, DEXA 2011, pp. 359–363 (2011)

2. Suchman, L.A.: Plans and Situated Actions: The Problem of Human-machine Communication. Cambridge University Press (1987)
3. Škuta, J.: Transferring data from mobile devices. In: Proceedings of the 12th International Carpathian Control Conference, ICCC 2011, May 25-28, pp. 389–392. VŠB-TU Ostrava, Velké Karlovice (2011), ISBN 978-1-61284-359-9. IEEE Catalogue Number: CFP1142L-CDR
4. Babiuch, M., Farana, R., Plandor, D.: Using NET Technology for Process Control and Monitoring. In: Proceedings of the 12th International Carpathian Control Conference, ICCC 2011, May 25-28, pp. 10–15. VŠB-TU Ostrava, Velké Karlovice (2011), ISBN 978-1-61284-359-9. IEEE Catalog Number: CFP1142L-CDR
5. Skripcak, T., Tanuska, P., Konrad, U., Schmeisser, N.: Toward Nonconventional Human - Machine Interfaces for Supervisory Plant Process Monitoring. In: IEEE Transactions on Human-Machine Systems, vol. 43 (2013) ISSN 2168-2291
6. Invensys Systems, Inc.: InTouch® HMI Alarms and Events Guide. Lake Forest, U.S.A (2007)
7. Kamarulzaman, W.A.W., Sufian, A.M., Embong, R., Shafik, M., Taha, M.: Substation alarm monitoring system. In: TENCON 2000, vol. 2, pp. 603–608 (2000)
8. Zhang, W., Li, J.: Research and Application of WCF Extensibility. In: 2010 International Conference on Web Information Systems and Mining (WISM), pp. 363–367 (2010)
9. Winkler, O., Valas, M., Osadník, P., Landryová, L.: Communication Standards Suitable for MES Systems Designed for SMEs. In: Transactions of the VŠB-Technical University of Ostrava. Mechanical Series, No. 2/2009, vol. LV, article No. 1710. VŠB-TU Ostrava, pp. 181–186 (2009), ISBN 978-80-248-2144-3. ISSN 1210-0471 (Print). ISSN 1804-0993 (Online). ISSN-L1210-0471
10. Zhang, W., Cheng, G.: A Service-Oriented Distributed Framework-WCF. In: International Conference on Web Information Systems and Mining, WISM 2009, pp. 363–367 (2010)
11. Jaloudi, S., Schmelter, A., Ortjohann, E., Sinsukthavorn, W., Wirasanti, P., Morton, D.: Employment of Programmable Controllers, Control Systems and Service Oriented Architecture for the Integration of Distributed Generators in the Grid
12. Daneels, A., Salter, W.: What is SCADA? In: National Conference on Accelerator and Large Experimental Physics Control Systems, Trieste, Italy (1999)

# Smartphone Robots

Daniel Lorenčík, Peter Sinčák, Martin Marek, and Jakub Tušan

Department of Cybernetics and Artificial Intelligence, Technical University of Košice
{daniel.lorencik,peter.sincak}@tuke.sk,
martin.marek@student.tuke.sk, jj.tusan@gmail.com

**Abstract.** In this chapter we aim to introduce the smartphone robots as a viable research and testing platform. As the smartphone gains on the popularity and market share, the creation of the applications that use artificial intelligence for these devices seems much more important. A smartphone robot is an incremental advancement of the smartphone and uses a smartphone for control functions (aka "the brain") and the chassis with actuators and sensors for moving and acting in an environment (aka "the brawl"). Although these robots have somewhat limited abilities, they offer interesting abilities, nonetheless. The easy development of programs working with sensory data (camera, gyro, accelerometers) or with services available (voice recognition service) means that we can focus on creating the smart applications for these robots.

## 1 Introduction

In recent years, smartphones have been on the rise. Currently, several major markets in developed countries are near the 50% penetration rate. That means that nearly (and in some countries already) 50% of mobile phones fall to the category "smartphone" [1].

The most recent smartphones can be compared in performance to the low to mid-level computers. They also boast a Wi-Fi or WAN (Wide Area Network in the form of 3G or 4G module) access to the network and an impressive array of sensors (accelerometer, gyroscope, magnetometer, camera and others). Smartphones can connect to other devices with the USB connector or the Bluetooth wireless communication, and the Software Development Kits are available for every major platform (Android, iOS, Windows). All these characteristics make the smartphones interesting for the consumer-oriented robotics.

Professor James Kuffner has used the term "cloud-enabled robots" in the interview robots with their heads in the clouds [2]. The core idea is: if we embrace cloud robotics, we can create "lighter, cheaper and smarter" robots. Cloud robotics is a new concept, the one which makes the idea of the remote brain [3] possible. At the core is the idea of offering the software of the robot as a cloud service. We have covered this topic in paper [4].

But if we combine the possibilities of today's smartphones with the services provided by cloud robotics, we can create a cheap, simple but nonetheless smart robot. There are several commercial projects utilizing the smartphone as a controller

© Springer International Publishing Switzerland 2015
P. Sinčák et al. (eds.), *Emergent Trends in Robotics and Intelligent Systems*,
Advances in Intelligent Systems and Computing 316, DOI: 10.1007/978-3-319-10783-7_15

for a wheeled chassis (Romo [5], SmartBot [6] and others) which can provide entertainment to the users. However, this area of robotics is also interesting for research for several reasons:

Robot-human interaction and how a robot can influence the human behaviour (for example when listening to music [7]).

Simulation of network delay when using the cloud services and testing the infrastructure [8].

Multiple robot control and interaction (especially for cloud robotics).Telepresence systems [9].

Using the smartphone controlled robotic platform to automatically map the area using complementary smartphone sensors [10].

Assistant services [11].

In robotics, smartphones can and are used as a brain of the robot. Actuators can be connected with the use of the micro USB, headphone jack or Bluetooth. The robot can have sensors and feed the data back to the smartphone to complement the smartphone sensors. The smartphone connection modules (Wi-Fi or WAN) provide the connection to the Internet, therefore allowing for remote servers or cloud services to provide required functionality.

This chapter is organized as follows: in section two is a brief description of the Android platform; in section 3 two projects incorporating smartphone and the Lego Mindstorm set are described; in section 4 we outline the future plans for use of the smartphone robots for testing the cloud robotics services; in section 5 we summarize this chapter.

## 2     Android Platform

Android Operating System is an open-source software stack for mobile devices [12]. It is based on the modified Linux kernel. The source code is released under the Apache License. Applications for Android can be written in a customized version of Java (with the addition of Android SDK) or in native C/C++ code (with the used Native Development Kit – NDK) for specific applications. The architecture of the Android OS is shown in Fig. 1.

Most of the Android smartphones are equipped with at least one camera, accelerometer, light sensor, GPS (Global Positioning System) module, Wi-Fi module, WAN module and Bluetooth module. All of these sensors can be accessed with the Android SDK and the data gained can be used in the application.

The applications are most commonly developed in Java programming language. The lifecycle of the application must be implemented by the developer, as the system postpones the GUI (Graphical User Interface) of the application (called activity) when the application is going to the background to conserve battery. If there is a need to have a longer running process, it is possible to run it as a service in the background process.

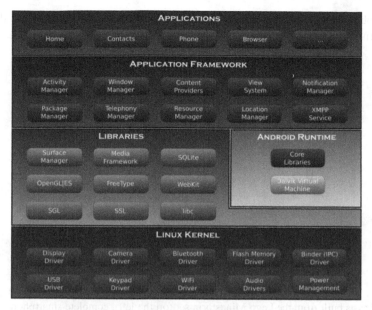

**Fig. 1.** Android system architecture overview

With the use of Android Application Framework, all developers can have access to the core services. The architecture allows for the easy reuse of components, and, most notably, every application can publish what type of action it can perform. This allows for loosely binding the components together without knowing the exact nature of them. Also, with the use of broadcast receivers, the application can receive and respond the system-wide broadcasts (an example can be the incoming phone call or message) or can respond to specific broadcasts (as a form of communicating between components of the application).

The availability of Software Development Kit, the excessive documentation and training manuals as well as broad and active community of developers backed by the Google Company allow for an easy adoption and learning of the Android applications development. The existence of Google Play Store [13] allows easy distribution of applications and their updates and the monetizing of apps. These features make Android the most used mobile operating system [14]. The problem with Android can be the high number of previous versions still active; therefore, the developers must test the application on these versions. Another problem is the number of different screen sizes and screen densities, so the user interface must be designed for each of them. However, the Android Application Framework offers several tools and methods to mitigate these problems.

# 3    Smartphone Robot Projects

We have worked on the problem of smartphone robots with two separate projects which will be described in detail in the subsections.

The robot consists of the smartphone and the wheeled chassis built from the Lego Mindstorm set. The robot system is shown in Fig. 2.

**Fig. 2.** Chassis built from the Lego Mindstorm set (on the left), complete smartphone robot (on the right)

The robot uses two motors for movement and the differential drive mode (to turn right, the right motor turns backward and the left forward, for the left turn the left motor turns backward and the right forward).

The smartphone controls the Lego NXT brick by sending specific signals via the Bluetooth communication channel. For this, code from the MINDdroid application [15] (available under Apache License) was adapted. Applications create the Bluetooth channel for communication and can send direct instructions for each motor and request the data from sensors. This does not require any sort of interface running on the NXT brick besides the operating system of the brick.

The orientation of smartphone can be reversed in case the smartphone without the front camera is used. Also, the applications can be run on a different chassis if the motors arrangement remains the same. It is possible for the developer of the application to provide the user with the option to specify the arrangement of motors (NXT brick has three output ports for motors).

## 3.1    Videoconferencing Robot

The task of the videoconferencing robot is to utilize the smartphone and wheeled chassis for communication over the internet. The main goal is to provide an easy setup and affordable system.

Although it is possible to use one of the free alternatives for videoconferencing (Skype, Google Hangouts and others), the possibility of controlling the camera and move with it is a step towards the telepresence applications. Although the use of a small Lego robotic vehicle is limiting the use of this application, it is a valuable hands-on practice.

The solution is based on the SpyDroid project [16]. It creates a streaming server from the smartphone and allows for accessing the device camera feed from the computer over the Internet. The code needed to control the robot and also the elements on the web interface which allow for controlling the robot movement over the internet were added to the application.

The application provides one-to-one communication and does not support the video streaming from the computer to the robot (due to the lack of support for the technologies used).

More detailed information can be found in [17].

## 3.2     Voice Controlled Robot

A second application created for the smartphone robot was the voice control. The goal of this application was to utilize existing software tools of the Android framework (Google Speech Recognition) and create an application for the voice control in several languages.

As there is an increasing number of services which provide much needed functionality, the task is to create applications with the use of them and to test their performance.

The application created is a simple control application which was built according to the Android developers guide. It provides the user with the ability to control the Lego robot by voice in several languages (Slovak, English, German, Russian, Spanish and French). The commands were previously specified. The application was tested by several users and also the reaction time was evaluated. The result of speech recognition was trimmed to the first five words and each of them was compared to the specified commands.

The results from the testing show that the use of a cloud based recognition service is viable in the non-critical applications as the robot was able to react in half of a second after the command was spoken. In the noisy environment, the recognition was 50% successful. The success rate of recognition was higher for the native and fluent speakers.

More detailed information can be found in [18].

# 4     Future Uses of Smartphone Robots

Our main research field is cloud robotics. However, smartphone robots can be very useful as a testing platform. Smartphone itself is a powerful computer with the attached sensors, and we can say it is ready for the cloud. That reduces the time needed to implement the cloud service on the device.

Another important factor is the price of the smartphone compared to the available robotics solutions. This allows for more robotic entities for the same price.

From the research point of view, the smartphone robots can be used as test devices for cloud services, to test the real network delay and the viability of using a cloud service for the robot control.

Besides the testing purposes, we will continue to employ smartphones and smartphone robots for small distinct tasks as human assistants. We will improve the videoconferencing application and allow for smartphone robots to be controlled by voice commands and also increase the level of control with the voice.

Another application will be the use of several smartphone robots as a group towards the common goal.

# 5     Conclusion

In this chapter, we have provided a brief introduction to the smartphone robots. There are several major operating systems on the market (Android, iOS, Windows), and we have provided an overview of Android OS. Then, we have described two projects we have done with the Android smartphone and the chassis from the Lego Mindstorm set.

The use of smartphone robots can ease the development process for applying the methods of artificial intelligence in real world applications, as many problems are alleviated by the operating system. These range from fast Wi-Fi connection, accurate and easy to obtain sensor and camera data, to the availability of certain services like voice recognition.

Another factor to consider is the affordability of smartphone robots. Not only for research purposes but also for the creation of real world applications. These can be used by general public, as the smartphones are already projected to excess more than 50% of all mobile phones in the developed countries.

**Acknowledgments.** Research supported by the "Centre of Competence of knowledge technologies for product system innovation in industry and service", with ITMS project number: 26220220155 for years 2012-2015.

# References

1. Smartphone AdoptionTips past 50% in Major markets Worldwide. Emarketer (May 29, 2013)
2. Guizzo, E.: Robots With Their Heads in the Clouds. In: IEEE Spectrum (February 28, 2011)
3. Inaba, M.: Remote-brained humanoid project. Advanced Robotics 11(6), 605–620 (1996)
4. Lorencik, D., Sincak, P.: Cloud Robotics: Current trends and possible use as a service. In: IEEE 11th International Symposium on Applied Machine Intelligence and Informatics, SAMI 2013, pp. 85–88 (2013)
5. Romo, http://romotive.com/ (accessed: July 25, 2013)
6. SmartBot, http://www.overdriverobotics.com/SmartBot/ (accessed: July 25, 2013)

7. Hoffman, G.: Dumb robots, smart phones: A case study of music listening companionship. In: 2012 IEEE RO-MAN: The 21st IEEE International Symposium on Robot and Human Interactive Communication, pp. 358–363 (2012)

8. Tsuchiya, R., Shimazaki, S., Sakai, T., Terada, S., Igarashi, K., Hanawa, D., Oguchi, K.: Simulation environment based on smartphones for Cloud computing robots. In: 2012 35th International Conference on Telecommunications and Signal Processing (TSP), pp. 96–100 (2012)

9. Choi, D., Song, T., Jeong, S., Jeon, J.W.: Design the Video Conferencing Robot for One-to-Many Communication using Smartphone. In: 1th International Conference on Control, Automation and Systems, pp. 671–675 (2011)

10. Klausner, A., Trachtenberg, A., Starobinski, D.: Phones and robots. In: Proceedings of the 9th ACM Conference on Embedded Networked Sensor Systems - SenSys 2011, p. 361 (2011)

11. Ho, Y., Sato-Shimokawara, E., Zhen, J., Yamaguchi, T.: An elderly people assistant system applying user model with robots and Smartphone. In: The 6th International Conference on Soft Computing and Intelligent Systems, and The 13th International Symposium on Advanced Intelligence Systems, pp. 410–415 (2012)

12. Android (2013), http://www.android.com/ (accessed: July 29, 2013)

13. Google play, https://play.google.com/store (accessed: July 27, 2013)

14. McCracken, H.: Who's Winning, iOS or Android? All the Numbers, All in One Place. Time (April 16, 2013)

15. LEGO MINDSTORMS MINDdroid source code, https://github.com/NXT/LEGO-MINDSTORMS-MINDdroid

16. SpyDroid IP Camera, https://code.google.com/p/spydroid-ipcamera/ (accessed: July 27, 2013)

17. Tušan, J.: The telepresence application for the mobile robotic platform. Technical University of Košice (2013) (in Slovak)

18. Marek, M.: The voice navigation for the mobile robotic platform. Technical University of Košice (2013) (in Slovak)

# Mathematical Model of Robot Melfa RV-2SDB

Peter Papcun and Ján Jadlovský

Department of Cybernetics and Artificial Intelligence (DCAI),
Faculty of Electrical Engineering and Informatics (FEEI), Technical University of Košice,
Letná 9, 042 00 Košice, Slovakia

**Abstract.** This article deals with the methodology of processing direct kinematics for a robotic arm. The paper describes basic matrices of direct kinematics and basic relations. A method of application and calculation of direct kinematics for industrial robot is also presented. The last part is the application of general inverse kinematics algorithm. This mathematical model is verified in the article.

## 1 Introduction

The article examines the mathematical model of the robotic arm MELFA RV-2SDB, it describes the general methodology of creating the robotic arm's mathematical model. The second chapter reminds basic matrices which are dedicated for calculation of the direct kinematics model. The next chapter describes the procedure of creating the mathematical model in steps. The fourth chapter describes the general algorithm for the inverse kinematics of the robotic arm. The fifth chapter compares the mathematical model and the real model of the robotic arm experimentally.

## 2 Direct Kinematics

This chapter reminds basic matrices for the calculation of direct kinematics. Mathematical descriptions of these matrices are described in [1]. This chapter summarizes all of the basic matrices only. Basic kinematics matrix:

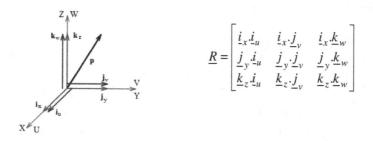

$$\underline{R} = \begin{bmatrix} \underline{i}_x \cdot \underline{i}_u & \underline{i}_x \cdot \underline{j}_v & \underline{i}_x \cdot \underline{k}_w \\ \underline{j}_y \cdot \underline{i}_u & \underline{j}_y \cdot \underline{j}_v & \underline{j}_y \cdot \underline{k}_w \\ \underline{k}_z \cdot \underline{i}_u & \underline{k}_z \cdot \underline{j}_v & \underline{k}_z \cdot \underline{k}_w \end{bmatrix}$$

**Fig. 1.** Basic kinematics

© Springer International Publishing Switzerland 2015

P. Sinčák et al. (eds.), *Emergent Trends in Robotics and Intelligent Systems*,

Advances in Intelligent Systems and Computing 316, DOI: 10.1007/978-3-319-10783-7_16

Rotation around axis X:

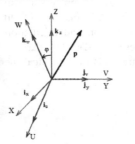

$$\underline{R}x, \alpha = \begin{bmatrix} 1 & 0 & 0 \\ 0 & \cos\alpha & -\sin\alpha \\ 0 & \sin\alpha & \cos\alpha \end{bmatrix}$$

**Fig. 2.** Rotation around axis X

Rotation around axis Y:

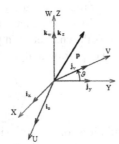

$$\underline{R}y, \varphi = \begin{bmatrix} \cos\varphi & 0 & \sin\varphi \\ 0 & 1 & 0 \\ -\sin\varphi & 0 & \cos\varphi \end{bmatrix}$$

**Fig. 3.** Rotation around axis Y

Rotation around axis Z:

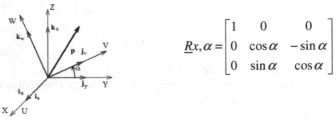

$$\underline{R}z, \vartheta = \begin{bmatrix} \cos\vartheta & -\sin\vartheta & 0 \\ \sin\vartheta & \cos\vartheta & 0 \\ 0 & 0 & 1 \end{bmatrix}$$

**Fig. 4.** Rotation around axis Z

Translation:

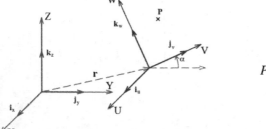

$$P = \begin{bmatrix} & & & | & r_x \\ & \underline{R} & & | & r_y \\ & & & | & r_z \\ - & - & - & - & - \\ 0 & 0 & 0 & | & 1 \end{bmatrix}$$

**Fig. 5.** Translation

# 3    Industrial Robot

This chapter describes the methodology of the creating and calculation of the direct kinematics mathematical model. This methodology is applied on the industrial robot MELFA RV-2SDB made by Mitsubishi corporation. We propose this methodology.

**The First Step: Present Technological Parameters of the Robotic Arm:**
Direct kinematics calculation needs to know accurate dimensions of the robotic arm and its maximal rotation:

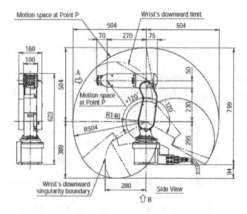

**Fig. 6.** Dimensions of robotic arm MELFA RV-2SDB

**The Second Step: Kinematics Structure of the Robotic Arm:**
Calculation also needs to know the kinematics structure of the robotic arm. This 3D kinematics structure is drawn with respect to the figure of the industrial robot and then the figure is removed and leaving only 3D kinematics structure of the robotic arm:

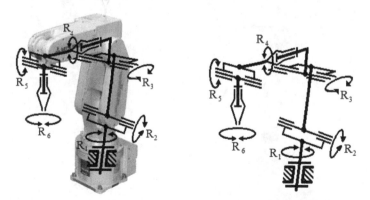

**Fig. 7.** 3D kinematics structure of industrial robot MELFA RV-2SDB

The 3D kinematics structure is redrawn to the classic kinematics structure for its simplification. The classic kinematics structure has arms marked with letters from $a$ to $f$ and joints $R_1$ - $R_6$ marked with Greeks letters (representing the angles) $\alpha$, $\beta$, $\gamma$, $\delta$, $\varepsilon$, $\varphi$ in the following figure:

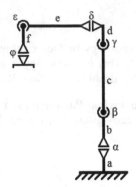

**Fig. 8.** Classic kinematics structure of the industrial robot MELFA RV-2SDB

Important information can be read from the scheme in Figure 8 and the dimensions of the robotic arm in Figure 6 (dimension a can be ignored because the nearest kinematics mechanism is rotary in axis z):

$$a = 0 \quad mm \qquad d = 50 \quad mm$$
$$b = 295 \ mm \qquad e = 270 \ mm$$
$$c = 270 \ mm \qquad f = 70 \quad mm$$

The next important information is the maximum range of joints:

$$\alpha \in < -240°; 240° > \qquad \delta \in < -200°; 200°>$$
$$\beta \in < -120°; 120° > \qquad \varepsilon \in < -120°; 120°>$$
$$\gamma \in < \quad 0°; 160°> \qquad \phi \in < -360°; 360°>$$

The third step: Determining the zero position of the robotic arm:

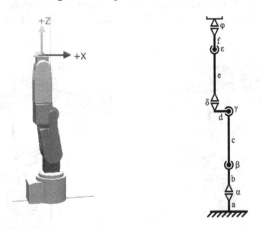

**Fig. 9.** Kinematics scheme of robotic arm if all angles have size 0°

**The Fourth Step: Transformation Matrices:**
Transformation matrices can be determined by the main information from the first three steps. They are created according to Figure 9 from the beginning of coordination system to the robot's end point.

Joint R1 is resolved to the first (because dimension a is ignored), this is the rotary motion around axis z (1), then the matrix calculation moves under joint R2 (2):

$$A = \begin{pmatrix} \cos\alpha & -\sin\alpha & 0 & 0 \\ \sin\alpha & \cos\alpha & 0 & 0 \\ 0 & 0 & 1 & 0 \\ 0 & 0 & 0 & 1 \end{pmatrix} \quad (1) \qquad B = \begin{pmatrix} 1 & 0 & 0 & 0 \\ 0 & 1 & 0 & 0 \\ 0 & 0 & 1 & b \\ 0 & 0 & 0 & 1 \end{pmatrix} \quad (2)$$

Joint $R_2$ causes rotary motion around axis y (3), then it moves under joint $R_3$ (4):

$$C = \begin{pmatrix} \cos\beta & 0 & \sin\beta & 0 \\ 0 & 1 & 0 & 0 \\ -\sin\beta & 0 & \cos\beta & 0 \\ 0 & 0 & 0 & 1 \end{pmatrix} \quad (3) \qquad D = \begin{pmatrix} 1 & 0 & 0 & 0 \\ 0 & 1 & 0 & 0 \\ 0 & 0 & 1 & c \\ 0 & 0 & 0 & 1 \end{pmatrix} \quad (4)$$

Joint $R_3$ rotates around axis y (5) and calculation moves under joint $R_4$ (6):

$$E = \begin{pmatrix} \cos\gamma & 0 & \sin\gamma & 0 \\ 0 & 1 & 0 & 0 \\ -\sin\gamma & 0 & \cos\gamma & 0 \\ 0 & 0 & 0 & 1 \end{pmatrix} \quad (5) \qquad F = \begin{pmatrix} 1 & 0 & 0 & -d \\ 0 & 1 & 0 & 0 \\ 0 & 0 & 1 & 0 \\ 0 & 0 & 0 & 1 \end{pmatrix} \quad (6)$$

Joint R4 rotates around axis z (7) and it moves under joint R5 (8):

$$G = \begin{pmatrix} \cos\delta & -\sin\delta & 0 & 0 \\ \sin\delta & \cos\delta & 0 & 0 \\ 0 & 0 & 1 & 0 \\ 0 & 0 & 0 & 1 \end{pmatrix} \quad (7) \qquad H = \begin{pmatrix} 1 & 0 & 0 & 0 \\ 0 & 1 & 0 & 0 \\ 0 & 0 & 1 & e \\ 0 & 0 & 0 & 1 \end{pmatrix} \quad (8)$$

Joint $R_5$ causes rotary motion around axis y (9), then it moves to last joint $R_6$ (10):

$$I = \begin{pmatrix} \cos\varepsilon & 0 & \sin\varepsilon & 0 \\ 0 & 1 & 0 & 0 \\ -\sin\varepsilon & 0 & \cos\varepsilon & 0 \\ 0 & 0 & 0 & 1 \end{pmatrix} \quad (9) \qquad J = \begin{pmatrix} 1 & 0 & 0 & 0 \\ 0 & 1 & 0 & 0 \\ 0 & 0 & 1 & f \\ 0 & 0 & 0 & 1 \end{pmatrix} \quad (10)$$

Last joint $R_6$ causes rotary motion around axis z (11):

$$K = \begin{pmatrix} \cos\varphi & -\sin\varphi & 0 & 0 \\ \sin\varphi & \cos\varphi & 0 & 0 \\ 0 & 0 & 1 & 0 \\ 0 & 0 & 0 & 1 \end{pmatrix} \quad (11)$$

The fifth step: The composite homogeneous transformation matrix:

The composite homogeneous transformation matrix is calculated from all transformations matrices, which is calculated in the fourth step:

$$T = A.B.C.D.E.F.G.H.I.J.K \tag{12}$$

Resultant composite homogeneous transformation matrix:

$$T = \begin{pmatrix} n_x & o_x & a_x & x \\ n_y & o_y & a_y & y \\ n_z & o_z & a_z & z \\ 0 & 0 & 0 & 1 \end{pmatrix} \tag{13}$$

Equations (14) are simplified $(\sin(x) \rightarrow sx)$:

$$x = f \left( c_\alpha c_\beta c_\gamma c_\delta s_\varepsilon - c_\alpha s_\beta s_\gamma c_\delta s_\varepsilon - s_\alpha s_\delta s_\varepsilon + c_\alpha c_\beta s_\gamma c_\varepsilon + c_\alpha s_\beta c_\gamma c_\varepsilon \right) +$$
$$+ e \left( c_\alpha c_\beta s_\gamma + c_\alpha s_\beta c_\gamma \right) - d \left( c_\alpha c_\beta c_\gamma - c_\alpha s_\beta s_\gamma \right) + c\, c_\alpha s_\beta$$

$$y = f \left( s_\alpha c_\beta c_\gamma c_\delta s_\varepsilon - s_\alpha s_\beta s_\gamma c_\delta s_\varepsilon + c_\alpha s_\delta s_\varepsilon + s_\alpha c_\beta s_\gamma c_\varepsilon + s_\alpha s_\beta c_\gamma c_\varepsilon \right) +$$
$$+ e \left( s_\alpha c_\beta s_\gamma + s_\alpha s_\beta c_\gamma \right) - d \left( s_\alpha c_\beta c_\gamma - s_\alpha s_\beta s_\gamma \right) + c\, s_\alpha s_\beta$$

$$z = f \left( c_\beta c_\gamma c_\varepsilon - s_\beta c_\gamma c_\delta s_\varepsilon - c_\beta s_\gamma c_\delta s_\varepsilon - s_\beta s_\gamma c_\varepsilon \right) + e \left( c_\beta c_\gamma - s_\beta s_\gamma \right) + d \left( s_\beta c_\gamma + c_\beta s_\gamma \right)$$
$$+ c\, c_\beta + b$$

$$n_x = c_\alpha c_\beta c_\gamma c_\delta c_\varepsilon c_\varphi - c_\alpha s_\beta s_\gamma c_\delta c_\varepsilon c_\varphi - s_\alpha s_\delta c_\varepsilon c_\varphi - c_\alpha c_\beta s_\gamma s_\varepsilon c_\varphi - c_\alpha s_\beta c_\gamma s_\varepsilon c_\varphi +$$
$$+ c_\alpha s_\beta s_\gamma s_\delta s_\varphi - c_\alpha c_\beta c_\gamma s_\delta s_\varphi - s_\alpha c_\delta s_\varphi$$

$$o_x = c_\alpha s_\beta s_\gamma c_\delta c_\varepsilon s_\varphi - c_\alpha c_\beta c_\gamma c_\delta c_\varepsilon s_\varphi + s_\alpha s_\delta c_\varepsilon s_\varphi + c_\alpha c_\beta s_\gamma s_\varepsilon s_\varphi + c_\alpha s_\beta c_\gamma s_\varepsilon s_\varphi +$$
$$+ c_\alpha s_\beta s_\gamma s_\delta c_\varphi - c_\alpha c_\beta c_\gamma s_\delta c_\varphi - s_\alpha c_\delta c_\varphi$$

$$a_x = c_\alpha c_\beta c_\gamma c_\delta s_\varepsilon - c_\alpha s_\beta s_\gamma c_\delta s_\varepsilon - s_\alpha s_\delta s_\varepsilon + c_\alpha c_\beta s_\gamma c_\varepsilon + c_\alpha s_\beta c_\gamma c_\varepsilon$$

$$n_y = s_\alpha c_\beta c_\gamma c_\delta c_\varepsilon c_\varphi - s_\alpha s_\beta s_\gamma c_\delta c_\varepsilon c_\varphi + c_\alpha s_\delta c_\varepsilon c_\varphi - s_\alpha c_\beta s_\gamma s_\varepsilon c_\varphi - s_\alpha s_\beta c_\gamma s_\varepsilon c_\varphi +$$
$$+ s_\alpha s_\beta s_\gamma s_\delta s_\varphi - s_\alpha c_\beta c_\gamma s_\delta s_\varphi + c_\alpha c_\delta s_\varphi$$

$$o_y = s_\alpha s_\beta s_\gamma c_\delta c_\varepsilon s_\varphi - s_\alpha c_\beta c_\gamma c_\delta c_\varepsilon s_\varphi - c_\alpha s_\delta c_\varepsilon s_\varphi + s_\alpha c_\beta s_\gamma s_\varepsilon s_\varphi + s_\alpha s_\beta c_\gamma s_\varepsilon s_\varphi +$$
$$+ s_\alpha s_\beta s_\gamma s_\delta c_\varphi - s_\alpha c_\beta c_\gamma s_\delta c_\varphi + c_\alpha c_\delta c_\varphi$$

$$a_y = s_\alpha c_\beta c_\gamma c_\delta s_\varepsilon - s_\alpha s_\beta s_\gamma c_\delta s_\varepsilon + c_\alpha s_\delta s_\varepsilon + s_\alpha c_\beta s_\gamma c_\varepsilon + s_\alpha s_\beta c_\gamma c_\varepsilon$$

$$n_z = s_\beta s_\gamma s_\varepsilon c_\varphi - s_\beta c_\gamma c_\delta c_\varepsilon c_\varphi - c_\beta s_\gamma c_\delta c_\varepsilon c_\varphi - c_\beta c_\gamma s_\varepsilon c_\varphi + s_\beta c_\gamma s_\delta s_\varphi + c_\beta s_\gamma s_\delta s_\varphi$$

$$o_z = s_\beta c_\gamma c_\delta c_\varepsilon s_\varphi + c_\beta s_\gamma c_\delta c_\varepsilon s_\varphi - s_\beta s_\gamma s_\varepsilon s_\varphi + c_\beta c_\gamma s_\varepsilon s_\varphi + s_\beta c_\gamma s_\delta c_\varphi + c_\beta s_\gamma s_\delta c_\varphi$$

$$a_z = c_\beta c_\gamma c_\varepsilon - s_\beta c_\gamma c_\delta s_\varepsilon - c_\beta s_\gamma c_\delta s_\varepsilon - s_\beta s_\gamma c_\varepsilon$$

$$\tag{14}$$

Matrix components x, y and z are the end point coordinates of the industrial robot. Rotation matrix is assembled from vectors n, o and a. The rotation of the coordination system is calculated from this rotation matrix. Presented below are the general rotation matrixes (15) around all axes:

$$R_C = \begin{pmatrix} \cos\tau & -\sin\tau & 0 \\ \sin\tau & \cos\tau & 0 \\ 0 & 0 & 1 \end{pmatrix} \cdot \begin{pmatrix} \cos\rho & 0 & \sin\rho \\ 0 & 1 & 0 \\ -\sin\rho & 0 & \cos\rho \end{pmatrix} \cdot \begin{pmatrix} 1 & 0 & 0 \\ 0 & \cos\omega & -\sin\omega \\ 0 & \sin\omega & \cos\omega \end{pmatrix} =$$

$$\begin{pmatrix} \cos\tau.\cos\rho & \cos\tau.\sin\rho.\sin\omega - \sin\tau.\cos\omega & \cos\tau.\sin\rho.\cos\omega + \sin\tau.\sin\omega \\ \sin\tau.\cos\rho & \sin\tau.\sin\rho.\sin\omega + \cos\tau.\cos\omega & \sin\tau.\sin\rho.\cos\omega - \cos\tau.\sin\omega \\ -\sin\rho & \cos\rho.\sin\omega & \cos\rho.\cos\omega \end{pmatrix} \tag{15}$$

$$\omega - \text{Rotation around axis } x$$
$$\rho - \text{Rotation around axis } y$$
$$\tau - \text{Rotation around axis } z$$

All axes rotations are deduced from equation (15):
Rotation around axis $y$, angle $\rho$:

$$n_z = -\sin\rho \tag{16}$$

$$\rho = \arcsin(-n_z) \tag{17}$$

Rotation around axis $z$, angle $\tau$:

$$n_y = \sin\tau.\cos\rho \tag{18}$$

$$\tau = \arcsin\left(\frac{n_y}{\cos(\arcsin(-n_z))}\right) \tag{19}$$

Rotation around axis $x$, angle $\omega$:

$$o_z = \cos\rho.\sin\omega \tag{20}$$

$$\omega = \arcsin\left(\frac{o_z}{\cos(\arcsin(-n_z))}\right) \tag{21}$$

# 4    Inverse Kinematics

Robot inverse kinematics is calculated by algorithm which we have proposed. This algorithm uses mathematical model of the robot's direct kinematics and Newton approximation method with the use of Jacoby matrix.

Inputs of the algorithm are an initial state, desired position and parameter δ. The initial state is the actual rotation state of the joints. The desired position consists of coordinates and axes rotations which the robotic arm has to be in. Parameter δ is described below. The shift has to be defined as zero before the start of the main loop of the algorithm.

The algorithm includes two functions. These functions are not shown in the flowchart for lack of space in this paper. These functions are AA (angle adjust) and DK (direct kinematics). AA function adjusts angles in the range from -360° to 360°. DK inputs are angles of the joints rotation and the output is vector with coordinates and axes rotation. The algorithms main loop includes two very important calculates: Jacoby matrix and shift. The classic formula for Jacoby matrix is (22):

$$J = \begin{pmatrix} \dfrac{df_1}{dx_1} & \dfrac{df_1}{dx_2} & \cdots & \dfrac{df_1}{dx_n} \\ \dfrac{df_2}{dx_1} & \dfrac{df_2}{dx_2} & \cdots & \dfrac{df_2}{dx_n} \\ \vdots & \vdots & \ddots & \vdots \\ \dfrac{df_m}{dx_1} & \dfrac{df_m}{dx_2} & \cdots & \dfrac{df_m}{dx_n} \end{pmatrix} \tag{22}$$

We modified the formula (22) to a discrete form. The discretization of derivation is different. Discrete Jacoby matrix consist of vectors which represent a difference between temporary positions and positions after minimal modification (δ) in the angle of one robotic joint. If the robot has 6 degrees of freedom and 6 joints, then the resultant Jacoby matrix has size 6x6. Discrete Jacoby matrix (Function DK is used in the formula for simplification):

$$J = [(DK(\alpha,\beta,\gamma,\delta,\varepsilon,\varphi))' \quad (DK(\alpha,\beta,\gamma,\delta,\varepsilon,\varphi))' \quad \cdots \quad (DK(\alpha+\Delta,\beta,\gamma,\delta,\varepsilon,\varphi))'] -$$
$$- [(DK(\alpha+\Delta,\beta,\gamma,\delta,\varepsilon,\varphi))' \quad (DK(\alpha,\beta+\Delta,\gamma,\delta,\varepsilon,\varphi))' \quad \cdots \quad (DK(\alpha,\beta,\gamma,\delta,\varepsilon,\varphi+\Delta))']$$

(23)

Inverse kinematics algorithm:

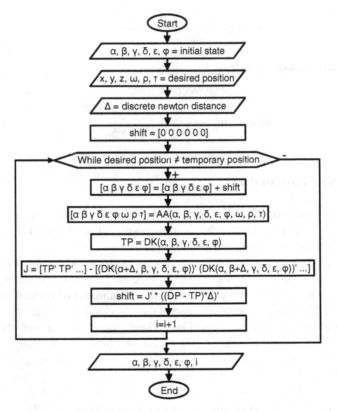

**Fig. 10.** Inverse kinematics algorithm

Shift is calculated through the basic formula:

$$dq = J^{-1}dp \tag{24}$$

Discretization of this formula (DP – vector of desired position, TP – vector of temporary position, s - shift):

$$s = J^{-1}(\Delta(DP-TP))' \tag{25}$$

The algorithm output is the angles vector of joint rotation and the number of loop iterations. It is important to point out that the robot's movement begins after the algorithm finishes; robot's movement never begins while an algorithm is running. Temporary position is not a real position, only an auxiliary position necessary for calculation.

## 5    Experimental Verification

These results were verified on the real industrial robot. Direct kinematics is verified by entering a random motor (joint) rotation to the real model and after the completion of the operation the actual position of the robot's endpoint is read from the robot controller. The same random joint rotation is calculated by composite homogeneous transformation matrix in MATLAB. These results are compared in Table 1:

Legend to table:

α, β, γ, δ, ε, φ:  joint rotation of industrial robot

X, Y, Z:          real position (coordinate) of robot's endpoint
A, B, C:          real rotation of coordination system of robot's endpoint
x, y, z:          calculated position (coordinate) of robot's endpoint
ω, ρ, τ:          calculated rotation of coordination system of robot's endpoint

**Table 1.** Verification of calculated direct kinematics – selected measurement

| č. | α[deg.]<br>X[mm]<br>x[mm] | β[deg.]<br>Y[mm]<br>y[mm] | γ[deg.]<br>Z[mm]<br>z[mm] | δ[deg.]<br>A[deg.]<br>ω[deg.] | ε[deg.]<br>B[deg.]<br>ρ[deg.] | φ[deg.]<br>C[deg.]<br>τ[deg.] |
|---|---|---|---|---|---|---|
| 1. | 87,47<br>-56,81<br>-56,8141 | -45,92<br>55,22<br>55,2198 | 98,93<br>676,16<br>676,1553 | 85,70<br>11,46<br>11,4617 | 58,00<br>-74,13<br>-74,127 | 121,00<br>-43,06<br>-43,057 |
| 2. | 225<br>-138,68<br>-138,6844 | 110<br>-237,68<br>-237,6793 | 70<br>-53,66<br>-53.6646 | -90<br>-90<br>-90 | -90<br>30<br>30 | 300<br>-45<br>-45 |
| 3. | 21,8<br>250<br>249,9950 | -1,32<br>100<br>99,9951 | 102,84<br>450<br>450,0048 | 0<br>180<br>180 | 78,48<br>0<br>0 | 21,8<br>180<br>180 |

On Table 1 it can be seen that the direct kinematics of an industrial robot MELFA RV-2SDB is calculated correctly. There is more verification but the results are same; this is the reason why the results are not in the table. Inverse kinematics is verified, too with excellent results (calculated joints rotation were the same as the real joints rotation).

# 6    Conclusion

This paper describes the process of creating a mathematical model of robotic arms. This mathematical model refers to the direct and inverse kinematics model. The process of creating the direct kinematics model was tested on 4 robotic arms (Mitsubishi MELFA RV-2SDB – model DCAI, Robkovia – model DCAI, SEF – Model DCAI and OWI 535 Robotic Arm). The algorithm of inverse kinematics was tested on 2 robotic arms (Mitsubishi MELFA RV-2SDB and SEF). We use the Newton approximation method for the inverse kinematics calculation but there are many other methods, for example Taylor expansion of the transformation matrix, Analytic solution, BFS (Broyden, Fletcher, Shanno) method and Vector method of inverse transformation. We use the mentioned method because we considered this method reliable and result needs maximally 8 iterations.

**Acknowledgment.** This work has been supported by the Scientific Grant Agency of Slovak Republic under project Vega No.1/0286/11 Dynamic Hybrid Architectures of the Multiagent Network Control Systems – 50%. This work has been supported by the Research and Development Operational Program for project: University Science Park Technicom for innovative applications with knowledge technology support, ITMS code 26220220182, co-financed by the ERDF – 50%.

# References

1. Krokavec, D.: Data processing in robotics (Spracovanie údajov v robotike), Košice (1985) ISBN 85-629-85
2. Mostýn, V., Krys, V.: Mechatronics of industrial robots (Mechatronika průmyslových robotů), Ostrava (2012) ISBN 978-80-248-2610-3
3. Božek, P., Barborák, O., Naščák, Ľ., Štollmann, V.: Specialized robotic systems (Špecializované robotické systémy), Bratislava (2011) ISBN 978-80-904766-8-4
4. Jadlovský, J., Papcun, P.: Optimizing industry robot for maximum speed with high accuracy. In: Procedia Engineering, Zemplínska Šírava, vol. 48, pp. 533–542 (2012) ISSN 1877-7058
5. Papcun, P., Čopík, M., Ilkovič, J.: Robot control integrated in flexible production system (Riadenie robota integrovaného v pružnom výrobnom systéme). In: Electro Scope, vol. 2012(2), Plzeň, pp. 1–9 (2012) ISSN 1802-4564

# Learning of Fuzzy Cognitive Maps by a PSO Algorithm for Movement Adjustment of Robots

Ján Vaščák and Roman Michna

Technical University of Košice, Letná 9, 042 00 Košice, Slovakia
jan.vascak@tuke.sk, romco.michna@gmail.com

**Abstract.** Motional stability and robustness play a very important role mainly in bipedal robots, especially if it is connected with a dynamic environment, where many motion changes are necessary to be done. Here, this problem is shown on kicking a ball in robotic soccer. The movement control leads to constructing movement trajectories which should secure stable behaviour. Some control approaches are oriented in creating smooth trajectories instead of complicated stability analyses. For such purposes the so-called Bézier curves are used. In this paper we use Fuzzy Cognitive Maps (FCMs) for determining parameters of Bézier curves as well as a Particle Swarm Optimization (PSO) algorithm for learning FCMs. The main advantages of PSO consist in their speed and necessity of a relatively small training set. Two types of a kicking system for generating smooth movement trajectories are proposed and compared in the paper, which is documented by performed experiments.

## 1 Introduction

The robustness of movement (gait, kicking, etc.) plays a key role in robotic control, especially in the case of bipedal robots. As conventional approaches for improving the movement stability which are based on in-depth system identification, encounter complexity problems, the research is becoming more oriented on skin-deep approaches as, e.g., smoothness of movement, where the only problem is to ensure differentiability of movement trajectories in each point, i.e. their continuity. In this paper we will focus our interest on the kicking problem as a part of the tasks to be solved in robotic soccer, but the proposed approach is usable for further kinds of movements, too.

Most kicking systems are more or less based on the so-called key-frames (see e.g. [2]), where a kicking motion is composed of several frames. Therefore, various types of kicking consist of different sequences of basic predefined key-frames. This method is based more intuitively than exactly utilizing a physical mathematical description. However, there are two basic disadvantages of such an approach. Firstly, there is available only a limited set of predefined kicking types which, more or less but not optimally, correspond to the needs of a real situation. Secondly, when a kicking movement starts, it cannot be changed during its processing because of the possible risk of discontinuities of such a movement which can lead to instabilities and, e.g., the

© Springer International Publishing Switzerland 2015
P. Sinčák et al. (eds.), *Emergent Trends in Robotics and Intelligent Systems*,
Advances in Intelligent Systems and Computing 316, DOI: 10.1007/978-3-319-10783-7_17

falling down of a player. Therefore, such a type of movement is defined as a static kick because of its inability to take into consideration dynamic changes of surroundings (e.g. moving players or ball). To minimize these drawbacks, several modifications have been designed where the main question is about how to define the key-frames and, mainly, how to define movements leading from one key-frame to another one. For instance, in [4] the so-called Zero Moment Point method is combined with the basic key-frames approach trying to incorporate the stability theory to minimize problems with instabilities. Key-frames are also used in the so-called Learning from Demonstration, where a human physically guides a robot to perform a skill. A hybrid method that combines trajectories and only some sparse key-frames performed by a human operator is proposed in [1]. On the contrary, the approach in [9] tries to substitute key-frames by a special adaptive motion controller using at least visual feedback and dividing the kick motion into four phases, but the basic principle of key-frames is still retained. We can see that despite the mentioned drawbacks, the key-frame approach represents a basis for more sophisticated methods of movement design.

The approach described in [10] belongs to the family of key-frame approaches too. However, it is neither purely intuitive nor purely physical-mathematical. It is based on a proposition that a smooth trajectory is the best guarantee for a stable behavior of a robot. Of course, it is a heuristics only, where the risk of instability is minimized but not fully eliminated. To obtain smooth trajectories the so-called Bézier curves are used which are well known because of their smoothness and this property has been already used in many other applications (e.g. computer graphics, economy, motion control, etc.). However, the main advantage of this method is that Bézier curves are recalculated in each motion phase depending on the definition of end points, i.e. key-frames which are defined by current circumstances and so any movement can be changed during its processing. In other words, a movement is able to immediately respond to sudden changes and in such a case we get a dynamic kick.

Until now, proposed kicking systems take into account only the position of the ball and prescribed direction of kicking. This data does not enable to determine the strength of kicking or to consider further influences like the position of obstacles or players (both opponents and teammates). In this paper we will describe our extension of the system designed in [10] which can be used for passing the ball between teammates or for a more realistic kicking comparable to human football. To achieve our goal we used the descriptions of the most significant factors in constructing the kicking process, the so-called Fuzzy Cognitive Maps (FCMs), as a user-friendly descriptive method. For constructing a proper FCM the well-known Particle Swarm Optimization (PSO) algorithm was used.

This paper is organized as follows: Sect. 2 introduces basic notions related to FCMs and PSO. The problem of constructing a dynamic kick is analysed and two new designs are presented in Sect. 3. In Sect. 4 results of some experiments are discussed and resulting concluding remarks are presented as well as recommendations for future research to accomplish the paper.

## 2    FCMs and PSO – Basic Notions

FCMs are basically oriented graphs, where nodes c represents notions in symbolic form and the connections are causal relations. Mostly, notions are states or conditions and connections are actions or transfer functions which transform a state in a node to another one in another node. FCM can be regarded as a set of rules too. In such a case, input nodes represent parts of a rule premise which are interconnected in the output node whose value corresponds to the consequent rule as we can see in Fig. 1. In contrast to conventional rule bases where the input and output variables are strictly divided in the case of an FCM, we can combine several rules and form implication chains or even closed loops. So the data flow is not forward as in conventional rule bases [7] but recurrent in general. Concerning implementation possibilities, FCMs are mostly designed as recurrent neural networks although the approach of a finite state machine is possible, too.

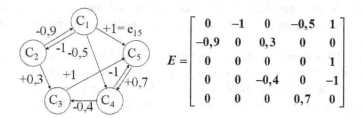

**Fig. 1.** An example of an FCM with its connection matrix E

Connections can be weighted by values eij from the interval [-1;1] (-1 because of negative connections), and so we can implement grades of membership into the inference process. If we define initial state values Ai in nodes Ci from the interval [0;1], then, using (1), we can calculate new state values for next time t+1:

$$A(t + 1) = L(A(t) + A(t) \cdot E),  \tag{1}$$

where A is the state vector of individual node values Ai, E is the connection matrix of connections eij and L is the limitation function to keep the values Ai in [0;1]. Thus we can simulate behaviour of an investigated system and then we can analyse its properties. Thus, FCMs can be very useful especially for prediction purposes. More detailed information about inference and applications of FCMs can be found, e.g., in [5] and [12].

FCMs learning is similar also in the case of neural networks either unsupervised or supervised. In the first case there are mostly various forms of Hebbian learning. However, for our purposes the unsupervised learning is necessary, where mainly genetic algorithms are used [11]. Regardless of the type of learning, we must always define nodes manually and the learning process is always restricted only to creating connections and setting up their weights under some other optimization criteria [14].

PSO algorithm belongs to the group of the so-called migration algorithms, where no new populations are generated, but original searchers (particles) are seeking better solutions. As they consider their ever best found solution as well as the ever best found solution of the swarm (the best particle), their behaviour is similar to a swarm of birds or fish [8]. If we consider a D-dimensional search space, then the position of a particle Xi will be described by a D-dimensional vector too, i.e. Xi = (xi1;xi2;...;xiD). Similarly, its velocity is defined as Vi = (vi1;vi2;...;viD). Besides, each particle knows also its best individual position reached at some time in its history Pi as well as the best position of the swarm at all, i.e. the best particle gi. The positions and velocities of particles can be changed in the following manner:

$$X_i(t + 1) = X_i(t) + V_i(t + 1), \tag{2}$$

$$V_i(t + 1) = \chi \cdot \left[ V_i(t) + C_1 \cdot R_1 \cdot \left( P_1(t) - X_i(t) \right) + C_2 \cdot R_2 \cdot \left( P_{g_i}(t) - X_i(t) \right) \right] \tag{3}$$

The parameters $\chi$;C1 and C2 are constants. Vectors R1 and R2 are random, which elements are from the interval [0;1], and they are analogous to mutations in genetic algorithms.

Generally, PSO algorithms do not need any additional knowledge about the system and it was proven they can also search large spaces very quickly but without any guarantee of finding the optimal solution. There are numbers of various PSO modifications, overview of which is summarized e.g. in [3]. PSO was used also for FCMs learning [13] which was a motivation for our research, too.

## 3     Implementation of FCMs in Dynamic Kick Control

A robot has a certain number of degrees of freedom which can form movement trajectories. In the case of a bipedal robot, these trajectories are realized by joints and some of them can be chained into one movement. For the Nao robot there are possibilities of up to eight movement trajectories that perform a movement: two for each leg and two for each arm (transitional and rotational).

In this paper we will solve the kicking movement only, where, e.g., the total movement of a leg can be divided into seven phases, i.e. key-frames, namely [10]: shifting the mass to the opposite leg, lifting the freed foot, striking it out, kicking the ball, taking the foot back, its lowering and shifting the mass of the robot back, respectively they are numbered in Fig. 2 (the last phase is not depicted). For each key-frame the required position of the foot tip is calculated and, using the inverse kinematics (for Nao robot see [6]), joint rotations are calculated. If we consider two coordinates X and Y, then they determine two movement trajectories for the transitional and rotational movement respectively. For simplicity reasons we will explain only the movement of one leg.

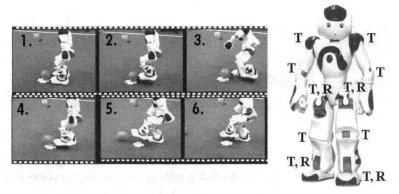

**Fig. 2.** The phases (key-frames) of the left leg movement from 1 to 6 [10] and movement types of joints for the robot Nao: T – transitional, R – rotational

To secure smoothness (differentiability) of the proposed movement trajectories, a cubic Bézier curve was used:

$$b(t) = \sum_{i=0}^{3} \binom{3}{i} t^i \cdot (1-t)^{3-i} \cdot P_i, \tag{4}$$

where $P_0,\ldots,P_3$ are the so-called control points, see Fig. 3. $P_0$ (t = 0) and $P_3$ (t = 1) are the start and end points, respectively. (In reality, the movement trajectory is $X_1 \to P_0 \to X_1 \to X_2$ because of the backward foot movement in the phase of striking the foot out , see Fig. 3.) $P_1$ and $P_2$ determine the form of such a curve. Just these last two control points are crucial for the trajectory smoothness because they determine the quality of the kick in frames 3 and 4, and they are the objects of our calculation.

The objective of FCMs learning is the connection matrix E. In our approach particles represent individual matrices so their dimensions are n2 if n is the number of nodes in an FCM. As the positions of start and end points of a given movement trajectory are limited, then the positions of P1 and P2 are also limited, i.e. $P_{x_i}^{min} \le P_{x_i} \le P_{x_i}^{max}$ and $P_{y_i}^{min} \le P_{y_i} \le P_{y_i}^{max}$ for coordinates X and Y, respectively, and i = 1; 2. The aim is to find the best compromise between the value limits, and so the fitness function F(E) is defined in the following manner [13]:

$$F(E) = \sum_{j=1}^{m} H\left(A_{P_j}^{min} - A_{P_j}\right) \cdot \left|A_{P_j}^{min} - A_{P_j}\right| + \sum_{j=1}^{m} H\left(A_{P_j} - A_{P_j}^{max}\right) \cdot \left|A_{P_j}^{max} - A_{P_j}\right| \tag{5}$$

APj are activation values of nodes, where the control points P1 and P2 are calculated. In our case there will be four such nodes for calculating Px1, Px2, Py1 and Py2 (m = 4). H is the well-known Heaviside function, where H(t)=0 if t <0, otherwise H(t)=1. The aim is to find the minimum of F(E) and corresponding elements for P1 and P2.

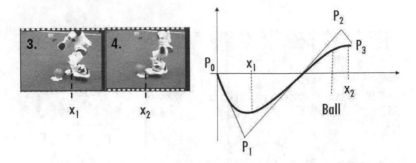

**Fig. 3.** Determination of control points depending on start ($x_1$) and end ($x_2$) points of the own kick

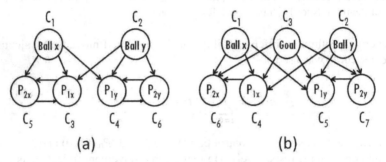

**Fig. 4.** FCMs for computing $P_1$ and $P_2$ – without (a) and with (b) considering the goal distance

Using FCMs for calculation of control points P1 and P2 (their elements Px1, Px2, Py1 and Py2), two schemes were proposed depicted in Fig. 4. The first scheme (Fig. 4(a)) is derived from the system described in [10]. It considers only the ball position related to the robot (Ballx and Bally) and, if the direction of kicking is given, it calculates movement trajectories. As this approach does not take into account further factors like the distance of an opponent's goal, the strength of kicking is constant (mostly the maximum one) which does not correspond to real human play, but, above all, such an approach is time consuming and threatens the stability of such a robot with a high risk of its falling down. Therefore, we proposed an improved version which considers also the goal distance, see Fig. 4(b).

## 4     Experiments and Conclusions

The main objective of our approach using FCMs was mainly to improve the kicking stability. The robot moved from various directions towards the ball at various speeds. We tested the number of falls and the time necessary for kicking the ball. Of course, we could observe potential staggering of the robot but it was not possible to measure it objectively. As expected, the first design led to almost constant kicking strengths regardless of the goal distance. Although the falling down was not considerably more frequent than in the second case (it was rare in general), the robot had large staggering problems. The second design showed more robustness at kicking at all and, of course, it produced various strengths of kicks depending on the goal distance which was the basic reason for improved behaviour.

The proposed kicking system preserves, like the system in [10], the ability of dynamic changes during kicking, too. After small modifications it can be used practically for all types of movement securing their smooth processing. Concretely, for purposes of robotic soccer, this system is especially suitable for tasks such as passing the ball between teammates which still is a great challenge in this research area . Concerning the learning quality, it is necessary to be aware of not only the precision but also speed. PSO is quick and needs less training data than, e.g., neural networks in general. However, it does not guarantee any satisfactory solution. This problem arose when we tried to include other factors like the presence of obstacles in the computation of control points. Therefore in the future our attention will be focused on improving FCMs learning.

**Acknowledgments.** Research supported by the "Center of Competence of Knowledge Technologies for Product System Innovation in Industry and Service", with ITMS project number: 26220220155 for years 20012–2015.

# References

1. Akgun, B., Cakmak, M., Yoo, J.W., Thomaz, A.L.: Trajectories and key frames for kinesthetic teaching: a human-robot interaction perspective. In: Proceedings of the Seventh Annual ACM/IEEE International Conference on Human-Robot Interaction, HRI 2012, pp. 391–398. ACM, New York (2012)
2. Antonelli, M., Dalla Libera, F., Menegatti, E., Minato, T., Ishiguro, H.: Intuitive humanoid motion generation joining user-defined key-frames and automatic learning. In: Iocchi, L., Matsubara, H., Weitzenfeld, A., Zhou, C. (eds.) RoboCup 2008. LNCS (LNAI), vol. 5399, pp. 13–24. Springer, Heidelberg (2009)
3. Clerc, M.: Particle Swarm Optimization. Wiley (2006)
4. Czarnetzki, S., Kerner, S., Klagges, D.: Combining key frame based motion design with controlled movement execution. In: Baltes, J., Lagoudakis, M.G., Naruse, T., Ghidary, S.S. (eds.) RoboCup 2009. LNCS (LNAI), vol. 5949, pp. 58–68. Springer, Heidelberg (2010)
5. Glykas, M. (ed.): Fuzzy Cognitive Maps. STUDFUZZ, vol. 247. Springer, Heidelberg (2010)
6. Graf, C., Härtl, A., Röfer, T., Laue, T.: A robust closed-loop gait for the standard platform league humanoid. In: Zhou, C., Pagello, E., Menegatti, E., Behnke, S., Röfer, T. (eds.) Proceedings of the Fourth Workshop on Humanoid Soccer Robots in Conjunction with the 2009 IEEE-RAS International Conference on Humanoid Robots, Paris, France, pp. 30–37 (2009)
7. Johanyák, Z.C., Kovács, S.: A brief survey and comparison on various interpolation-based fuzzy reasoning methods. Acta Polytechnica Hungarica 3(1), 91–105 (2006)
8. Kennedy, J., Eberhart, R.C., Shi, Y.: Swarm intelligence. Morgan Kaufmann series in evolutionary computation. Morgan Kaufman Publishers, San Francisco (2001)

9.  Mellmann, H., Xu, Y.: Adaptive motion control with visual feedback for a humanoid robot. In: Proc. of IEEE/RSJ International Conference on Intelligent Robots and Systems (IROS), pp. 3169–3174 (2010)
10. Müller, J., Laue, T., Röfer, T.: Kicking a ball – modelling complex dynamic motions for humanoid robots. In: Ruiz-del-Solar, J. (ed.) RoboCup 2010. LNCS (LNAI), vol. 6556, pp. 109–120. Springer, Heidelberg (2010)
11. Papageorgiou, E.: Learning algorithms for fuzzy cognitive maps: A review study. IEEE Transactions on Systems, Man, and Cybernetics, Part C: Applications and Reviews 42(2), 150–163 (2012)
12. Papageorgiou, E., Salmeron, J.: A review of fuzzy cognitive maps research during the last decade. IEEE Transactions on Fuzzy Systems 21(1), 66–79 (2013)
13. Papageorgiou, E.I., Parsopoulos, K.E., Stylios, C.D., Groumpos, P.P., Vrahatis, M.N.: Fuzzy cognitive maps learning using particle swarm optimization. International Journal of Intelligent Information Systems 25(1), 95–121 (2005)
14. Precup, R.E., Preitl, S.: Optimisation criteria in development of fuzzy controllers with dynamics. Engineering Applications of Artificial Intelligence 17(6), 661–674 (2004)

# Higher Speed of Data between Computers and Mobile Robots Based on Increase in the Number of Transmitters Robosoccer

Marek Sukop, Mikulas Hajduk, and Jaromir Jezny

Department of Production Systems and Robotics, Faculty of Mechanical Engineering,
Technical University of Kosice, 042-00 Kosice, st. Nemcovej 32
{marek.sukop,mikulas.hajduk}@tuke.sk

**Abstract.** A very important feature in the management of robotic soccer is the response time of the system. This article is one of the possible ways of reducing the time and using concatenation transmitters. This way to the first team (under MiroSot) used by FME TUKE Robotics World Championship 2009 in Korea. The first section describes the conventional method of data transmission as it said it applied. The second part describes the hardware and software method by which transmitters and finally compare responses with conventional manner.

**Keywords:** robot soccer, transmitter, receiver, mobile robot.

## 1   Introduction

Robot soccer is an application for testing multi-agent systems consisting of multiple mobile robots. Each team, consisting of five robots, has a camera attached above the playground in hight of 2 to 2.5 meters. Camera is connected to the control computer in which image processing algorithms are being recalculated, and then algorithms strategy. Then the robot player instructions are sent through the transmitter. In this article is described how to decrease the response by adjusting the transmitting part.

## 2   The Standard Configuration the Transmitting Device

The system is very sensitive to the speed of response (time from shooting scenes to perform interventions in the scene) and the number of hits per unit of time (usually a specified frequency framing cameras around 50-100 fps). The speed of response in general is mainly dependent on the system parameters: speed image acquisition (camera shutter, the speed of data transfer to a computer), the speed of image processing (to obtain required positional data on all participating entities), speed over the strategic calculations (to generate commands for robots), the data rate to transmit

© Springer International Publishing Switzerland 2015                                             163
P. Sinčák et al. (eds.), *Emergent Trends in Robotics and Intelligent Systems*,
Advances in Intelligent Systems and Computing 316, DOI: 10.1007/978-3-319-10783-7_18

module (eg via RS232 or USB), the speed of data transfer between the RF transmitter module and robots (eg 433MHz, ZigBee, Bluetooth), the processing of the data microcontroller robot and finally the change of the speed parameters.

One of the ways to positively affect the rate of response was applied to non-standard transmitting data in category MiroSot. The standard configuration transmitter link to the transmitter requires a computer with RS232 or USB interface (Fig.1).

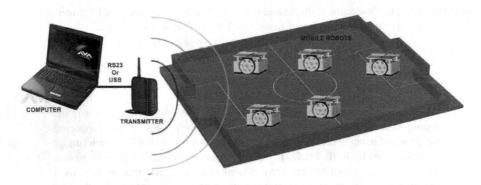

**Fig. 1.** The standard configuration of devices in the transmit chain

As a final member to use modules with free frequencies 433MHz (418MHz in some countries), or 915MHz. The disadvantage of these modules is normally quite low bit rate of RF parts. Modules can transfer data up to 115.2 kbit / s. Data is sent with the sequence to all mobile robots. Each robot receives data with a delay of:

$$t_n = \frac{n * BM}{TR} * 10 \qquad (1)$$

n...ID of robot (n=1,2,3,4,5, large league n=1,2,...,11)
BM...number of bytes in message for one robot
TR...transfer rate v bit/s

When using the maximum possible transmission speed 115.2 kBaud number of bytes being transferred and 4 for each robot (speed 16 bit, 16 bit angular velocity), the resulting delay complete information for the last robot in the chain of 1.74 ms. But it delays the ideal situation is that using conventional transmitter can not be achieved. Transmitters operating with a carrier frequency hundred MHz are limited data transfer by the ratio of ones and zeros in a short time should it be 1:1, 1:3 in the worst case.

SjF TUKE Robotics team was due to technological limitations, and with the help of using coding delay amounted to around 870us per robot, which is 4.35 ms for the fifth robot.

## 3    The Use of Multiple Transmitters

Philosophy arrangement of several transmitters in the transmit chain is on Fig.2.

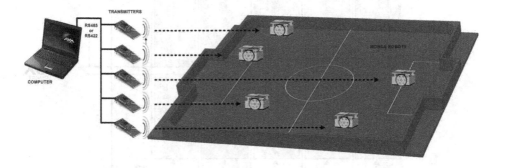

**Fig. 2.** Non-standard configuration of devices in the transmit chain

To shorten the time of data transmission modules were used, based on the nRF24L01 chip. Its advantage is the data transfer speed of up to 2 Mbit / s. Communication with the control processor runs over SPI at up to 8MHz. Minimum configuration is sent bytes: 1 byte preamble, 3 byte address, data transferred 1-32bytov, 1 CRC byte. The non-standard configuration, we achieved a delay of sending data to capture all the data about the last robot. 300-320us. There was room for repetition level data microcontroller, which ensures repeat transmission to reception of new data. Using the principle of that problem is eliminated accidental data loss. If the robot does not capture the first data pack, then catch the second resp. third etc. Of course there is a longer delay, but that would not be still, so long as the standard broadcast. Time display data transfer from PC to microcontroller and RF module is then to Fig.3.

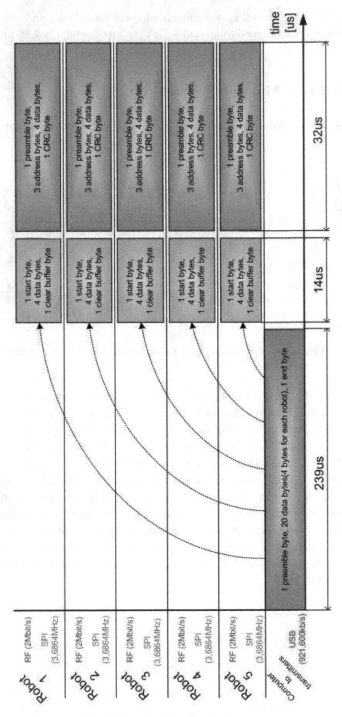

**Fig. 3.** Time display data transfer

The big advantage is visible in the graph is constant delay for all robots. The delay does not depend on the number of robots, but only from faults in transmission. In the earlier arrangement, however, there has been a complete failure in the data processor for the image, if there is a failure of transmission between the transmitter robot. The complete block diagram of the implemented data transmission between the control computer and the mobile robot is shown on Fig.4.

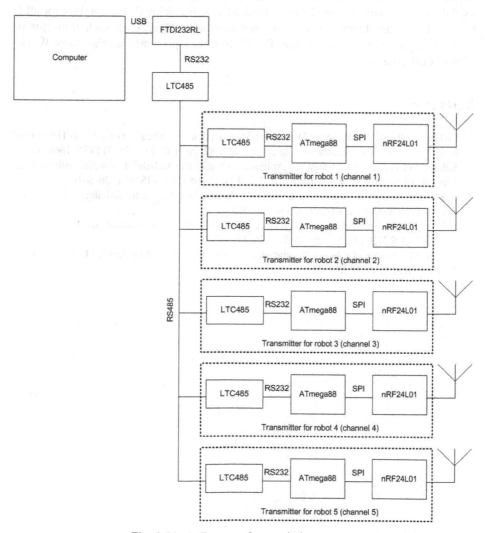

**Fig. 4.** Block diagram of transmission system

Part of the algorithm in the control computer, which is responsible for transferring data from a software module strategic calculations to the transmitter via USB, remained unchanged. After filling the system as shown in Fig.4, it was necessary to change the algorithms in microcontrollers upstream transmitter module and also part of the algorithms in the robot which processes data from the receiving module.

## 4    Conclusions

The advantage of applying the method described in shortening the response time of mobile robots to load images from a camera positioned above the playground. For optimal transmission, we reduced the time required for data transmission speed and the desired yaw rate compared to our old system of approx. 4ms, so we reduced the total delay in the control loop by more than 20%. It is a delay that is constant for all 5 robots. This latter feature is another important advantage. With such transmitters arranged in a chain, our team SjF TUKE Robotics presented at the 2009 World Championships in Korea.

## References

1. Fedák, V., Bačík, J.: Hardware Design for State Vector Identification of a Small Helicopter Model. In: Aplied Mechanics and Materials, vol. 282, pp. 107–115 (2013) ISSN 1660-9336
2. Sukop, M., Hajduk, M., Varga, J.: Aplikácia robotického futbalu: 1. Použité softvérové a hardvérové moduly. ATP Journal PLUS 1(2013), 66–68 (2013) ISSN 1336-5010
3. Sukop, M., Varga, J., Jánoš, R., Svetlík, J.: Aplikácia robotického futbalu: 2. Robot ako hráč. ATP Journal PLUS 1(2013), 69–71 (2013)
4. Sukop, M., Páchniková, L.: Aplikácia robotického futbalu: 3. Spracovanie obrazu. ATP Journal PLUS 1(2013), 72–75 (2013) ISSN 1336-5010
5. Sukop, M.: Aplikácia robotického futbalu: 4. Modul stratégií. ATP Journal PLUS 1(2013), 76–82 (2013) ISSN 1336-5010

# Increasing of Robot's Autonomy in Nuclear Application

Ladislav Vargovcik and Roland Holcer

ZTS VVU KOSICE a.s., Juzna trieda 95, 041 24 Kosice, Slovakia
{vargovcik,holcerr}@ztsvvu.eu

**Abstract.** The article describes development of robotic system for application in nuclear power plants (NPP), particularly example of heavy water evaporator fragmentation. Its control system started on Master-Slave level and based on requests of users it was extended to robotic system and then improved with autonomous sub-functions.

**Keywords:** robotic system, autonomy, fragmentation, nuclear equipment.

## 1 Introduction

As an example of the nuclear robotic application will be described liquidation of the heavy water evaporator for which has been applied a mobile robotic system MT 80, which had been developed, designed and constructed as a general-purpose decommissioning equipment.

The heavy water evaporator as a part of the NPP heavy water system is located inside the room 220 of the main production unit building of NPP A1 in Jaslovske Bohunice where the inner surface contamination is from 101 Bq/cm2 to the level of 103 Bq/cm2, dose rate up to 1.5 mGy/h and the feeding pipeline contained LRAW with high tritium content.

The first step to solve the fragmentation of the evaporator was the development of special tooling for application with the robot, such as hydraulic shears, circular saw, reciprocating saw, circular pipe cutter and a system for quick tool-change without direct intervention of the operators. All operations are remotely controlled on basis of visual information from four cameras, with consistent radiation protection of the operators.

Also robotic arm MT80 was extended to mobile version using mobile platform with additional two arms with the aim to increase working space for technological operations.

## 2 Robotic Manipulator MT 80

Manipulator MT 80 is a remote controlled handling device with hydraulic drives enabling handling of loads of up to 80 kg total weight. It is intended for work in the industry and liquidation of industrial accidents. The application is also

© Springer International Publishing Switzerland 2015

P. Sinčák et al. (eds.), *Emergent Trends in Robotics and Intelligent Systems*,

Advances in Intelligent Systems and Computing 316, DOI: 10.1007/978-3-319-10783-7_19

accommodated for use in environment with ionising radiation and ambient temperatures between -5°C to + 30°C. Its waterproof construction enables rinsing and spraying of the surface with decontaminating media.

The configuration of manipulator MT 80 with 6 degrees of freedom consists of the following five function units:

- manipulating arm
- control panel
- regulation module
- hydraulic power unit
- interconnecting cables and pressure hoses

## Technical Specification:

Main dimensions:

- length 1 938 mm
- height 537 mm
- width 320 mm
- the arm can be inserted through opening size min. 360 x 400 mm

Kinematical parameters:

- maximum reach of manipulator 1 800 mm
- No. of degrees of freedom 6
- maximum circumferential velocity for manipulation 0,2 m/s
- angles of rotation:
  - base -130° to +130°
  - arm 0° to +125°
  - forearm -135° to +85°
  - wrist angle -88° to +88°
  - wrist deflection - 88° to +88
- jaws rotation - 102° to +102°
- jaws opening - 0 - 104 (102 - 206)mm

Weight and load capacity:

- maximum load capacity 80 kg
- weight 150 kg
- hydraulic system operational pressure 17 MPa

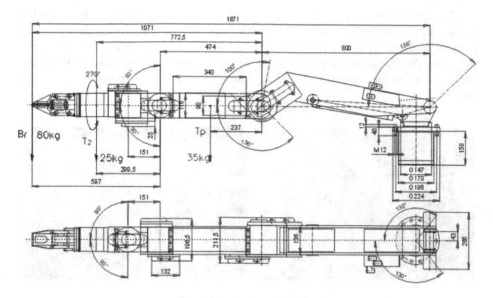

**Fig. 1.** Manipulator MT 80

## 3    Tooling Head with Connections and Quick Couplings for Utilities Transfer

The tooling head is located at the end element of the manipulator MT80 arm. It consists of a felloe and a guiding section. It is fixed in place by means of a gripping joint and is secured against axial movement by a bolt. The front part of the head is provided with two centring pins the purpose of which is to secure reliable centring during sliding in of the guiding system holding the tool (circular saw, hydraulic shears, reciprocating saw, etc.).

The felloe body carries one part of the quick coupling with the relevant connections of the hydraulic medium. The second part of the quick coupling is situated on the guiding bed body of the guiding system. The connection is used for feeding hydraulic power when using the hydraulic shears.

In the rear section of the head (the guiding section) there is a hydraulic lock with connection hoses which provides interlock with the guiding system and the relevant tool.

## 4    Quick Tool Change Stands

The stands are used as storage places for the particular tools. There are two stands altogether. The first stand is used for storing the following tools: handling effector, hydraulic shears and reciprocating saw. The second stand is used for the electromagnetic effector, circular saw and the sampling device.

The stand consists of an outer frame made from rectangular tube. In its bottom section there is a pair of wheels and ion the opposite side there is a handle enabling transfer of the stand. In the top part of the stand there is a separate frame for each of the tools. The frames are suspended from the bottom section to facilitate picking up of the tools by means of the MT80 manipulator. All stands are made from stainless steel.

**Fig. 2.** Hydraulic shears                    **Fig. 3.** Pipes fragmentation

**Fig. 4.** Manipulator handling via control panel

## 5     Evaporator Dismantling and Fragmentation

The parts that had to undergo fragmentation first were those that presented an obstacle and limited the manoeuvring possibilities of the manipulator in the area. These were the connecting rods for distant control of the armatures.

Prior to starting with the fragmentation of the piping it was necessary to take samples of the medium inside it in order to decide about the manner of the subsequent decontamination and handling of the fragments. The next step was the fragmentation of the insulated piping situated between the front wall (with the technological opening) and the evaporator by cutting it with a reciprocating saw into lengths of approx. 400 mm. Some of pipes were cut by hydraulic shears. [2]

When the connecting rods and pipes were removed from the space between the wall with the technological opening and the evaporator and the space was empty, enabling better manoeuvring possibilities for the MT80, dismantling and fragmentation of the evaporator could begin, starting from the front (accessible) section in the following order:

- Insulation sheathing – front section
- Cover and jacket – front section
- Wash water and pulp piping –interior section
- Cover and jacket – rear section
- Insulation sheathing – rear section
- Pedestal – location of the heating element
- Heating element

**Fig. 5.** Evaporator body fragmentation

For fragmentation of highly positioned pipes mobile robotic system MT 15 with a special tool based on reciprocating saw was used (see Fig. 6).

With pure teleoperatory control of the arm, high demand for experience of the operator was manifested, especially so in cases of manipulation in a small constrained space, supported only by a camera system with limited view, accompanied by frequent switching of cameras and remote adjustment of their views. It is for this reason that the control algorithm was supplemented by further partially autonomous

**Fig. 6.** Mobile robot MT 15 and its predecessor

functions providing for correct positioning and determination of the correct cutting path, in order that the tool position would be perpendicular to the cutting surface and thus enable the subsequent autonomous execution of the cutting operation. The 3D model of the fragmented body was firstly obtained by contact of the effector in five arbitrary points and assigning the geometrical type of body (cylinder, sphere, etc.).

For speeding up of the process there are technically elaborated solutions using 3D scanning by means of MESA 400 sensor supported by processing of images from the cameras. This type of scanning can provide a map of the space as well as of the object to be disposed. The first approach can be supplemented by precise contact measuring in three points. The data and the 3D graphic representation of the workspace thus obtained provide conditions for application of partially autonomous modes also for generally shaped bodies.

The higher degree of intelligence in exploiting the obtained data further lies in automatic determination of cross-sections of the object being disposed and their optimal distribution into zones depending on the handling reach of the arm and also for determination of positions of the mobile platform of the manipulator for these zones. The robot is subsequently able to autonomously perform the individual cutting operations and move to new positions for the particular zones. Also optimisation of cutting conditions and the tool´s leaving the cutting zone with subsequent replacement of the tool for a technologically more suitable one for the given task can be solved as autonomous operations. Here the role of the operator is reduced to visual checking of operations and possibly correction of the cutting plan.

# 6    Summary

During the design and construction phase, long-term experience in decommissioning was put to good use. Specific requirements had to be fulfilled:

- adjustment of robotic equipment for use in environment with intensive ionising radiation
- its waterproof construction suitable for proper decontamination
- nonstandard reach of arm ( max 1 800 mm horizontal, 4 200mm vertical)
- use of powerful tooling with quick exchange of individual tools during the work process
- extended intelligence of particular autonomous regimes helping to increase safety and efficiency of robot´s work

The example of the heavy water evaporator demonstrated typical procedure for decommissioning of contaminated technological equipment by remotely controlled manipulators - planning of decommissioning tasks, preparatory tasks, modification of applied tools and design of specific supporting construction for manipulator and finally decontamination and dismantling themselves.

Due to the particularly demanding conditions in strongly contaminated A1 NPP, a team of experts with special know-how in the field of decommissioning has grown up, and unique technological equipment enabling effective and safe work in environments with a high radiation level has been developed.

# References

1. Medved, J., Vargovcik, L.: Decommissioning of the A1 NPP Long Term Storage Facility. In: ICEM Conference, Liverpool, GB (2009)
2. Daniska, V., et al.: The plan of the A1 NPP decommissioning second phase. In: DECONTA (2008) (in Slovak)
3. Bohacik, B., et al.: Fragmentation cell application on evaporator equipment in rooms No. 220, 219 - working program. ZTS VVU Kosice (2009) (in Slovak)

# Part II
# Intelligent Systems

# An Emerging Trend in Ambient Intelligence: Large-Scale Outdoor Applications

Peter Mikulecky

University of Hradec Kralove, Rokitanskeho 62, 50003 Hradec Kralove, Czech Republic
peter.mikulecky@uhk.cz

**Abstract.** The area of Ambient Intelligence recently seems to be matured enough and widely accepted as one of the basic supporting technologies and approaches for human living environment enhancement. There are many interesting applications of wireless sensor networks coupled with ambient intelligence applications, mainly for the creation of indoor intelligent spaces. However, there are just a few of them aiming at the enhancement of outdoor environments in order to support human activities typical for wide-area spaces in open natural environments. There are also a number of useful applications of wireless sensor networks aiming at environmental monitoring, however, just a few of them, if any, are taking advantage of recent achievements of Ambient Intelligence. In our paper, we intend to go further with the ideas of how the Ambient Intelligence used in a wide manner in connection with wireless sensor networks throughout the open natural environment could be beneficial not only for early warning in case of possible disasters, but also as an important supporting tool for people located in outdoor areas due to various reasons, e.g. hitchhikers, hikers on difficult mountains tracks, or even workers in exacting outdoor workplaces (i.e., a coal mine). An idea of a large-scale Ambient Intelligence system based on a wide area wireless sensor network is presented, where the system should be able to monitor the environment, evaluate the collected data and, if necessary, inform the workers in the environment about possible threats and also give some hints for their rescue, using various communication devices.

## 1    Introduction

The environmentally-oriented wireless sensor networks [5] have recently matured enough to become a basis for more complex support of various outdoor activities. Wireless sensor networks are more and more seen as a solution to wide area tracking and monitoring applications, but these networks are usually designed to serve a single application and collected information is commonly available to one authority, usually to the owner of the sensor network. According to [2], the vision for the future generation of wireless sensor networks is of a world where sensing infrastructure is a shared resource that can be dynamically re-purposed and reprogrammed in order to support multiple desirable applications. Furthermore, multiple sensor networks (possibly owned by different authorities) can be combined in a federated fashion in order to create a more

© Springer International Publishing Switzerland 2015                                              179
P. Sinčák et al. (eds.), *Emergent Trends in Robotics and Intelligent Systems*,
Advances in Intelligent Systems and Computing 316, DOI: 10.1007/978-3-319-10783-7_20

complete picture of the world. This idea can be largely exploited for the purposes of a "large-scale Ambient Intelligence" (see, e.g., [8], [15], [16], or [29]).

The first ideas of the Ambient Intelligence used in a large-scale manner appeared in [8]. In addition to incorporating intelligence into sensor nodes within a wide-area wireless sensor network, the authors of [8] proposed to upgrade this vision to the next level where these geographically distributed intelligent sensor networks would become intelligent sensor resources accessible to the users anytime and anywhere. In our earlier papers [15] and [16], we started with some contemplations related to possibilities of using large-scale Ambient Intelligence approaches and applications in a number of environmental problems.

In our paper, we intend to go further on with the ideas of how the Ambient Intelligence used in wide manner in connection with wireless sensor networks throughout the open natural environment could be beneficial not only for early warning in case of possible disasters, but also as a supporting tool for people located in outdoor areas due to various reasons, e.g. hitchhikers, hikers on difficult mountains tracks, or even workers in exacting outdoor workplaces (i.e., a coal mine). An idea of a large-scale Ambient Intelligence system based on a wide-area wireless sensor network is presented, where the system could be able to monitor the environment, evaluate the collected data and, if necessary, inform the workers in the environment about possible threats and possibly give some hints for their rescue, using their mobile devices or other communication means. Such "intelligent outdoor spaces" could use already existing and matured technology of wireless sensor networks, intelligent sensor grids and the Ambient Intelligence with a clear benefit for people situated in such spaces.

## 2    Wireless Sensor Networks Properties

A wireless sensor network is usually a network of distributed autonomous devices that can sense or monitor physical or environmental conditions in mutual cooperation. They consist of a large number of small, inexpensive, disposable and autonomous sensor nodes that are generally deployed in an ad hoc manner in large geographical areas primarily for remote operations. Sensor nodes are severely constrained in terms of storage resources, computational capabilities, communication bandwidth and power supply, as they should be small and inexpensive.

Typically, the sensor nodes are grouped in clusters, and each cluster has a node that acts as a cluster head. All nodes forward their sensor data to the cluster head which, in turn, routes it to a specialized node called sink node (or base station) through a multi-hop wireless communication (cf. [9]). From basic stations, the collected sensor data flows further on to subsequent analysis, usually provided by human specialists. If necessary, they will take a decision related to the circumstances in monitored environment and eventually start an action (e.g., a rescue action).

Kulkarni et al. [9] have identified the following four main properties of wireless sensor networks that seem to be most important for any further contemplation about their large-scale utilization:

- wireless ad hoc nature
- mobility and topology changes
- energy limitations
- physical distribution.

It means that there is no fixed communication infrastructure in wireless sensor networks. New problems, like unreliable and asymmetric links, can appear as a consequence. An advantage of this can be found in the fact that a packet transmitted by a node to another is received by all neighbours of the transmitting node. In such a case a faulty node could be easily substituted by one of its neighbours and communication could continue.

The mobility and topology changes cause that wireless sensor networks may involve dynamic scenarios of their topology composition. New nodes may join the network and the existing nodes may either move through the network or out of it. Nodes may stop their functioning and other nodes may go in or out of transmission radii of other nodes. Environmental wireless sensor networks applications have to be resistant to node failure and dynamic topology. This feature designed them to become a good and reliable basis for the outdoor environments early warning systems (cf. [14], [17], or [20]).

Recently, it is quite often that nodes in most wireless sensor networks have limited energy resources. Usual scenario considers a topology of sensor nodes with a limited number of more powerful base stations. The maximum power available to the sensor nodes is consumed by communication tasks. The maintenance or recharging of batteries on the sensor nodes is not possible after the deployment, therefore new energy sources are desirable. The most promising seem to be solar powered batteries, but some further research in this direction is highly appreciated.

Each node in a wireless sensor network is an autonomous computational unit that communicates with its neighbours via messages. Data is distributed throughout the nodes in the network and can be gathered at a central station at high communication costs only. Consequently, algorithms that require global information from the entire network become very expensive [9]. However, this physical distribution can be naturally modelled by multi-agent systems, and various agent-based approaches can be utilized.

## 3    Related Works

The outdoor-oriented applications of Ambient Intelligence approaches are driven by recent achievements in the area of wide-area wireless sensor networks. Let us shortly describe some related works focused on this field.

Considerable effort has recently been devoted to the area of sensor networks and their important applications, as mentioned in [11]. There is a number of publications dealing with technical possibilities and properties of sensor networks. The book [11] lists interesting results oriented on context-awareness of sensors and sensor networks. The idea of employing context-awareness is that if sensors knew more about the context in which they are acting, then they could adapt their behaviour and function only when needed and to the extent adequate to current circumstances. This aspect can be important also for energy consumption of the sensor. A lot of related results can be found in [1], [3] or [7], or in a useful book on sensor networks [21].

Among a number of recent interesting environmental applications, the FieldServer Project [22], and the Live E! Project [12] can be mentioned.

The FieldServer Project is oriented on development and networked applications of so-called Field Servers. A Field Server [22] is a wireless sensor network that will enhance the monitoring of environmental factors by allowing the sensing nodes to be located at precise locations on fields, reducing overhead installation costs and allowing for real-time data collection. For instance, in Japan, Field Servers were developed for applications at farms. They produce real-time images for security guards and environmental data for farming. Agronomists, physiologists and ecologists can exploit high-resolution real-time images in order to react to any specific situation that could appear in the environment. Many types of Field Servers have been developed up to now, however, still without implementing any approaches related to Ambient Intelligence area.

The second example, the Live E! Project [12], is an open research consortium with the aim to explore a platform sharing digital information related to the living environment. Using a low cost weather sensor nodes with Internet connectivity, a nationwide sensor network was deployed [12]. The network has accommodated more than 100 stations. The application of this weather station network is intended for disaster protection/reduction/recovery and also as educational material for students. Nevertheless, proper function of the whole network is also strongly dependent on human supervisors evaluating the collected data in related dispatching centres.

One of the most significant drivers for the research of wireless sensor networks is environmental monitoring. Its potential will not only enable scientists to measure properties that have not been observable before but also, by ubiquitous monitoring of the environment and supplying the related data to relevant supervising bodies, to create a basis of early warning systems for various environmental disastrous situations and their management. As [10] points out, the relatively low cost of the wireless sensor network devices allow the installation of a dense population of nodes that can adequately represent the variability present in the environment. They can provide various risk assessment information, for example alerting farmers at the onset of frost damage. Wireless sensor network-based fire surveillance systems were designed and implemented as well. They can measure temperature and humidity, and detect smoke following by early warning information broadcasting [10]. Sensors are able to consider certain dynamic and static variables such as humidity, type of fuel, slope of the land, direction and speed of the wind, smoke, etc. They also allow to determine the direction and possible evolution of the fire edge.

In literature, there is only a little works oriented on a kind of a service to the potentially endangered persons in an open-door environment, e.g., in natural or urban environment. For instance, there are some attempts of preventing children from potentially dangerous situations in urban environment. Probably the first ubiquitous system to assist the outdoor safety care of schools kids in real world is described in [24].

A number of papers are devoted to various solutions for tourist assistance, mainly oriented on context-aware tourist navigation on their routes. The usual approach is in the deployment of intelligent agents which collectively determine the user context and retrieve and assemble a kind of simple information up to multi-media presentations that are wirelessly transmitted and displayed on a Personal Digital Assistant (PDA). However, these tourism-oriented applications are usually deployed for the navigational purposes without having capabilities to warn the user from potentially dangerous situations that can appear on their routes.

As an example of, in a sense, similar system, we refer to [28]. The deployed sensor network aimed to assist the geophysics community and, in contrast to the volcanic data acquisition equipment existing at that time, the nodes used in the sensor network were smaller, lighter and consumed less power. The resulting spatial distribution greatly facilitated scientific studies of wave propagation phenomena and volcanic source mechanisms. Certainly, we can imagine a number of potentially dangerous situations that may endanger people working closely to a volcano. Enhancing the purely geophysical sensor networks by the features mentioned above could improve the safety of works near a volcano.

One of the rare outdoor health-care applications is described in [26]. The application aims to support patients suffering of dementia when being in outdoor environment. However, this application is based more on wearable sensors approach than on a wireless sensor network installed independently from monitored humans.

Another example belongs also to the area of potentially dangerous workplaces. The result of [27] seems to be one of those attempts that aimed directly at developing a sensor network for monitoring possibly dangerous situations (gas explosion) in a large yet closed environment - a coal mine in China. Localization of miners present in a coal mine is also implemented in this system, however, there is no possibility to start any rescue actions by the system itself, automatically by processing the results supplied by the network nodes. Nevertheless, the experience with this sensor network in a Chinese coal mine is good and inspirational, according to [27].

## 4 Proposed Approach

In order to support personal outdoor activities, her/his geographic location must be identified as an important contextual information that can be used in a variety of scenarios as disaster relief, directional assistance, context-based advertisements, or early warning of whether the particular person is in any potentially dangerous situation. GPS provides accurate localization outdoors, although it is not very useful inside buildings.

Based on the above mentioned positive properties of wireless sensor networks, we propose a complex ambient intelligence system covering a wide area with wireless sensor network implemented in a potentially risky natural environment (mountain areas, surface mines, seashores, river basins or water reservoirs, forests, etc.) that will be able to perform following tasks:

- monitoring usual hydro-meteorological parameters of the environment (air pressure, temperature, humidity, soil moisture, etc.),
- monitoring indications of possible threats (seismo-acoustic signals, smoke, water on unusual places, etc.),
- monitoring appearance and movement of human beings (and possibly also animals, vehicles, etc.) in an area,
- evaluating data collected from the sensor network and identifying possibly dangerous situations,
- identification of possibly endangered human beings in under monitored area,
- attempting to contact persons in danger possibly via their mobile devices and starting to provide all the necessary information and knowledge support aiming to help them to escape from the dangerous situation (including eventual alarming of a rescue squad).

An AmI system working over such a sensor network will be able to evaluate all collected data and decide about possible active intervention in the environment aiming to help the people located at outdoor spaces in danger.

Based on ideas presented by Iqbal and others in [8], we can think about a Large-scale Ambient Intelligence application as a large set of geographically widely distributed intelligent sensor resources with the main purpose to significantly increase the intelligence of various segments of real nature. By a smart sensor resource we mean a kind of ambient artefact, namely a combination of an advanced sensor with ubiquitous computing and a communicating processor integrated in the sensor. They purpose will be given by their main tasks, so that a number of their specific types could be possible.

Multi-agent architectures seem to be applicable here very well, as it is common in the case of large networks of sensors. We can tract various types of intelligent sensors and guards as agents with appropriate level of intelligence, recent dispatchers or even dispatching centres can be modelled as supervising agents (e.g., river basin management dispatching centres or fire brigades dispatching centres, etc.), see [19].

Personal outdoor activity support should provide relevant and reliable information to users often engaged in other activities and not aware of some hazardous situations that he or she could possibly encounter. There are only a small number of attempts to solve the related dangerous situations that can be described using the following scenario:

A user appears in an open-door natural environment performing her/his working mission, a kind of leisure time activity (hiking tour, mountaineering, cycling, etc.), or simply being an inhabitant of the area. A sudden catastrophic situation (storm, flash flood, landslide, forest fire, etc.) could put the person in a risky if not a life endangering situation. The "intelligent outdoor space", based on a federated wireless sensor network, is ubiquitously monitoring the area and estimating any possible appearance of a dangerous situation. If necessary, the network will proactively broadcast an early warning message to the user, offering her/him related navigation services supporting escape from the dangerous situation. This message can be delivered via mobile device of the user or broadcasted to the area by other suitable means (radio, speakers installed on suitable places, etc.).

In order to design some solution for outdoor spaces, we can imagine a number of sensors acting as various kinds of guards. Let us present some examples which are technologically feasible and frequently used in large-scale wireless sensor networks:

- water level guards, monitoring surface water level or even groundwater level and watching over potentially dangerous or at least unusual situations.
- water quality guards, monitoring surface and groundwater quality and watching over possible contaminations or pollutions.
- air pollution guards, monitoring air quality, watching over possible pollutions.
- wind velocity sensors, monitoring wind velocity and watching over potentially dangerous situations.
- soil moisture sensors, measuring level of soil humidity, e.g. in forests or in a river watershed, aiming at monitoring the degree up to which the land segment is saturated by water and measuring the capacity of further possible saturation.

Of course, other kinds of intelligent sensors integrating ubiquitous monitoring (computing) of measured parameters with ubiquitous communication with other sensors – agents – in the area are possible as well.

We believe that the main application area for large-scale ambient intelligence will be any kind of prevention, connected with early warning facilities. Such areas as fire prevention, water floods prevention and early warning or accident prevention in urban traffic could be clear candidates.

In water floods prevention we can imagine the usage of the following agents:

- water level monitoring agents;
- land segments saturation (moisture) guards;
- water reservoir handlers;
- supervising agents.

The concept of our solution of a problem e.g., water floods, could, etc., consists of the following steps:

- Establishing a large-scale wireless sensor network consisting of, e.g., water level guards, completed by a number of sensor clusters composed of soil moisture guards situated in the land segments that are already known as critical from the soil saturation point of view.
- The established large scale wireless sensor network will be embedded in a multi-agent architecture where the particular sensor clusters of various types will play roles of group of agents in the multi-agent architecture.
- Special roles are assigned to manipulating agents as, e.g., water reservoir handlers or river weirs manipulators.
- The whole system can be designed as hierarchical, as there could be a number of concentrators (agents collecting the data) as well as messages from the groups of agents (sensor clusters) defined in the previous steps. These concentrators then communicate mutually as well as with the supervisor which is an agent with the task of evaluating the data as well as messages.

- Further on, the supervising agent will evaluate the messages from the lower level agents and, after judging the level of their importance, it will start a respective action or a whole sequence of actions adequate to the situation.
- The supervising agent will communicate also with localization agents which are responsible for keeping information about the people localized in the monitored area. If the supervising agent evaluates the whole situation as dangerous for localized people, it will then send a request to the communicating agents to send an urgent message to PDAs of the monitored people with a hint what to do in order to escape from the danger.
- The crowdsourcing approach can likely be utilized here for monitoring and collecting actual messages or responses from people appearing throughout the monitored area and communicating wirelessly with the underlying system or with other users (cf. [4] or [6]).

Similar solutions can be imaged also for other dangerous situations that are likely to appear in an outdoor space. These will be elaborated in a detail further on in other publications. Some our work in this direction has been already published in [13], [16], [18] or [19].

## 5    Conclusions

The ideas about the application of sensor networks in open natural environments are not too old, first relevant applications are dated to early 2000s (see, e.g., [25]). As the technology of wireless sensor networks is evolving rapidly and the advancements in the area is tremendous, it is time not only to deploy interesting wireless sensor network applications for monitoring open natural environments but also to implement more ambitious projects reflecting recent advancements in the area of Ambient Intelligence. In the paper, after a short analysis of various recently used approaches, we presented an idea (or a concept) of a large-scale ambient intelligence application over a wide environmental sensor network aiming at monitoring and, possibly, early warning in case of a threat in the monitored outdoor environment. The system should be primarily focused on outdoor workplaces and workers acting there as well as it could be beneficial for all the people located in the monitored outdoor space. Their mobile devices can be naturally used for communicating with the monitoring system with a possibility of prescription of the desired messaging or warning service. The idea of the described outdoor environments enhanced by large-scale ambient intelligence (we shall call them "intelligent outdoor spaces", see [14]) is in further development. We hope that such intelligent outdoor spaces could become a reality in a short time, thus contributing to saving many lives of potentially endangered people.

**Acknowledgments.** This research has been partially supported by the Czech Science Foundation project No. P403/10/1310 "SMEW – Smart Environment at Workplaces" as well as the FIM UHK Excellence Project "Agent-based models and Social Simulation".

# References

1. Cardell-Oliver, R., Smettem, K., Kranz, M., Mayer, K.: A reactive soil moisture sensor network: Design and field evaluation. Int. Journal of Distributed Sensor Networks 1, 149–162 (2005)
2. Efstratiou, C.: Challenges in supporting federation of sensor networks. In: NSF/FIRE Workshop on Federating Computing Resources (2010)
3. Elnahrawy, E., Nath, B.: Context-aware sensors. In: Karl, H., Wolisz, A., Willig, A. (eds.) EWSN 2004. LNCS, vol. 2920, pp. 77–93. Springer, Heidelberg (2004)
4. Goodchild, M.F., Glennon, J.A.: Crowdsourcing geographic information for disaster response: a research frontier. International Journal of Digital Earth 3(3), 231–241 (2010)
5. Hart, J.K., Martinez, K.: Environmental sensor networks: A revolution in the earth system science? Earth-Science Reviews 78, 177–191 (2006)
6. Heinzelman, J., Waters, C.: Crowdsourcing Crisis Information in Disaster-Affected Haiti. In Special Report 252, United States Institute of Peace (October 2010)
7. Huaifeng, Q., Xingshe, Z.: Context-aware Sensornet. In: Proc. 3rd International Workshop on Middleware for Pervasive and Ad-Hoc Computing, pp. 1–7. ACM Press, Grenoble (2005)
8. Iqbal, M., et al.: A sensor grid infrastructure for large-scale ambient intelligence. In: 2008 Ninth International Conference on Parallel and Distributed Computing, Applications and Technologies, pp. 468–473. IEEE (2008)
9. Kulkarni, R.V., Förster, A., Venayagamoorthy, G.M.: Computational Intelligence in Wireless Sensor Networks: A Survey. IEEE Communications Surveys and Tutorials 13(1), 68–96 (2011)
10. Lloret, J., Garcia, M., Bri, D., Sendra, S.: A wireless sensor network deployment for rural and forest fire detection and verification. Sensors 9, 8722–8747 (2009)
11. Loke, S.: Context-Aware Pervasive Systems. Auerbach Publications, Boca Raton (2007)
12. Matsuura, S., et al.: LiveE! Project: Establishment of infrastructure sharing environmental information. In: 2007 International Symposium on Applications and the Internet Workshops (SAINTW 2007), p. 67. IEEE (2007)
13. Mikulecky, P.: Ambient intelligence and smart spaces for managerial work support. In: 3rd IET International Conference on Intelligent Environments, IE 2007, September 24-25, pp. 560–563 (2007)
14. Mikulecky, P.: Intelligent Outdoor Spaces. In: Botía, J.A., Charitos, D. (eds.) Workshop Proceedings of the 9th International Conference on Intelligent Environments, pp. 377–385. IOS Press, Amsterdam (2013)
15. Mikulecky, P.: Large-scale ambient intelligence. In: Mastorakis, N., Mladenov, V. (eds.) Advances in Data Networks, Communications, Computers, Proc. of the 9th WSEAS Conference on Data Networks, Communications, Computers, p. 12. WSEAS Press (2010)
16. Mikulecky, P.: Large-scale ambient intelligence – Possibilities for environmental applications. In: Čech, P., Bureš, V., Nerudová, L. (eds.) Ambient Intelligence Perspectives II. Ambient Intelligence and Smart Environments, vol. 5, pp. 3–10. IOS Press (2010)
17. Mikulecky, P.: User Adaptivity in Smart Workplaces. In: Pan, J.-S., Chen, S.-M., Nguyen, N.T. (eds.) ACIIDS 2012, Part II. LNCS (LNAI), vol. 7197, pp. 401–410. Springer, Heidelberg (2012)
18. Mikulecký, P., Olševičová, K., Cimler, R.: Outdoor Large-scale Ambient Intelligence. In: IBIMA (2012)

19. Mikulecky, P., Olsevicova, K., Ponce, D.: Knowledge-based approaches for river basin management. Hydrol. Earth Syst. Sci. Discuss 4, 1999–2033 (2007)
20. Mikulecky, P., Tucnik, P.: Ambient Intelligence for Outdoor Activities Support: Possibilities for Large-Scale Wireless Sensor Networks Applications. In: ICSNC 2012: The Seventh International Conference on Systems and Networks Communications, pp. 213–217. IARIA, Lisbon (2012)
21. Misra, S., Woundgang, I., Misra, S.C. (eds.): Guide to Wireless Sensor Networks. Springer, London (2009)
22. Ninomiya, S., Kiura, T., Yamakawa, A., Fukatsu, T., Tanaka, K., Meng, H., Hirafuji, M.: Seamless integration of sensor network and legacy weather databases by MetBroker. In: 2007 International Symposium on Applications and the Internet Workshops (SAINTW 2007), p. 68. IEEE (2007)
23. Ruiz-Garcia, L., Lunadei, L., Barreiro, P., Robla, J.I.: A review of wireless sensor technologies and applications in agriculture and food industry: State of the art and current trends. Sensors 9, 4728–4750 (2009)
24. Takata, K., Shina, Y., Komuro, H., Tanaka, M., Ide, M., Ma, J.: Designing a context-aware system to detect dangerous situations in school routes for kids outdoor safety care. In: Yang, L.T., Amamiya, M., Liu, Z., Guo, M., Rammig, F.J. (eds.) EUC 2005. LNCS, vol. 3824, pp. 1016–1025. Springer, Heidelberg (2005)
25. Tanenbaum, A.S., Gamage, C., Crispo, B.: Taking Sensor Networks from the Lab to the Jungle. Computer 2006, 98–100 (2006)
26. Wan, J., Byrne, C., O'Hare, G.M.P., O'Grady, M.J.: OutCare: Supporting Dementia Patients in Outdoor Scenarios. In: Setchi, R., Jordanov, I., Howlett, R.J., Jain, L.C. (eds.) KES 2010, Part IV. LNCS (LNAI), vol. 6279, pp. 365–374. Springer, Heidelberg (2010)
27. Wang, X., Zhao, X., Liang, Z., Tan, M.: Deploying a wireless sensor network on the coal mines. In: Proceedings of the 2007 IEEE International Conference on Networking, Sensing and Control, London, UK, pp. 324–328 (2007)
28. Werner-Allen, G., Lorincz, K., Welsh, M., et al.: Deploying a wireless sensor network on an active volcano. IEEE Internet Computing 10(2), 18–25 (2006)
29. Yeong, Y.-S., et al.: Large-Scale Middleware for Ubiquitous Sensor Networks. IEEE Intelligent Systems, 2–13 (March/April 2010)

# Interactive Evolutionary Computation
# for Analyzing Human Characteristics

Hideyuki Takagi

Kyusyu University, Fukuoka, Japan
http://www.design.kyushu-u.ac.jp/~takagi/

**Abstract.** We emphasize that interactive evolutionary computation (IEC) can be used not only to optimize a target system based on an IEC user's subjective evaluations but also to analyze the characteristics of the IEC user. We introduce four research works as concrete examples of this new research direction: measuring a perceived range for emotional expressions, finding unknown auditory knowledge through hearing-aid fitting and cochlear implant fitting, and modeling of human awareness mechanism.

## 1    Introduction

Interactive evolutionary computation (IEC) is a framework for optimizing a target system based on human subjective evaluations. There are many tasks which performances are hard to be measured or almost impossible but can be evaluated by human beings. IEC shown in Fig. 1 (a) can optimize such tasks by involving a human user in an optimization loop.

The first direction of IEC research is to expand IEC application areas. IEC has been applied to wide variety of areas. They are roughly categorized into three: (1) artistic applications such as creating computer graphics (CG), music, editorial design, and industrial design, (2) engineering applications such as acoustic or image processing, robotics control, data mining, generating programming code, and media database retrieval, and (3) others such as educations, games, and geological simulation. See these perspectives in [8].

The second direction of IEC research is to reduce IEC user fatigue and make IEC practical. IEC users must repeat evaluations many times and feedback them to a tireless computer. This nature causes IEC user fatigue, and especially, it becomes a serious problem for practical use when end-users use IEC.

Accelerating IEC search is one of solutions for this fatigue problem, and developing new IEC framework with less human fatigue is other solution. We introduce our trials in this paper. Besides them, there are several other approaches to overcome the fatigue problem such as improving IEC user interface, allowing an IEC user to intervene in an EC search, introducing a user model made by machine learning, and others. See these works in [8].

© Springer International Publishing Switzerland 2015
P. Sinčák et al. (eds.), *Emergent Trends in Robotics and Intelligent Systems,*
Advances in Intelligent Systems and Computing 316, DOI: 10.1007/978-3-319-10783-7_21

The third direction of IEC research is to use the IEC as a tool to analyze human characteristics; see Fig. 1(b). This is a new and unique approach, and there are few related works so far. We focus on this third research direction in this paper.

Since an IEC target system is optimized based on a human psychological evaluation scale, we may know the scale indirectly by analyzing the optimized target system. It somehow has similarity to reverse engineering in software engineering. Other explanation of this approach is that IEC is a tool to visualize impressions or images in mind. Artists have skills for expressing them by drawing pictures, playing musical instruments, programming computer graphics and writing in poems, for example. However, it is hard for many ordinary people who have no such skills to express the impressions or mental images. IEC helps those who have less skill to express the mental images using IEC-based systems.

Thanks to this kind of IEC use, we may be able to analyze human characteristics by analyzing obtained optimized systems and their system outputs. Through the analysis, we are looking forward to finding out new psychological or physiological unknown facts.

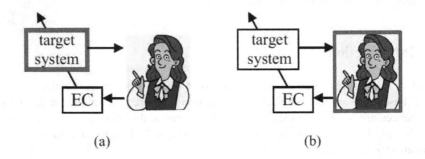

(a)                                    (b)

**Fig. 1.** IEC frameworks (a) for optimization and (b) for human science

## 2    Measuring a Perceived Range for Emotional Expressions

We applied IEC to measure a happy–sad range in human mind and compared the ranges of schizophrenics and mental healthy people. Some therapists feel that face emotional impressions of schizophrenic patients are fewer than those of mental healthy people through their experiences. However, there was no way to measure the range.

We asked 3 schizophrenics and 5 mental healthy students to design 3-D CG lighting of the happy impression and the sad impression using our IEC-based 3-D CG lighting design support system [1, 2] and asked 33 human subjects to evaluate 28 pairs (=(3+5) C2 ) of designed lighting images. Fig. 2 is the psychological scale of happy constructed using the Scheffe´'s method of paired comparison.

The happy–sad ranges (Fig. 3) obtained from the experimental results imply that it is hard for schizophrenic patients to identify especially a happy impression in lighting. It is expected that this IEC approach may provide new data that are helpful for psychiatric diagnostics [10].

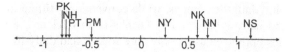

**Fig. 2.** Psychological scale constructed using the Scheffe´'s method of paired comparison and impression levels of the eight best lightings designed by three schizophrenics (PK, PT, and PM) and five mental healthy students (NH, NY, NK, NN, and ND). The bigger measure values, the higher evaluation of happy.

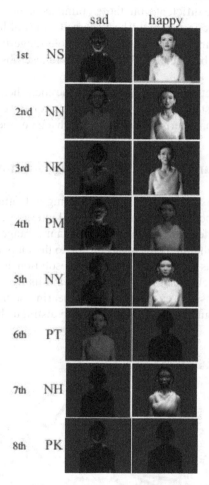

**Fig. 3.** Rank order of happy–sad expression range obtained as a difference between two happy and sad scales. See the subject ID's in the caption of Fig. 2.

## 3    Hearing-Aid Fitting and Finding Unknown Knowledge

IEC is the best way for hearing-aid fitting because sound qualities for a certain hearing-aid user cannot be measured. Other advantage is that it allows us to fit a hearing aid using any daily-life sounds, while conventional fitting method has to use only pure tones and narrow band noise. Thanks to this feature, we could compare fitting characteristics optimized using only pure sounds and daily-life sounds and find several unknown facts [11].

They are: the characteristics of hearing aids optimized using speech sounds were different from those optimized using pure tones or band pass noise; those optimized using speech sounds of speaker i with/without noise were almost similar to those optimized using speech sounds of speaker j with/without noise (i = j); those optimized using speech sounds were different from those optimized using music.

Nobody had known the facts that the best characteristics of hearing-aids depend on sound types used for fitting. It implies that an audible range in human sense level is not the final cue for the best hearing. We could find these observations thanks to an IEC technique.

From the obtained facts, we can imagine that the ideal hearing-aids in the future would have multiple best hearing-aid characteristics for different acoustic environments and switch their characteristics based on the change of the environments.

## 4    Cochlea Implant Fitting and Finding Unknown Knowledge

Cochlea-implant fitting is a similar task with hearing-aid fitting and has been conducted based on two hypotheses for better fitting: (1) the more electric channels of a cochlea-implant, the better and (2) the wider dynamic range of each channel, the better. As frequency resolution increases according to the number of electric channels, the hypothesis (1) means that higher frequency resolution helps to distinguish the difference of frequency characteristics of phonemes; this hypothesis sounds natural. The hypothesis (2) means that enabling a user to hearing sounds from the minimum level to the maximum comfortable level is helpful to distinguish sounds; this hypothesis also sounds natural.

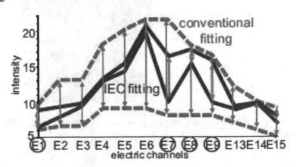

**Fig. 4.** Fitting characteristics of conventional cochlear implant fitting and that used IEC. Horizontal axis means electric channels and a vertical axis means electric voltage of each channel. This figure was remade based on an image in [6].

Interactive genetic algorithms (IGA) was used to tune the fitting parameters of cochlea implants [6]. Their experimental result was that dynamic ranges of all 15 channels were almost 0 except 3 or 4 channels, and the dynamic ranges of the exceptional 3 or 4 channels were narrower than the maximum ranges (see Fig. 4). Nevertheless, its recognition rate with IGA fitting was higher than that of manual fitting.

This result did not match to the mentioned two hypotheses. It implies that there must be unknown audio-psychophysiological facts. We are conducting this cochlea-implant fitting using paired comparison-based interactive differential evolution (IDE) [12] together with medical departments of two universities and trying to find the unknown facts.

## 5     IEC for Awareness Scence

Although there are many application papers on awareness computing, such as content awareness, location awareness, power awareness, and others, there are few papers that focus on basic research for awareness science such as analyzing awareness mechanisms and making awareness models. Once computer has an awareness model, it must be useful for communication with human users and support human awareness. IEC can be a tool for this research.

There must be several types of human awareness at different psychological levels including a sensory level, a perceptual level, and a cognitive level. To start this basic research, let us define awareness as realize latent variables that explain the relationship between inputs to a human and outputs from the human.

Let us consider an example case that someone finds a preferred pot but cannot explain why he or she prefers it. This case means that the relationship between a visual image of the pot (input x) and the evaluation (output z) is complex. After thinking for a while, he or she may become aware that the vertical-horizontal ratio of the pot (r) and the curvature of design pattern (c) are the points underpinning their evaluation and become to be able to explain the reason for their preference as Fig. 5. Then, we may say that he or she is aware of the two hidden variables of the vertical-horizontal ration of the pot and the curvature of design pattern.

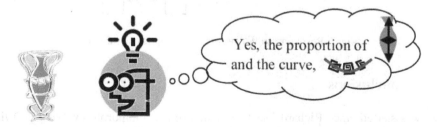

**Fig. 5.** He or she is aware of the vertical-horizontal ratio of the pot (r) and the curvature of design pattern (c) as key points of his or her preference of the pot

That is, we can say that " f () for z = f (x) was so complex that he or she could not explain the input-output relationship at the first glance. However, he or she found the latent variables, r and c, and could interpret them as $z = f(x) = g1(r, c)$, where $r = g2(x)$ and $c = g3(x)$." In other words, we can say that he or she became to explain a complex relationship between z and x explicitly thanks to r and c.

One approach to model an awareness mechanism is to make an IEC user model using machine learning (Fig. 6) and resolve the obtained user model (Fig. 7). Now we are considering some approaches for obtaining latent variables: (a) structure analysis of a structured NN-FS model, (b) introducing statistical methods for finding latent variables, (c) making a learning model by a math equation using genetic programming and analyzing the obtained equation, and (d) others.

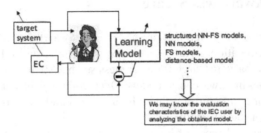

**Fig. 6.** Leaning inputs and outputs to/from an IEC user and making its user model

IEC for Analyzing Human Characteristics

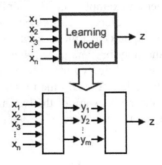

**Fig. 7.** Resolving an IEC user model using latent variables

# 6    Conclusions

IEC was started since Richard Dawkins demonstrated biomorph, evolving 2-D line figures, in 1986, and its research increased from 1990's. IEC applications were spread from CG applications and engineering, edutainment, and many other areas. Now the number of IEC papers became 835 (in Scopus database as of July, 2013).

Still now, majority of IEC research is on optimizing target systems. Although expanding IEC to application areas that are hard for conventional optimization methods are useful and important, IEC has other potentials. To demonstrate it, we introduced unique our IEC research, i.e. IEC for human science, as the third direction of IEC research. We hope that many people realize that the capability of IEC is not only optimization from these efforts and IEC can con-tribute quite widely.

**Acknowledgments.** This work was supported in part by Grant-in-Aid for Scientific Research (23500279).

# References

1. Aoki, K., Takagi, H.: 3-D CG lighting with an interactive GA. In: 1st Int. Conf. on Conventional and Knowledge-based Intelligent Electronic Systems (KES 1997), Adelaide, Australia, pp. 296–301 (May 1997)
2. Aoki, K., Takagi, H.: Interactive GA-based design support system for lighting design in 3-D computer graphics. Trans. of IEICE J81-DII(7), 1601–1608 (1998) (in Japanese)
3. Dawkins, R.: The blind watchmaker. W. W. Norton & Company, Inc., New York (1986)
4. Ingu, T., Takagi, H.: Accelerating a GA convergence by fitting a single-peak function. In: IEEE Int. Conf. on Fuzzy Systems (FUZZ-IEEE 1999), Seoul, Korea, pp. 1415–1420 (August 5, 1999)
5. Jin, Y.: A comprehensive survey of fitness approximation in evolutionary computation. Soft Computing 9(1), 3–12 (2005)
6. Legrand, P., Bourgeois-Republique, C., Peán, V., et al.: Interactive evolution for cochlear implants fitting. Genetic Programming and Evolvable Machines 8(4), 319–354 (2007)
7. Pei, P., Takagi, H.: Fourier analysis of the fitness landscape for evoltionary search acceleration. In: IEEE Congress on Evolutionary Computation (CEC 2012), Brisbane, Australia, pp. 1–7 (June 2012)
8. Takagi, H.: Interactive evolutionary computation: fusion of the capabilities of EC optimization and human evaluation. Proceedings of the IEEE 89(9), 1275–1296 (2001)
9. Takagi, H., Ingu, T., Ohnishi, K.: Accelerating a GA convergence by fitting a single- peak function. J. of Japan Society for Fuzzy Theory and Intelligent Informatics 15(2), 219–229 (2003) (in Japanese)
10. Takagi, H., Takahashi, T., Aoki, K.: Applicability of interactive evolutionary computation to mental health measurement. In: IEEE Int. Conf. on Systems, Man, and Cybernetics (SMC 2004), The Hague, the Netherlands, pp. 5714–5718 (October 2004)
11. Takagi, H., Ohsaki, M.: Interactive evolutionary computation-based hearing-aid fitting. IEEE Trans. on Evolutionary Computation 11(3), 414–427 (2007)
12. Takagi, H., Pallez, D.: Paired comparison-based interactive differential evolution. In: 1st World Congress on Nature and Biologically Inspired Computing (NaBIC 2009), Coimbatore, India, pp. 375–480 (December 2009)

# Robust TS Fuzzy Fault Detection Filters Design

Anna Filasová, Vratislav Hladký, and Dušan Krokavec

Department of Cybernetics and Artificial Intelligence, Letná 9/B, 042 00 Košice, Slovakia
{anna.filasova,vratislav.hladky,dusan.krokavec}@tuke.sk

**Abstract.** One principle for designing the robust Takagi-Sugeno fuzzy fault detection filter, dedicated to a class of continuous-time nonlinear MIMO system,
is treated in this paper. The problem addressed can be designated as an approach exploiting the fuzzy reference model to reflect the problem as an $H_\infty$ optimization task, guaranteeing the fault detection performance and the state observer stability. The conditions are outlined in the terms of linear matrix ine-
qualities to possess a stable structure closest to optimal asymptotic properties.

## 1  Introduction

The fault detection filters, usually relying on the use of any type of state observers, are
mostly used to produce the fault residuals in fault tolerant control systems. Because it is
generally not possible to decouple totally fault effects from the perturbation influence in
residuals, the $H_\infty$ approach is used to tackle this conflict [3], [6] in part. Since faults are
detected usually by setting a threshold on the residual signal, determination of actual
threshold is often formulated as an adaptive task [4], [5].

The nonlinear system theory has emerged as a method of use in the state observer
based residual generator design for nonlinear systems [10], [18], [21], although
Lipschitz conditions may be strongly restrictive in many cases [22]. An alternative is
the Takagi–Sugeno (TS) fuzzy approach [16] which avails local system dynamics
approximation techniques and gives descriptions permitting utilization of the system
state space representation. In the light of the above, TS fuzzy fault detection filters
(FDF) have attracted interest in fault detection (see, e.g., [9]). Different TS fuzzy
observers [13], [17] as well as FDF structures, were designed [2], [19], [20], usually
exploiting the linear matrix inequality (LMI) approach. The new trends reduce the
robust FDF design to a standard $H_\infty$ model-matching problem with the main goal to
discriminate the effect between the fault and the disturbances in FDF signals.

The main contribution of the paper is to present an extension principle for designing robust TS fuzzy FDFs. Following the idea concerning the residual reference models [1], [7], the new specification of TS fuzzy reference residual models (RRM) by a
cross-bonds matrix is introduced in the paper. Using RRM, the robust TS fuzzy FDF
design problem is formulated as an $H_\infty$ optimization task, considering both the robustness against disturbances and the sensitivity to faults, as well as guaranteeing the
fault detection performance and the observer stability.

© Springer International Publishing Switzerland 2015

P. Sinčák et al. (eds.), *Emergent Trends in Robotics and Intelligent Systems,*

Advances in Intelligent Systems and Computing 316, DOI: 10.1007/978-3-319-10783-7_22

## 2     TS System Model

The systems under consideration devolve to a class of multi-input multi-output (MIMO) nonlinear dynamic continuous-time systems, described as follows:

$$\dot{q}(t) = \sum_{i=1}^{s} h_i\left(\boldsymbol{\theta}(t)\right)\left(A_i q(t) + B_i u(t) + B_{fi} f(t) + B_{di} d(t)\right), \tag{1}$$

$$y(t) = Cq(t), \tag{2}$$

where $q(t) \in \mathbb{R}^n$, $u(t) \in \mathbb{R}^r$, $y(t) \in \mathbb{R}^m$ are vectors of the state, input and output variables, respectively, $C \in \mathbb{R}^{m \times n}$, $A_i \in \mathbb{R}^{n \times n}$, $B_i \in \mathbb{R}^{n \times r}$, $B_{fi} \in \mathbb{R}^{n \times r_f}$, $B_{di} \in \mathbb{R}^{n \times r_d}$, $i = 1,2, \dots s$ are known real matrices and $d(t) \in \mathbb{R}^{r_d}$ is the disturbance input that belongs to $L_2\langle 0, +\infty)$. It is considered that a fault $f(t)$ may occur at an uncertain time, the size of the fault is unknown but bounded, and all pairs $(A_i, C)$, $i = 1,2, \dots s$ are observable. The membership function $h_i(\boldsymbol{\theta}(t))$ is the averaging weight for the $i$-th fuzzy rule, satisfying, by definition, the following properties

$$0 \le h_i\left(\boldsymbol{\theta}(t)\right) \le 1, \quad \sum_{i=1}^{s} h_i\left(\boldsymbol{\theta}(t)\right) = 1 \text{ for all } i \in \langle 1, s\rangle. \tag{3}$$

Assuming that nonlinear terms in the nonlinear system description are bounded in associated sectors, will operate within the system, the number of these nonlinear terms is $p$ and the number of sector functions varies from 2 to k, then the number of linear sub-models is $s \in \langle 2^k, p^k \rangle$. Note, $h_i(\boldsymbol{\theta}(t))$, $i = 1,2, \dots s$ are calculated from all combinations of the sector functions. The vector $\boldsymbol{\theta}(t) \in \mathbb{R}^o$ of the structure

$$\boldsymbol{\theta}(t) = [\theta_1(t) \quad \theta_2(t) \quad \cdots \quad \theta_o(t)] \tag{4}$$

is the vector of premise variables. A premise variable generally represents any measurable system variable occurring in the sector nonlinear terms. It is supposed in the following that all premise variables are measurable and none of them are a function of the input variables defined in $u(t)$. More details can be found, e.g., in [12], [17].

## 3     Reference Model Design

For the purpose of residual generation, the TS fault detection filter (FDF) is used in a standard structure, designed together with the residual signal equation as

$$\dot{q}_e(t) = \sum_{i=1}^{s} h_i\left(\boldsymbol{\theta}(t)\right)\left(A_i q_e(t) + B_i u(t) + J_i\left(y(t) - y_e(t)\right)\right) \tag{5}$$

$$y_e(t) = Cq_e(t), \tag{6}$$

$$r(t) = \sum_{i=1}^{s} h_i\left(\boldsymbol{\theta}(t)\right)V_i C\left(y(t) - y_e(t)\right), \tag{7}$$

where $q_e(t) \in \mathbb{R}^n$ is the estimation of the system state, $y_e(t) \in \mathbb{R}^m$ is the observed output vector, $r(t) \in \mathbb{R}^{m_r}$ is the residual signal and $J_i \in \mathbb{R}^{n \times m}$, $V_i \in \mathbb{R}^{m_r \times m}$, $i = 1,2, \ldots s$, are the set of the observer and residual generator gain matrices.

Introducing the observer state error $e(t) = q(t) - q_e(t)$ and taking the time derivative of $e(t)$, the dynamics of FDF can be expressed by

$$\dot{e}(t) = \sum_{i=1}^{s} h_i \left( \theta(t) \right) \left( A_{ei} e(t) + B_{fi} f(t) + B_{di} d(t) \right), \tag{8}$$

$$r(t) = \sum_{i=1}^{s} h_i \left( \theta(t) \right) V_i C e(t) = \sum_{i=1}^{s} h_i \left( \theta(t) \right) H_i e(t), \tag{9}$$

where $A_{ei} = A_i - J_i C$, $H_i = V_i C$, $H_i \in \mathbb{R}^{m_r \times n}$, $V_i \in \mathbb{R}^{m_r \times m}$. Then (8) can be written as

$$\dot{e}(t) = \sum_{i=1}^{s} h_i \left( \theta(t) \right) \left( (A_i - J_i C) e(t) + [B_{di} \quad -B_{fi}] \begin{bmatrix} d(t) \\ -f(t) \end{bmatrix} \right). \tag{10}$$

Considering, for the sake of simplicity, $r_f = r_d = r_g$ and, using the equivalent observer structure with the same cross-bonds between $d(t)$ and $f(t)$, it can be set

$$\dot{e}^{\circ}(t) = \sum_{i=1}^{s} h_i \left( \theta(t) \right) \left( A_{ei}^{\circ} e^{\circ}(t) + [B_{di} \quad -B_{fi}] T^{\circ} \begin{bmatrix} d(t) \\ -f(t) \end{bmatrix} \right), \tag{11}$$

where $A_{ei}^{\circ} = A_i - J_i^{\circ} C$ and the cross-bonds matrix $T^{\circ}$ is selected as

$$T^{\circ} = \begin{bmatrix} I_{r_g} & I_{r_g} \\ I_{r_g} & I_{r_g} \end{bmatrix} = \begin{bmatrix} I_{r_g} \\ I_{r_g} \end{bmatrix} [I_{r_g} \quad I_{r_g}] = N^{\circ} N^{\circ T}, \quad N^{\circ T} = [I_{r_g} \quad I_{r_g}] \tag{12}$$

Applying (12), the reference model is defined in the next form

$$\dot{e}^{\circ}(t) = \sum_{i=1}^{s} h_i \left( \theta(t) \right) \left( (A_i - J_i^{\circ} C) e^{\circ}(t) + G_i^{\circ} g^{\circ}(t) \right) \tag{13}$$

where, with $G_i^{\circ} \in \mathbb{R}^{n \times r_g}$, $g^{\circ}(t) \in \mathbb{R}^{r_g}$,

$$G_i^{\circ} = [B_{di} \quad -B_{fi}] N^{\circ}, \quad g^{\circ}(t) = d(t) - f(t). \tag{14}$$

The FDF parameters $J_i \in \mathbb{R}^{n \times m}$, $V_i \in \mathbb{R}^{m_r \times m}$, $i = 1,2, \ldots s$, have to be such that

$$\| r^{\circ}(t) \|_{\infty}^2 \leq \gamma^{\circ} \| g^{\circ}(t) \|_{\infty}^2, \tag{15}$$

where the square of the $H_{\infty}$ norm $\gamma^{\circ} > 0$, $\gamma^{\circ} \in \mathbb{R}$ is as small as possible.

According to (9), the formulation of the optimization criterion means that the double summation through membership function occurs in the calculation of the product $r^T(t) r(t)$ in (16). Since $\sum_{i=1}^{s} h_i \left( \theta(t) \right) = 1$, the next approximation can be applied (for the proof see, e.g., in [11])

$$r^T(t) r(t) = = e^{\circ T}(t) \sum_{i=1}^{s} \sum_{j=1}^{s} h_i(\theta(t)) h_j(\theta(t)) H_i^{\circ T} H_j^{\circ} e^{\circ}(t) \leq e^{\circ T}(t) \sum_{i=1}^{s} h_i(\theta(t)) \\ H_i^{\circ T} H_i^{\circ} e^{\circ}(t). \tag{16}$$

The following design conditions are now proposed for the design of the set of the reference model parameters.

**Theorem 1.** The reference model (13), (14) is asymptotically stable with the quadratic performance $\| \mathbf{R}^\circ(\mathbf{s}) \|_\infty < \sqrt{\gamma^\circ}$ if there exists a symmetric positive definite matrix $\mathbf{P}^\circ > \mathbf{0}$, $\mathbf{P}^\circ \in \mathbb{R}^{n \times n}$, matrices $\mathbf{Y}_i^\circ \in \mathbb{R}^{n \times m}$, $\mathbf{V}_i^\circ \in \mathbb{R}^{m_r \times n}$, $i = 1, 2, \dots s$ and a positive scalar $\gamma^\circ > 0$, $\gamma^\circ \in \mathbb{R}$ such that for all $i$

$$\mathbf{P}^\circ = \mathbf{P}^{\circ T} > 0, \quad \gamma^\circ > 0,$$

$$\begin{bmatrix} \mathbf{P}^\circ \mathbf{A}_i + \mathbf{A}_i^T \mathbf{P}^\circ - \mathbf{Y}_i^\circ \mathbf{C} - \mathbf{C}^T \mathbf{Y}_i^{\circ T} & * & * \\ \mathbf{G}_i^{\circ T} \mathbf{P}^\circ & -\gamma^\circ \mathbf{I}_{r_g} & * \\ \mathbf{V}_i^\circ \mathbf{C} & \mathbf{0} & -\mathbf{I}_{m_r} \end{bmatrix} < 0. \tag{17}$$

When the *above conditions hold, the reference model gain matrices are given as*

$$\mathbf{J}_i^\circ = \mathbf{P}^{\circ -1} \mathbf{Y}_i^\circ, \qquad \mathbf{H}_i^\circ = \mathbf{V}_i^\circ \mathbf{C} \quad \text{for all } i. \tag{18}$$

Here and hereinafter, $*$ denotes a symmetric item in a symmetric matrix.

**Proof.** Defining the Lyapunov function candidate of the form

$$v(e^\circ(t)) = e^{\circ T}(t) \mathbf{P}^\circ e^\circ(t) + \int_0^t (r^{\circ T}(x) r^\circ(x) - \gamma^\circ g^{\circ T}(x) g^\circ(x)) \, dx, \tag{19}$$

then, after the evaluation of the derivative of (21), it is obtained

$$\dot{v}(e^\circ(t)) = \dot{e}^{\circ T}(t) \mathbf{P}^\circ e^\circ(t) + e^{\circ T}(t) \mathbf{P}^\circ \dot{e}^\circ(t) + r^{\circ T}(t) r^\circ(t) - \gamma^\circ g^{\circ T}(t) g^\circ(t) < 0 \tag{20}$$

Substitution of (8) in (22) gives

$$\dot{v}(e^\circ(t)) \le$$

$$\le -\gamma^\circ g^{\circ T}(t) g^\circ(t)) + \sum_{i=1}^s h_i(\theta(t)) e^{\circ T}(t) (\mathbf{A}_{ei}^{\circ T} \mathbf{P}^\circ + \mathbf{P}^\circ \mathbf{A}_{ei}^\circ + \mathbf{H}_i^{\circ T} \mathbf{H}_i^\circ) e^\circ(t) + \tag{21}$$

$$+ \sum_{i=1}^s h_i(\theta(t)) \left( e^{\circ T}(t) \mathbf{P}^\circ \mathbf{G}_i^\circ g^\circ(t) + g^{\circ T}(t) \mathbf{G}_i^{\circ T} \mathbf{P}^\circ e^\circ(t) \right) < 0.$$

Thus, defining the composite vector

$$e_c^{\circ T}(t) = [e^{\circ T}(t) \quad g^{\circ T}(t)], \tag{22}$$

can be rewritten as

$$\dot{v}(e^\circ(t)) \le \sum_{i=1}^s h_i(\theta(t)) e_c^{\circ T}(t) \mathbf{P}_{ci}^\circ e_c^\circ(t) < 0,$$

$$\mathbf{P}_{ci}^\circ = \begin{bmatrix} \mathbf{P}^\circ(\mathbf{A}_i - \mathbf{J}_i^\circ \mathbf{C}) + (\mathbf{A}_i - \mathbf{J}_i^\circ \mathbf{C})^T \mathbf{P}^\circ + \mathbf{H}_i^{\circ T} \mathbf{H}_i^\circ & * \\ \mathbf{G}_i^{\circ T} \mathbf{P}^\circ & -\gamma^\circ \mathbf{I}_{r_g} \end{bmatrix} < 0. \tag{23}$$

Using the notation

$$Y_i^\circ = P^\circ J_i^\circ, \tag{24}$$

and Schur complement property, (26) implies (19). This concludes the proof.

## 4    Fault Detection Filter Design

Since residuals generated by (13), (14) are, in general, not totally decoupled from the unknown input $d(t)$, then setting $T^\circ = I_{2r_g}$ and using the obtained parameters (20), the model (11), (14) can be interpreted as the ideal reference.

Thus, inserting (20) into (8), (9) leads to the ideal reference model equations

$$\dot{e}(t) = \sum_{i=1}^{s} h_i\,(\theta(t))((A_i^\circ - J_i^\circ C)e(t) + B_{fi}f(t) + B_{di}d(t)), \tag{25}$$

$$r^\circ(t) = \sum_{i=1}^{s} h_i\,(\theta(t))H_i^\circ e(t). \tag{26}$$

To design the generated residual $r(t)$ as closely as possible to a reference model, the overall FDF model, incorporating (8), (9) and (28), (29) can be expressed as

$$\dot{e}^\bullet(t) = \sum_{i=1}^{s} h_i\,(\theta(t))\big((A_i^\bullet - J_i^\bullet C^\bullet)e^\bullet(t) + G_i^\bullet g^\bullet(t)\big), \tag{27}$$

$$r^\bullet(t) = \sum_{i=1}^{s} h_i\,(\theta(t))(V_i^\bullet - V_i^\star)Ce^\bullet(t), \tag{28}$$

where emphasizing structured LMI matrix variables are used

$$e^\bullet(t) = \begin{bmatrix} e(t) \\ e^\circ(t) \end{bmatrix}, g^\bullet(t) = \begin{bmatrix} f(t) \\ d(t) \end{bmatrix}, G_i^\bullet = \begin{bmatrix} B_{fi} & B_{di} \\ B_{fi} & B_{di} \end{bmatrix} \tag{29}$$

$$A_i^\bullet = \mathrm{diag}[A_i \quad A_i], C^\bullet = \mathrm{diag}[C \quad J^\circ C], J_i^\bullet = \mathrm{diag}[J_i \quad I_n], \tag{30}$$

$$V_i^\bullet = [V_i \quad 0], V_i^\star = [0 \quad V_i^\circ], H_i^\bullet = V_i^\bullet C_r, H_i^\star = V_i^\star C_r, C_r^T = [C^T \quad C^T],$$

$e^\bullet(t) \in \mathbb{R}^{2n}, \; g^\bullet(t) \in \mathbb{R}^{r_f + r_d}, \; G_i^\bullet \in \mathbb{R}^{2n \times (r_f + r_d)}, \; A_i^\bullet \in \mathbb{R}^{2n \times 2n}, \; C^\bullet \in \mathbb{R}^{(m+n) \times 2n}, \; J_i^\bullet \in \mathbb{R}^{2n \times (m+n)}, V_i^\bullet, V_i^\star \in \mathbb{R}^{m_r \times m}, H_i^\bullet, H_i^\star \in \mathbb{R}^{m_r \times 2n}.$

The design conditions are formulated in the sense of existence of a robust FDF of the type (5)-(7) which achieves the asymptotic stability as well as the $H_\infty$ performance condition simultaneously.

**Theorem 2.** The residual filter (5)-(7), associated with the reference model (1), (2), is stable with the quadratic performance $\| R^\bullet(s) \|_\infty < \sqrt{\gamma^\bullet}$ if there exist symmetric positive definite matrices $P_1^\bullet > 0$, $P_2^\bullet > 0$, $P_1^\bullet, P_2^\bullet \in \mathbb{R}^{n \times n}$, matrices $Y_{1i}^\bullet \in \mathbb{R}^{n \times m}$, $V_{1i}^\bullet \in \mathbb{R}^{m_r \times m}$, $i = 1, 2, ..., s$ and a scalar $\gamma^\bullet > 0$, $\gamma^\bullet \in \mathbb{R}$ such that for all $i$

$$P_1^\bullet = P_1^{\bullet T} > 0, \quad P_2^\bullet = P_2^{\bullet T} > 0, \quad \gamma^\bullet > 0,$$

$$
\begin{bmatrix}
P^{\bullet}A_i^{\bullet} + A_i^{\bullet T}P^{\bullet} - Y_i^{\bullet}C^{\bullet} - C^{\bullet T}Y_i^{\bullet T} & * & * \\
G_i^{\bullet T}P^{\bullet} & -\gamma^{\bullet}I_{r_f+r_d} & * \\
(V_i^{\bullet} - V_i^{*})C_r & 0 & -I_{m_r}
\end{bmatrix} < 0,
\qquad (31)
$$

where

$$
P^{\bullet} = \mathrm{diag}[P_1^{\bullet} \quad P_2^{\bullet}], \quad Y_i^{\bullet} = \mathrm{diag}[Y_{1i}^{\bullet} \quad P_2^{\bullet}], \quad V_i^{\bullet} = [V_i \quad 0]
$$

and $P^{\bullet} \in \mathbb{R}^{2n \times 2n}$, $Y_i^{\bullet} \in \mathbb{R}^{2n \times (m+n)}$ *are structured matrix variables.*
*When the above conditions hold, then*

$$
J_i = P_1^{\bullet -1}Y_{1i}^{\bullet}, \quad V_i = V_i^{\bullet}[I_m \quad 0]^T \text{ for all } t.
$$

**Proof.** Now, the Lyapunov function candidate is defined as

$$
v(e^{\bullet}(t)) = e^{\bullet T}(t)P^{\bullet}e^{\bullet}(t) + \int_0^t (r^{\bullet T}(x)r^{\bullet}(x) - \gamma^{\bullet}g^{\bullet T}(x)g^{\bullet}(x))dx \qquad (32)
$$

and its time derivative is

$$
\dot{v}(e^{\bullet}(t)) = \dot{e}^{\bullet T}(t)P^{\bullet}e^{\bullet}(t) + e^{\bullet T}(t)P^{\bullet}\dot{e}^{\bullet}(t) + r^{\bullet T}(t)r^{\bullet}(t) - \gamma^{\bullet}g^{\bullet T}(t)g^{\bullet}(t) < 0,
\qquad (33)
$$

where the structure of $A_i^{\bullet}$, $C^{\bullet}$ implies the structure of $P^{\bullet}$. As well as (2) and (13) have the same structure, then, evidently,

$$
\dot{v}(e^{\circ}(t)) \le \sum_{i=1}^s h_i\,(\theta(t))e_c^{\bullet T}(t)P_{ci}^{\bullet}e_c^{\bullet}(t) < 0 \qquad (34)
$$

$$
P_{ci}^{\bullet} =
\begin{bmatrix}
P^{\bullet}(A_i^{\bullet} - J_i^{\bullet}C^{\bullet}) + (A_i^{\bullet} - J_i^{\bullet}C)^TP^{\bullet} + H_i^{\bullet T}H_i^{\bullet} & * \\
G_i^{\bullet T}P^{\bullet} & -\gamma^{\bullet}I_{r_f+r_d}
\end{bmatrix} < 0 \qquad (35)
$$

and with the notations

$$
P^{\bullet}J_i^{\bullet} = \mathrm{diag}[P_1^{\bullet}J_i \quad P_2^{\bullet}] = \mathrm{diag}[Y_{1i}^{\bullet} \quad P_2^{\bullet}], \quad Y_{1i}^{\bullet} = P_1^{\bullet}J_i,
$$

and Schur complement property, (42) implies (36). This concludes the proof.

## 5     Ilustrative Example

The nonlinear fourth order state-space model of the hydrostatic transmission was taken from [8] and is used in the design and simulations. Its form is

$$
\begin{aligned}
\dot{q}_1(t) &= -a_{11}q_1(t) + b_{11}u_1(t) + b_{d1}d(t) \\
\dot{q}_2(t) &= -a_{22}q_2(t) + b_{22}u_2(t) + b_{d2}d(t) \\
\dot{q}_3(t) &= a_{31}q_1(t)p(t) - a_{33}q_3(t) - a_{34}q_2(t)q_4(t) + b_{d3}d(t) \\
\dot{q}_4(t) &= a_{43}q_2(t)q_3(t) - a_{44}q_4(t) + b_{d4}d(t)
\end{aligned}
\qquad (36)
$$

where $q_1(t)$ is the normalized hydraulic pump angle, $q_2(t)$ is the normalized hydrau-lic motor angle, $q_3(t)$ is the pressure difference [bar], $q_4(t)$ is the hydraulic motor speed [rad/s], $p(t)$ is the speed of hydraulic pump [rad/s], $u_1(t)$ is the normalized control signal of the hydraulic pump, and $u_2(t)$ is the normalized control signal of the hydraulic motor. It is supposed that the external variable $p(t)$, as well as the second state variable $q_2(t)$, are measurable. In given working point, it yields

$$\begin{aligned}
&a_{11} = 7.6923, &&a_{22} = 4.5455, &&a_{33} = 7.6054.10^{-4}, \\
&a_{31} = 0.7877, &&a_{34} = 0.9235, &&b_{11} = 1.8590.10^3, &&\mu_d = 0, \\
&a_{43} = 12.1967, &&a_{44} = 0.4143, &&b_{22} = 1.2879.10^3, &&\sigma_d^2 = 10^{-5}, \\
&b_{d1} = 1.00.10^3, &&b_{d2} = 0.80.10^3, &&b_{d3} = 0.07.10^3 &&b_{d4} = 0.01.10^3.
\end{aligned}$$

Since the variables $p(t) \in \langle c_1, c_2 \rangle = \langle 105, 300 \rangle$ and $q_2(t) \in \langle d_1, d_2 \rangle = \langle 0.0001, 1 \rangle$ are bounded on the prescribed sectors, then the premise variables were chosen as $\theta_1(t) = q_2(t)$ and $\theta_2(t) = p(t)$. Thus, the set of nonlinear sector functions

$$\begin{aligned}
w_{11}(q_2(t)) &= \frac{d_1 - q_2(t)}{d_1 - d_2}, &w_{12}(q_2(t)) &= 1 - w_{11}(q_2(t)), \\
w_{21}(p(t)) &= \frac{c_1 - p(t)}{c_1 - c_2}, &w_{22}(p(t)) &= 1 - w_{21}(p(t)),
\end{aligned} \tag{37}$$

implies the set of normalized membership functions with the next associations

$$\begin{aligned}
h_1(t) &= w_{11}(q_2(t))w_{21}(p(t)), &h_2(t) &= w_{11}(q_2(t))w_{22}(p(t)), \\
h_3(t) &= w_{12}(q_2(t))w_{21}(p(t)), &h_4(t) &= w_{12}(q_2(t))w_{22}(p(t))
\end{aligned} \tag{38}$$

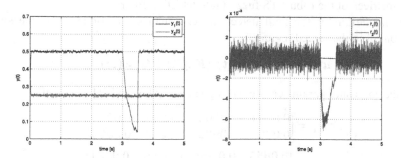

**Fig. 1.** a) System output response, b) residual output response of the system in a fault regime

$$i = 1 \leftarrow (lk = 11) \quad i = 2 \leftarrow (lk = 12) \quad i = 3 \leftarrow (lk = 21) \quad i = 4 \leftarrow (lk = 22)$$

and the conversion of the nonlinear system equations into TS fuzzy system gives

$$A_i = \begin{bmatrix} -a_{11} & 0 & 0 & 0 \\ 0 & -a_{22} & 0 & 0 \\ a_{31}c_k & 0 & -a_{31} & -a_{34}d_l \\ 0 & 0 & a_{43}d_l & -a_{44} \end{bmatrix}, B = \begin{bmatrix} a_{11} & 0 \\ 0 & b_{22} \\ 0 & 0 \\ 0 & 0 \end{bmatrix}, B_d = \begin{bmatrix} b_{d1} \\ b_{d2} \\ b_{d3} \\ b_{d4} \end{bmatrix}, C^T = \begin{bmatrix} 0 & 1 \\ 1 & 0 \\ 0 & 1 \\ 0 & 0 \end{bmatrix},$$

Note the necessary but not sufficient condition that the system is observable and the state variable $q_2(t)$ is measurable. Considering this condition, the system output matrix $C$ was chosen as above.

Supposing that faults are affecting the second actuator, i.e., $B_f = B_2$ and choosing the maximal rank of residual model $m_r = m = 2$ then, exploiting Self–Dual–Minimization (SeDuMi) package for Matlab [15], the design problems (18), (19) and subsequently (35), (36), were feasible with the filter parameters $\gamma^* = 2.0736$,

$$J_1 = 10^3 \begin{bmatrix} 0.8007 & 1.4516 \\ 2.6256 & 0.8342 \\ 0.0546 & 0.1365 \\ 0.0080 & 0.0145 \end{bmatrix}, J_2 = 10^3 \begin{bmatrix} 0.8424 & 1.4521 \\ 2.7065 & 0.8520 \\ 0.0572 & 0.1368 \\ 0.0085 & 0.0204 \end{bmatrix}$$

$$J_3 = 10^3 \begin{bmatrix} 0.7919 & 2.6609 \\ 2.6103 & 1.5111 \\ 0.0540 & 0.2564 \\ 0.0079 & 0.0282 \end{bmatrix}, J_4 = 10^3 \begin{bmatrix} 0.7531 & 2.6401 \\ 2.5283 & 1.4794 \\ 0.0515 & 0.2553 \\ 0.0075 & 0.0339 \end{bmatrix}$$

$$V_1 = 10^{-3} \begin{bmatrix} 0.0041 & 0.0013 \\ 0.1706 & 0.0486 \end{bmatrix}, V_2 = 10^{-7} \begin{bmatrix} 0.8116 & 0.2329 \\ 0.0955 & 0.0247 \end{bmatrix}$$

$$V_3 = 10^{-6} \begin{bmatrix} 0.0100 & 0.0030 \\ 0.3349 & 0.0916 \end{bmatrix}, V_4^\circ = 10^{-8} \begin{bmatrix} -0.3292 & 0.3438 \\ 0.0006 & -0.0062 \end{bmatrix}$$

As can be seen, the most interesting, however, is a substantial difference in the weight matrices of the robust TS fuzzy fault detection filter.

The simulation for the fault detection performance was done under the system fuzzy control in the forced regime, i.e.,

$$u(t) = \sum_{j=1}^{s} h_j(\theta(t))(K_j q(t) + W_j w(t)) \tag{39}$$

where the controller parameters were designed according to [11] as follows

$$K_1 = \begin{bmatrix} 0.0834 & 0.0000 & 0.0844 & 0.0001 \\ 0.0000 & 0.0902 & 0.0000 & 0.0000 \end{bmatrix},$$

$$K_2 = \begin{bmatrix} 0.0855 & 0.0000 & 0.0869 & 0.0001 \\ 0.0000 & 0.0898 & 0.0000 & 0.0000 \end{bmatrix},$$

$$K_3 = \begin{bmatrix} 0.1353 & 0.0000 & 0.1933 & 0.0004 \\ 0.0000 & 0.0907 & 0.0000 & 0.0000 \end{bmatrix},$$

$$K_4 = \begin{bmatrix} 0.1350 & 0.0000 & 0.1926 & 0.0004 \\ 0.0000 & 0.0903 & 0.0000 & 0.0000 \end{bmatrix},$$

$$W_1 = \begin{bmatrix} 0.0000 & 0.0861 \\ 0.0937 & 0.0000 \end{bmatrix}, W_2 = \begin{bmatrix} 0.0000 & 0.0884 \\ 0.0933 & 0.0000 \end{bmatrix}$$

$$W_3 = \begin{bmatrix} 0.0000 & 0.1936 \\ 0.0942 & 0.0000 \end{bmatrix}, W_4 = \begin{bmatrix} 0.0000 & 0.1927 \\ 0.0938 & 0.0000 \end{bmatrix}$$

The working point was set by $p(t) = 105$ and the desired output $w^T(t) =$ [0.50   0.25], the fault was modeled as the short-circuit outage of the second actuator during $t \in \langle 3.0, 3.5 \rangle$ [s]. Fig. 1 shows the system output response to the fault (a), the corresponding residual signals are presented in (b).

## 6    Concluding Remarks

Newly introduced robust TS fuzzy fault detection filter design method, as augmentation of the residual generators synthesis for one class of nonlinear systems, is presented in the paper. This is achieved by the application of TS fuzzy reasoning, relating to multi-model approximation, as in the observer structure as well as in the residual signals frame, and is supported by the TS fuzzy residual reference model. Design conditions are derived in terms of optimization over LMI constraints using standard numerical optimization procedures to achieve, simultaneously, the fuzzy observer asymptotic stability and the optimal residual signal $H_\infty$ performance with respect to the unknown disturbances. Since the TS fuzzy fault detection filter design task is generally singular for a non-square system output matrix $C$, to obtain more regular conditions any further extensions can be included in the design conditions.

**Acknowledgments.** The work presented in this paper was supported by VEGA, the Grant Agency of the Ministry of Education and the Academy of Science of Slovak Republic under Grant No. 1/0348/14. This support is very gratefully acknowledged.

## References

1. Bai, L., Tian, Z., Shi, S.: Robust fault detection for a class of nonlinear time-delay systems. J. Franklin Institute 344(6), 873–888 (2007)
2. Chadli, M.: State and an LMI approach to design observer for unknown inputs Takagi-Sugeno fuzzy models. Asian J. Control. 12(4), 524–530 (2010)
3. Chen, W., Saif, M.: Observer-based strategies for actuator fault detection, isolation and estimation for certain class of uncertain nonlinear systems. IET Control Theory Appl. 1(6), 1672–1680 (2007)
4. Ding, S.: Model-Based Fault Diagnosis Techniques. Design Schemes, Algorithms, and Tools. Springer, Berlin (2008)
5. El-ghatwary, M.G., Ding, S.X., Gao, Z.: Robust fuzzy fault detection for continuous-time state-delayed nonlinear dynamic systems. In: Proc. 6th IFAC Symp. Fault Detection, Supervision and Safety of Technical Processes, Beijing, China, pp. 240–245 (2006)
6. Guo, J., Huang, X., Cui, Y.: Design and analysis of robust fault detection filter using LMI tools. Computers and Math. with Appl. 57(11-12), 1743–1747 (2009)

7. Gao, Z., Jiang, B.: Delay-dependent robust fault detection for a class of nonlinear time-delay systems. In: Proc. 2nd Int. Symp. Systems and Control in Aerospace and Astronautics ISSCAA 2008, Shenzhen, China, pp. 1–6 (2008)
8. Gerland, P., Gross, D., Schulte, N., Kroll, A.: Robust adaptive fault detection using global state information and application to mobile working machines. In: Proc. Conf. Control and Fault Tolerant Systems SysTol 2010, Nice, France, pp. 813–818 (2010)
9. Ichalal, D., Marx, B., Maquin, D., Ragot, J.: Observer design and fault tolerant control of Takagi-Sugeno nonlinear systems with unmeasurable premise variables. In: Rigatos, G.G. (ed.) Fault Diagnosis in Robotic and Industrial Systems, ch. 5, 21p. CreateSpace, Seattle (2012)
10. Koshkouei, A.J., Zinober, A.S.I.: Partial Lipschitz nonlinear sliding mode observers. In: Proc. 7th Mediterranean Conf. Control and Automation, MED 1999, Haifa, Israel, pp. 2350–2359 (1999)
11. Krokavec, D., Filasová, A.: Optimal fuzzy control for a class of nonlinear systems. Mathematical Problems in Engineering 29, ID 481942 (2012)
12. Krokavec, D., Filasová, A.: Actuator faults reconstruction using reduced-order fuzzy observer structures. In: Proc. 12th European Control Conf., ECC 2013, Zürich, Switzerland, pp. 4299–4304 (2013)
13. Lendek, Z., Guerra, T.M., Babuška, R., De Schutter, B.: Stability Analysis and Nonlinear Observer Design Using Takagi-Sugeno Fuzzy Models. Springer, Berlin (2010)
14. Passino, K.M., Yurkovich, S.: Fuzzy Control. Addison-Wesley Longman, Berkeley (1998)
15. Peaucelle, D., Henrion, D., Labit, Y., Taitz, K.: User's Guide for SeDuMi Interface 1.04. LAAS-CNRS, Toulouse (2002)
16. Takagi, T., Sugeno, M.: Fuzzy identification of systems and its applications to modelling and control. IEEE Trans. Syst. Man, and Cyber. 15(1), 116–132 (1985)
17. Tanaka, K., Wang, H.O.: Fuzzy Control Systems Design and Analysis. A Linear Matrix Inequality Approach. John Wiley & Sons, New York (2001)
18. Thau, F.E.: Observing the state of nonlinear dynamical systems. Int. J. Control 17(3), 471–479 (1973)
19. Xu, D., Jiang, B., Shi, P.: Nonlinear actuator fault estimation observer. An inverse system approach via a T–S fuzzy model. Int. J. Appl. Math. Comput. Sci. 22(1), 183–196 (2012)
20. Yang, F., Li, Y.: Set-membership fuzzy filtering for nonlinear discrete-time systems. IEEE Trans. Syst. Man, and Cyber. Part B: Cybernetics 40(1), 116–124 (2010)
21. Zhang, X., Polycarpou, M.M., Parisini, T.: Fault diagnosis of a class of nonlinear uncertain systems with Lipschitz nonlinearities using adaptive estimation. Automatica 46(2), 290–299 (2010)
22. Zolghadri, A., Henry, D., Monsion, M.: Design of nonlinear observers for fault diagnosis. A case study. Control Engineering Practice 4(11), 1535–1544 (1996)

# Advantage of Parallel Simulated Annealing Optimization by Solving Sudoku Puzzle

Ladislav Clementis

Institute of Applied Informatics, Faculty of Informatics and Information Technologies,
Slovak University of Technology, Ilkovičova 2, 842 16 Bratislava, Slovakia
clementis@fiit.stuba.sk

**Abstract.** Simulated Annealing is one of the leading stochastic optimization algorithms. Parallel Simulated annealing, as a modification of Simulated annealing, if properly adapted, is able to solve the problem of selected global optimization issues. An example of a linearly dependent representative of problems in which we are looking for just one solution or global optimum is publicly known as a Sudoku puzzle. The object of this project is to study behaviour of Simulated annealing and Parallel simulated annealing by finding solutions to the Sudoku puzzle, evaluate and verify the success and efficiency of these algorithms, compared to the Backtracking algorithm.

## 1 Introduction

Finding solutions to specific comprehensive problems is one of the toughest challenges in the domain of stochastic optimization algorithms. Finding solutions to the tasks such as solving Sudoku puzzles is a subset of classic optimization tasks.

Unlike traditional optimization problems, finding solutions to such a problem is not only a local or global optimum approximation method but also finding the extreme, often global.

When we know the Sudoku puzzle assignment, we know nothing about the solution until we find it. If the assignment of the Sudoku puzzle is adequate, it has just one solution. This solution is a vector of numbers, putting entries into the empty boxes so that each number from 1 to 9 in all the relevant subsets of the problem space - in rows, columns and nine subfields - is found just once.

Therefore, if we search for the solution of the Sudoku puzzle assignment using stochastic optimization algorithms, we search for global optimum - the vector of these numbers - which, if inserted into the problem space, satisfies the linear boundaries, and that each of the relevant subsets of the problem space is a permutation of the numbers in range from 1 to 9.

Simulated annealing, Parallel simulated annealing, as well as many different genetic and evolutionary algorithms, are also involved in finding solutions to problems, such as the NP complete problems as the Travelling Salesman Problem, polynomial complex problems but also finding solutions to problems such as a Sudoku puzzle.

© Springer International Publishing Switzerland 2015

P. Sinčák et al. (eds.), *Emergent Trends in Robotics and Intelligent Systems*,

Advances in Intelligent Systems and Computing 316, DOI: 10.1007/978-3-319-10783-7_23

## 2     Sudoku as an Optimization Problem

The Sudoku puzzle is likely known to anyone who reads a daily or weekly press, though they don't need to be enthusiasts of logic games, puzzles and anagrams.

The Sudoku puzzle is interesting not only because of its global expansion and popularity but also in the domain of mathematics, computer science, informatics and optimization. Resolved, Sudoku puzzle assignment is a Latin square of size 9, while also satisfying the condition of the uniqueness of each digit in each of the nine subfields. A correct Sudoku puzzle assignment also has the property of having just one solution which is the global optimum in terms of optimizing.

One of the algorithms guaranteed to find a solution to the Sudoku puzzle assignment is the Backtracking algorithm. This algorithm, when adapted to search for Sudoku puzzle assignment solutions, works in the way that the empty boxes are gradually filled with numbers that do not cause conflicts. If there is no number that could be placed in a box without causing conflicts, the algorithm continues with previous number after the initialization and, therefore, in the Sudoku puzzle assignment has an empty box. Otherwise, once the last number has been placed into the last empty box without causing conflicts, the Sudoku puzzle assignment is algorithmically solved. Techniques on how to choose the following blanks or choice of number insertion order into the empty boxes are, of course, dependent on the design, implementation or algorithm parameters.

In solving a Sudoku puzzle using Backtracking algorithm, it is sufficient to decide on a conflict-causing or non-conflict-causing number insertion into the empty box sign whether the given number already is in the given row, column or subfield. In other algorithms, such as genetic algorithms or Simulated annealing, there is a need of the given state assessment. For this a fitness function is used which will return the value of Fitness which can be used to evaluate the state, and, hence, to compare several states to each other. The fitness function also depends on the particular design and implementation. It can be the sum of numbers that cause conflicts, the number of conflicts between numbers and the like.

Genetic algorithms use the fitness function for evaluating individuals in the population. Based on the value of Fitness are then individuals selected for crossover, mutation or progress to the next generation. There are many selection techniques, such as roulette, for example. Choice of technique depends on the particular design and implementation.

Since the correct Sudoku puzzle assignment has just one solution, it is an interesting model representing the complex problem of linear dependence, and the problem space is characterized by its unique solution. In other words, in solving a Sudoku puzzle there are no better or worse current states, there is only a solution. As in solving Sudoku puzzles using genetic algorithms, Simulated annealing and similar methods, we use a fitness function and modify the current state in a way that is not only an approximate solution but a complete one. Despite the pessimistic estimates of the analysis, Parallel simulated annealing has proven being effective in solving this problem.

# 3    Sudoku Puzzle and Simulated Annealing

Simulated annealing [1-5] is effective in finding good solutions of problems such as the Travelling salesman problem and others. Many of these tasks and their problem spaces have their local minimums in which the Simulated annealing algorithm is able to get out due to the slow reduction of temperature.

A Sudoku puzzle assignment problem space is matrix of size 9 times 9 boxes, some of which are filled - contain a number - and other are empty. The box containing the number in the assignment are given fixed, i.e. when running, the algorithm will remain constant. Empty fields are modifiable - it is possible to fill them with numbers from 1 to 9. Problem space initialization can be made, for example, that for each of the nine subfields we identify a set of numbers that are not already present in it. Blanks for each subfield are then filled with randomly arranged permutation made of missing numbers from the set. The state initialized this way meets the requirement of a solution which is that a permutation of numbers from 1 to 9 must appear in each of the nine subfields.

Because we do not want to lose an advantage of this initialized state, we use the type of mutation - a sufficiently small symmetric change - that only exchanges the position of two numbers in the two modifiable boxes in one of the nine subfields. An example of such a small symmetric change is shown in Fig. 1, where the underlined bold numbers 3 and 9 mutually exchange their positions in the boxes in the upper middle subfield. Fields with a black background are defined by assignment, so they are given fixed. Since this type of mutation changes only the position of numbers within the subfields which, after initialization, contain the complete permutations of numbers 1 to 9, it will not cause the presence of a number within a subfield more than once, or a number in a subfield missing. This also helps us for easier implementation of the fitness function which, in this case, does not need to identify the conflicts of numbers within the subfields because these types of conflicts do not occur and it is sufficient for us to identify conflicts in the rows and columns. Therefore, computational complexity of the fitness function has decreased approximately by one third.

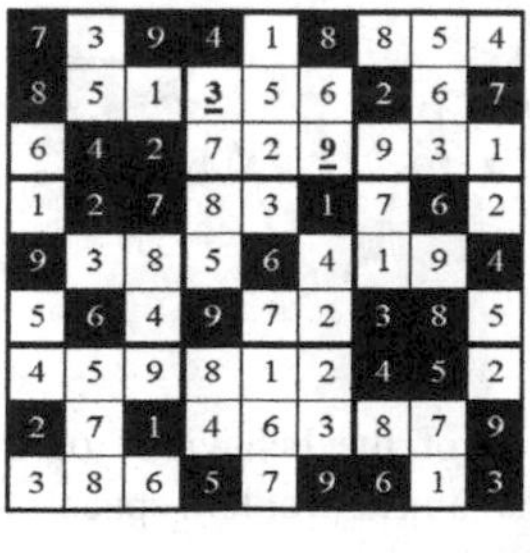

**Fig. 1.** Visual representation of Sudoku puzzle mutation as a small symmetric change

Taking into account the fitness function which gives us the number of conflicts of numbers in given state as an indicator in evaluating this state and for comparing the two states, it is obvious that if the function returns the number zero (meaning no conflicts), we have found a solution. It is easy to verify that a current state is the solution. However, while this state (the solution) is not known explicitly, we have little information about how it looks, and that information is the assignment itself, from which is difficult to read out additional information without additional calculations.

Although by the Simulated Annealing algorithm we gradually modify the state during the run so that the value of Fitness is lowest, this does not always mean that we are converging this state to a solution. This is due to the nature of the Sudoku puzzle assignment problem space which is revealed in the simulations. The nature of this problem space is outlined using an abstract metaphor in Fig. 2.

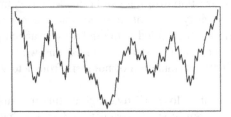

**Fig. 2.** Metaphorical representation of the Sudoku puzzles assignment problem space

As it is evident from Fig. 2, this problem space metaphorically depicted as a function, have many local minimums which are very "deep". Global minimum in this problem space is just one and its surrounding is not so very different from the surroundings of local minimums. Therefore, there is a high probability that even if the fitness function indicates us a low number of conflicts of numbers in rows and columns, the current state and the solution are very different. How does an instance of a Sudoku puzzle behave, solved by Simulated Annealing in this problem space, is metaphorically depicted in Fig. 3.

Simulated Annealing algorithm in the Sudoku assignment problem space handles shallow local minimums of which it can jump out using the Metropolis criterion and probability function 1.

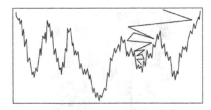

**Fig. 3.** Metaphorical representation of the behaviour of the Parallel Simulated Annealing algorithm in the Sudoku puzzles assignment problem space

$$Pr(perturbed \leftarrow current) = min(1, exp(-\Delta E/kT))$$

However, in case of searching for global optimum, namely the Sudoku puzzle assignment solution, the Simulated annealing is very little successful. Simulated annealing runs were able to find the Sudoku puzzle assignment solution configurations with 12% success rate. By changing the simulation parameters, such as slowing down the lowering of temperature and thereby slowing the simulations and the like, the success rate of Simulated annealing algorithm by solving Sudoku puzzle improved only marginally but computing the duration of the simulations has enormously increased.

It is important to note that if the Simulated annealing algorithm with the appropriately chosen parameters finds the solution to the Sudoku puzzle assignment, it finds it relatively quickly. In figures, by using the parameters modified by the number of filled boxes in the assignment, the number of calculations or the number of fitness function calls ranged from 7400 to 20720. For comparison, in finding the Sudoku puzzle assignment solutions by the Backtracking algorithm, the number of fitness function calls ranged from 3809 to 2565290. Compared with genetic algorithms where population sizes reach hundreds and sometimes thousands, not to mention the number of generations, even such a low success rate of Simulated annealing by finding Sudoku puzzle solutions is compensated by computing efficiency.

As shown by simulations, as the success rate of Simulated annealing in finding solutions is insufficient due to the nature of the Sudoku assignment problem space, it is appropriate to use a more robust approach in solving problems of this type which is Parallel simulated annealing which, unlike Simulated annealing, uses the power of the crossover subjects from the population and the population itself while retaining the strength of Simulated annealing.

## 4     Sudoku Puzzle and Parallel Simulated Annealing

Parallel simulated annealing [2] is a combination of classic Simulated annealing and genetic algorithms. It takes advantage of the population that has more independent instances of the Simulated annealing involved in finding solutions to the problem. As a genetic algorithm, it performs operations in the selection of subjects, crossover of subjects and mutation of subject. In Parallel simulated annealing, a mutation is a sufficiently small symmetric change, as in simulated annealing.

There are of course many techniques of possible crossovers of two individuals. If we use the initialization technique, after which any of the nine subfields contains permutation of the numbers 1 to 9, it is appropriate to use a type of crossover in which there are given subfields with numbers in them maintained as separate entities. Suitable simulated annealing instance crossover technique in finding the Sudoku puzzle assignment solution is to change randomly chosen subfields between the two instances, respectively each of the nine subfields between the two instances exchange with a probability of $0.5. The results of the crossover of two individuals are two new individuals.

The decision about the inclusion of a new individual in the population is made by the Metropolis criterion. Since the temperature during the run of the algorithm decreases, and the average incidence of subjects with a lower Fitness value rises, the probability of the inclusion of an individual to a population which is created by crossover is gradually decreasing. For these reasons, crossover was successfully applied in particular in the first third of the algorithm computational run time, as was evident from the simulations. The advantage of this parallel run of multiple instances of Simulated annealing, compared to the serial run of one instance, is that individuals were evenly placed in the problem space in the first third of the algorithm run time. This is significantly increasing the probability that at least one instance - the individual - will find the Sudoku puzzle assignment solution even at lower sizes of the population. Another advantage of parallel running of multiple instances is the possibility to use custom hardware to run parallel computing or distributed processing. How individuals of Parallel simulated annealing population in the Sudoku puzzle assignment problem space behave is abstractly shown in Fig. 4.

**Fig. 4.** Metaphorical representation of the behaviour of the Parallel Simulated Annealing algorithm in the Sudoku puzzles assignment problem space

The larger is the size of Sudoku puzzle problem space, the larger is the number of blank boxes present in the assignment, or a smaller number of filled ones. Population size in this case was sufficient in the range of 16 to 34 depending on the number of filled fields in the assignment. Of course, compared to classic Simulated annealing, the number of fitness function calls was increased 16 to 34 times, from 118400 to 704480. The success of the algorithm in simulations, however, statistically increased from the aforementioned 12% to 96%, which is comparatively better algorithm statistic. This is an empirical verification that the Parallel simulated annealing is effective in finding solutions to Sudoku puzzle assignments, especially in the case of more blank boxes where Backtracking algorithm was ineffective. These numerical results suggest that the complexity of Parallel simulated annealing, respectively the number of calculations with an increasing number of empty boxes, does not increase as fast as in the case of the Backtracking algorithm.

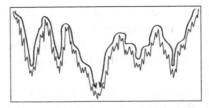

**Fig. 5.** Metaphorical representation of the behaviour of the Backtracking algorithm in the Sudoku puzzles assignment problem space

# 5     Conclusion

Addressing the complex issue often depends on the efficiency of the process of finding solutions. Many times we decide for a solution which is good enough, even if not the best one but very cheap. For these types of problems the stochastic optimization algorithms are often appropriate. We can, however, find out that many of these algorithms, when we choose the appropriate candidate from among them, adapted to the problem domain and with suitably chosen parameters, have the potential of even global optimization which is relatively quite effective compared to the brute force approach as in the Backtracking algorithm.

# References

1. Kirkpatrick, S., Gelatt, C.D., Vecchi, M.P.: Optimization by simulated annealing. Science 220, 671–680 (1983)
2. Kvasnička, V., Pospíchal, J., Tiňo, P.: Evolučné algoritmy. Edícia vysokoškolských učebníc / Slovenská technická univerzita. Slovenská technická univerzita (2000), http://books.google.sk/books?id=ambEAAAACAAJ
3. Metropolis, N., Rosenbluth, A.W., Rosenbluth, M.N., Teller, A.H., Teller, E.: Equation of State Calculations by Fast Computing Machines. The Journal of Chemical Physics 21(6), 1087–1092 (1953)
4. Černý, V.: Thermodynamic approach to the traveling salesman problem: An efficient simulation algorithm. Journal of Optimization Theory and Applications 45(1), 41–51 (1985)
5. Weise, T.: Global optimization algorithms theory and application (2008)

# Extended Self Organizing Maps
# for Structured Domain: Models and Learning

Gabriela Andrejková and Jozef Oravec

Institute of Computer Science, P.J. Šafárik University in Košice, Jesenná 5, Košice, Slovakia
gabriela.andrejkova@upjs.sk, jozi.oravec@gmail.com

**Abstract.** Modelling of neural networks is still a very interesting and important field in the area of computing models. If input domains of neural networks are data structures represented by graphs and output domain is expected in a similar form, it is necessary to consider it in the process of neural network modelling. We propose four models of extended Self Organizing Maps (SOM) that can be applied to graph data structures as input and output domains together with learning algorithms. Extensions of the SOM model are based on the idea to remember information of connections in data structures (using some context neurons).

With regards to the evaluation of developed models and trained structures, we used data from the study programs of the Faculty of Science, P. J. Šafárik University in Košice. We evaluated the ability of models to enumerate output descendants of a node in a graph structure and the ability to interpret structures by developed neural networks. We also evaluated the quality of the developed networks in a learning process.

## 1    Introduction

Kohonen's Self-Organizing Map (SOM) is one of the most popular neural network models. The model was motivated by the retina-cortex mapping and it was developed in 1982 for an associative memory [12]. It uses an unsupervised learning algorithm based on a similarity of input to information in the associative memory. In general, the basic model was developed for problems with input presented by vectors. But now, SOM networks are used in solving many problems working with data structures. Basic knowledge on SOM networks can be found in [11].

The SOM networks are associated with the nodes of regular (sometimes hexagonal), usually two-dimensional grids. The classical SOM algorithm constructs a model in the following way [13]:

More similar data structures will be associated with nodes that are closer in the grid, whereas less similar data structures will be situated gradually further away in the grid.

The central idea of learning principles and mathematics of the SOM can be illustrate in the following form [13]:

Every input data item shall select the grid element that matches best with the input item and this element, as well as a subset of its spatial neighbours in the grid, shall be modified for better matching.

P. Sinčák et al. (eds.), *Emergent Trends in Robotics and Intelligent Systems*,
Advances in Intelligent Systems and Computing 316, DOI: 10.1007/978-3-319-10783-7_24

Many versions of SOM algorithms have been suggested over the years. For example, a model of recurrent SOM was developed in [4][14]. The main idea is in a computation of neuron output using (modified) values of output from previous steps and it has an influence on how long a neuron will have a capability to be a winner.

In [18], Merge SOM was developed, (MSOM) based on classical SOM using the context of neurons with information about previous winners. In the computation of a winner, it is necessary to use a distance of weight vectors from an input vector and an influence of the context.

In [10], we can find an extension of MSOM for tree structures. Contexts are computed for descendants of a node in a tree and they have an influence on the computation of a winner. The next type is a recursive SOM, published in [20].

The SOM model of the neural network was used for sequence processing. The structure processing by neural networks can be found in [9].

SOM neural networks have been proposed by many authors [1][5-8][15] [16][17][19] as models which are very good for learning graph data structures. It means the graph structures in connection to neural networks should be represented in some special representation (a trained neural network) and the neural networks can be trained to graph structures. In the training, it is possible to prepare input data of the graph, in some parts, it is not necessary to put a full graph as one input data to the network. In the papers [2][3], we have prepared theoretical points of view to the presented application of using SOM neural networks to acyclic data structures. When we follow applications of SOM models to solving many problems, we can find the common idea: some function is put to the classical SOM which follows some properties of the network (plasticity).

**Preliminaries:**

A directed graph or digraph is a pair $G = (V, E)$, where $V$ is a set nodes or vertices, $E$ is a set of ordered pairs of nodes, called directed edges. A Directed Acyclic Graph (DAG) is a directed graph with no directed cycles. DAGs will be used as the theoretical model for the data structures representation. A supersource (supernode) of DAG is a node from which all nodes are reachable; if the graph has not a supernode, we have to add it. A node labelled graph has some information in nodes. $\mathcal{U}$ will be a domain of node labels, $\mathcal{U} \#$ will be a domain of graph data structures with labels in $U$. A usual set of labels are the set of real numbers R.

If we choose some nodes from DAG together with the oriented edges among them, then we get a set of subgraphs (one or more subgraphs). The subgraphs are data structures which can be used as input (output) structures to neural networks and neural networks are trained to them. $T, T_B \subseteq \mathcal{U}^\#$ are supervised ($T$) and unsupervised ($T_B$) training sets of data structures. If $\mathbf{D}, \mathbf{D} \in T_B$, then $_v\mathbf{D}$ is the subgraph constructed by the node $v$ and by all descendants of the node $v$; $ch_k[v]$ is the $k$-th descendant of $v$.

In the following text we describe four systematically extended SOM networks of neural networks for the processing of data structures together with the learning algorithms. In the sections 2–5, the developed networks are described. The following section contains criteria of network evaluations and the evaluation of networks. In the conclusion, the achieved results are summarized and a description of the plan for the following work.

## 2    Counterpropagation SOM for Data Structures - CP SOM SD

The first proposed network is CounterPropagation SOM for Structured Data (CP SOM SD) that comes out from a self-organizing map (SOM) for data structures extended by an output layer of Grossberg's neurons. Some examples of SOM SD network are given in Fig. 1. The developed network is capable to work with a training set prepared by pairs of input and output prepared for Grosberg's neurons.

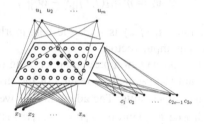

**Fig. 1.** Counterpropagation SOM for data structures - CP SOM SD neural network. The network has n input neurons, 2*o context neurons, o is the maximal number of descendants in all data structures and the output layer has m neurons, the grid is 2-dimensional.

The structure of the network can be split into three parts: an input layer $x_1, ..., x_n$, SOM SD part and an output layer of Grossberg's neurons. SOM SD part has $q$ dimensional SOM grid, $n_i$ is the size of $i$-th dimension, $N = n1 \times ... \times nq$ is the number of neurons in the grid and context neurons $c_1, ..., c_{q*o}$. SOM grid cooperates with the input layer in a classical way. Context neurons are used in the learning of connections between nodes in trained data structures. If $o = max\{o_D; \mathbf{D} \in T_B\}$, $o_D$ is the number of descendants of a node in the structure $\mathbf{D}$, then the number of context neurons is $q * o$. The weight $w_{N+k,i}$ is the weight between the $k$-th context neuron, $k = 1, ... q * o$, and $i$-th grid neuron, $i = 1, ..., N$.

The learning algorithm works in two steps for one training example $(\mathbf{D}, \mathbf{Y})$:

1. The learning of SOM SD means the learning of the mapping $\mathcal{M}: \mathcal{U}^{\#} \to \mathcal{A}$, where $\mathcal{U}^{\#}$ is a set of labelled graph data structures and $\mathcal{A}$ is an output domain represented by grid neurons results. The given data structure $\mathbf{D}$ has to be processed in the following recursive way:

$$\mathcal{M}(\mathbf{D}) = \begin{cases} nil_{\mathcal{A}}, & \text{if } \mathbf{D} = \xi, \\ \mathcal{M}_{node}\left(\mathbf{D}_s, \mathcal{M}\left(_{ch_1[s]}\mathbf{D}\right), ..., \mathcal{M}\left(_{ch_o[s]}\mathbf{D}\right)\right), & \text{otherwise,} \end{cases} \quad (1)$$

where $nil_{\mathcal{A}}$ is the empty element in $\mathcal{A}$, $\xi$ is the empty data structure, $s$ is a supernode of $D$, $ch_k[s]D$, $k = 1, \ldots, o$ are subgraphs in which supernodes are descendants of $ch_k[v]$ of the supernode $s$. $\mathcal{M}_{node}$ is defined by (2).

$$\mathcal{M}_{node}(v) = \arg \min_{l=1,\ldots,N} \|L(x - w_l)\| = i^*, \tag{2}$$

where $i^*$ is the number of the winner neuron for the data structure $_vD$ of the training structure $D$ and $v$ is the current processed node. In (2), $L = (\lambda_{ij})$ is the diagonal matrix of type $(n + oq) \times (n + oq)$. The matrix $L$ is used to balance the weights, $\lambda_{1,1}, \ldots, \lambda_{n,n}$ represent an influence of a node evaluation, usually set to values $\alpha \in \langle 0, 1 \rangle$. Values $\lambda_{n+1,n+1}, \ldots, \lambda_{n+oq,n+oq}$ represent an influence of a context, usually set to $\beta \in \langle 0, 1 \rangle$. The modification of weights are done by (3)

$$\Delta w_j^t = \eta^t h_{i^*}(j,t)\|x - w_j\|^2, \tag{3}$$

where $\eta t$ is a learning rate, $h_{i^*}(j,t)$ is a classic neighborhood function, $t$ is the iteration number and $x$ is an input vector.

2. The learning of the counterpropagation part is the learning of the mapping $\mathcal{C} : \mathcal{A} \to \mathcal{V}$, where $\mathcal{A}$ is the domain represented by results of the grid neuron and $\mathcal{V}$ is the domain of Grosberg's neurons results. The adaptation of weights $q_{ri}$ from the grid $(i, i = 1, \ldots, N)$ to the Grosberg's neurons $(r, r = 1, \ldots, m)$ is the following

$$q_{ri}^{(t)} = q_{ri}^{(t-1)} + \eta(t)\left(-q_{ri}^{(t-1)} + Y_r^{(t)}\right)z_i , \tag{4}$$

where $Y_r$ is an evaluation in the node $r$ of an expected structure $Y$ from the current training example, $z_i \in \{0,1\}, z_i = 1$ for winner neuron only. The vector represented by the structure $Y$ corresponds to the input structure $D$, it means it corresponds to the supernode of the input structure.

# 3    Neural Network with Lateral Weights – SLW

The second network is a network with lateral weights (SLW) modified by using a special method (5). It comes out from a self-organizing network, neurons are connected through lateral weights, an illustration is given in Fig. 2, but a learning algorithm for lateral weights uses a different function than in [11]. This network is able to topologically interpret trained structures. In the unsupervised process of learning, there are modified weights between an input layer and a grid in the same way as in SOM networks. Lateral weights are modified according to the training data structure, the weights between neurons corresponding to nodes in the data structure are reinforced and other weights are inhibited. The output of the learning algorithm is an interpreted structure; its topology should correspond to the input data structure.

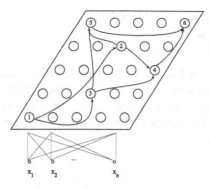

**Fig. 2.** Neural network with lateral weights - SLW. Nodes of data structures correspond to the winner neurons and edges correspond to excited lateral weights. The disadvantage of this network is that, after the training of multiple structures, they may "merge" and the neural network may interpret also parts of other structures.

The learning algorithm: The modification of the lateral weight between the $i$-th and the $j$-th neuron in the grid is done by the formula (5).

$$\Delta w_{ji}^{t} = \eta^{t} h_{l}\left(i^{*}, i, j, x_{j}^{t} - w_{ji}^{t-1}, t\right)\left(x_{j}^{t} - w_{ji}^{t-1}\right) \qquad (5)$$

where the neighbourhood function $h_l$ has the following special shape:

$$h_l(i^*, i, j, p, t) = \begin{cases} exp\left(-\dfrac{\|i^* - i\|}{2\delta^2(t)}\right) & 1 \le i \le N, 1 \le j \le n, \\ \dfrac{1}{1 + e^{-\alpha p}} & p \ge 0, i = i^* \\ & 1 \le j - n \le N, l = j - n, j \ne i, \\ \dfrac{1}{1 + e^{-\beta p}} & p < 0, i \ne i^* \\ 0 & \text{otherwise}. \end{cases} \qquad (6)$$

The interpreted (computed) data structure can have a different structure because of the process of adaptation, the reinforced weight in next steps can be inhibited and the association between the neurons in the grid could be lost. After learning, the network works with an input structure in the following way: the network computes an output structure in the process of recursive activated neurons from the first node put on input in a direction of excited lateral weights sequentially. We used a parameter "limit of association $(L_a)$" to distinguish excited from nonexcited lateral weights. Lateral weights $> L_a$ are excited and other ones are inhibited.

## 4    Neural Network with Lateral Weights in Levels – SLWL

We propose an extension of the SLW network by levels of lateral weights. It is an improvement of the network SLW in which interpreted structures could contain parts of other structures. To avoid the risk, the network SLW was extended by (1) a number of lateral weight levels and (2) some switch among the levels, as shown in Fig. 3. The switch is a real SOM SD with a 1-dimensional grid.

The learning algorithm works in two phases:

- The learning of the switch using the classical learning rules for SOM SD.
- The learning of lateral weights in levels. Using the switch, there is a chosen level for the learning process of a current processed data structure and then lateral weights are modified in the chosen level of lateral weights in the analogous way as in the network SLW.

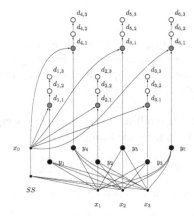

**Fig. 3.** Neural network with lateral weights in levels - SLWL. $SS$ is a switch of the network, $x_1$, $x_2$, $x_3$ are input values, $d_{i,1}, \ldots, d_{i,l}$ are levels of lateral weight to the **i**-th neuron in the grid. The levels of lateral weights offer better results in the interpretation of networks.

The work of the trained network will be similar to the work of SLW, but the network has to know information about an interpreted structure from the starting node. The computed structure is given by the mapping (7).

$$\mathcal{M}_{LW} \colon \mathcal{U}^{\#} \times \mathcal{U} \to \mathcal{J}^{\#}, \tag{7}$$

where $\mathcal{U}^{\#}$ is the set of structures in the domain $\mathcal{U}$, $\mathcal{U} \subseteq R^n$ is a label of an input node and $\mathcal{J}^{\#}$ is the output domain of structures in the domain $\mathcal{J} = R \times \mathcal{A}$.

## 5    Neural Network SLW with a Supervised Learning – SLWT

The last proposed network is based on the network SLW and a supervised learning where a hidden layer and an output layer is added to each neuron in a grid, as shown in Fig. 4. This network is trained by pairs of data structures where the input structure corresponds to a statement and the output structure corresponds to intermediate results of the step-wise evaluation of the statement. A context layer of values is added to the network. The output layer corresponds to the output value of a partial computation of an expression. The computation of an expression is a recursive process following the substructures. It is necessary to add a context layer of values which will contain partial results of computations for a substructure connected to a current processed node.

The learning algorithm is done for the "part" of the network SLW in the first step and then other weights are modified.

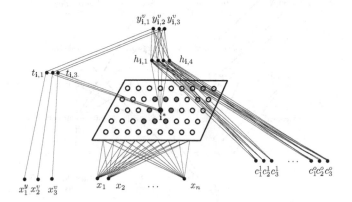

**Fig. 4.** Neural network SLWT with a supervised learning. The network is prepared for the learning of expressions together with the corresponding computations. A hidden layer and an output layer is added to each neuron in the grid.

The work of the trained network is done by the mapping (8).

$$\mathcal{M}_T : \mathcal{U}^{\#} \times (\mathcal{U} \times \mathcal{Y})^p \to \mathcal{Y}^{\#}, \tag{8}$$

where $\mathcal{U}^{\#}$ is the set of structures (arithmetic or logic expressions) in the domain $\mathcal{U}$, $\mathcal{Y}$ contains evaluations in nodes and values of variables in an expression. A part of the network SLWT interprets a structure and the interpreted structure is a base for a recursive computation of an expression value.

# 6     Encoding of Data Structures and the Training Sets

We used DAG to define a study program, an example is given in Fig. 5.

Let K = |V| be the number of subjects in all prepared acyclic data structure of the study program. The subjects are encoded in binary {0,1}, it means that each subject $p^i = [p_1^i, p_2^i, ..., p_b^i] \in \{0,1\}^b, 1 \le i \le K$, where $b = log_2 (2 * K + 10000) + 1$ is the number of bits used to the encoding of each subject.

The preparation of training data structures —a substructure with all prerequisites on the first and second level is prepared to each subject. The prepared training structures $T, T_B$ were prepared by random choice, $|T|, |T_B|$.

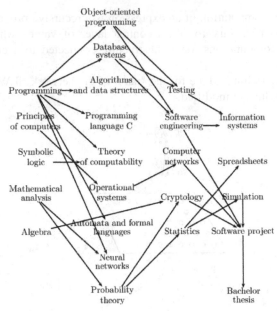

**Fig. 5.** The part of the study program "Informatics". The subjects are in the nodes and the oriented edges represent prerequisites. If the subject A is the prerequisite of the subject B, then in DAG there is the oriented edge from A to B. The subject A is called the direct prerequisite or the first level prerequisite to B. If C is the right prerequisite to A, the C is the second level prerequisite to B.

# 7     Results in the Applications

From theoretical point of view, we are interested in the quality of the learning process of neural networks. The size of a grid in SOM neural network is a very important parameter. The differentiation of winners refers to the spreading of winners in the grid and it is dependent on its size. Confidence of right answers refers to how good the network constructs the output data structure.

## 7.1     Criteria of the Evaluation

- **Winner** differentiation **(WD).** It describes the ratio between the number of all winner neurons and the number of different inputs.

$$WD = |\{j;\ \exists t : j = i^*(t)\}|/N, \tag{9}$$

If $WD < 1$, the same winners are computed for some different inputs. The situation is the best if $WD = 1$.

- **Confidence of right answers.** (Quality of an interpretation. Let $p^S$ be the sum of all the numbers of predecessors of the node $v_D$ together with the subject in the training set and $p^N$ be the sum of all the numbers of predecessors of the node $v_D$ together with the subject computed by a neural network, then the confidence of right answer is

$$Confid = p^N/p^S. \tag{10}$$

- **Quality of a learning** of logical expressions representing prerequisites. It will be evaluated by Mean Squared Error (M. S. E).

## 7.2    Evaluation of all Developed Networks

The networks in the experiments were evaluated for grids from $15 \times 15$ to $54 \times 54$ and parameters $\alpha, \beta = 0,6; 0,75; 0,9$. We concentrate on the evaluation of descendants naming to the current input node. In the evaluation, we observe that the grid size has a principal influence on the evaluated criteria. Some results are described in Table 1.

**Table 1.** Results of the winner differentiation criterion and the confidence

| Network | $WD_{min}$ | $WD_{max}$ | $Confid_{min}$ | $Confid_{max}$ |
|---|---|---|---|---|
| **CP SOM SD** | 0,1303 | 0,8145 | 0,2872 | 0,9456 |
| # grid | 15 x 15 | 51 x 51 | 15 x 15 | 51 x 51 |
| **SLW** | 0,259 | 0,8280 | 0,017 | 0,556 |
| # grid | 15 x 15 | 54 x 54 | 15 x 15 | 54 x 54 |
| **SLWL**: #levels | 0,287 : 9 | 0,9970 : 6 | 0,022 : 1 | 0,788 : 4 |
| # grid | 15 x 15 | 54 x 54 | 15 x 15 | 54 x 54 |
| **SLWT**: #hidNeur | 0,511 : 25 | 0,930 : 20 | 0,328 : 5 | 0,831 : 20 |
| # grid | 15 x 15 | 15 x 15 | 12 x 12 | 25 x 25 |

- **CP SOM SD** - The number of right answers depends on the grid size, the smallest value is 285 for the grid $15 \times 15$ and the best value 427 has the network with the $42 \times 42$ grid .
- **SLW** - The values of the winner differentiation in SLW were higher than in the network CP SOM SD because of the omitted context in SLW.

We have some interesting results for $Confid$. It was evaluated for all prepared data structures and for the data structures with some maximal depth. The capability of a right structure interpretation improves with an increasing size of the grid. We can follow the same observation in the case of confidence according to levels of data structures. But the confidence gets worse if the depth of the structure increases. For example, the best values of the confidence for levels 1,2,3 and 4 of data structure and the grid $54 \times 54$ were 0,664; 0,388; 0,184 and 0,500, but for the grid $30 \times 30$ were 0,387; 0,155; 0,027 and 0,000. The worst number of answers, 400, gave the network with the grid 15,18 and 24, and the best number of answers, 439, gave network with the grid $54 \times 54$.

- **SLWL** - We present the example of the best values for the grid $30 \times 30$ with 10 levels of lateral weights - 0,592; 0,353; 0,161 and 0,000, the values are better than the results in SLW. The worst number of answers, 399, gave the network with the grid $18 \times 18$ and with 9 levels of lateral weights, and the best number of answers, 498, gave the network with the grid $42 \times 42$ and 9 levels of lateral weights.

- **SLWT** - The evaluation of $WD$ and $Confid$ can be found in Table 1. The evaluation of the mean square error (M.S.E) is presented in Table 2. We observed that the grid size has quite a big influence to the capability of a network to compute the good result of an expression. A very important observation is that the quality of results for the increasing depth of an expression must not be dependent from the grid size. And the number of neurons in the hidden layer has a small influence to a quality of network. The biggest computed values of maximal M.S.E were computed in the level $4 - 0$.

**Table 2.** Mean squared errors $M.S.E.$ according to levels x – y, $x$ is the highest number of levels in used structures, $y$, $y=0, 1, \dots, x-1$ is current processed level. $Tr$ is the training set, $Tst$ is the testing set. $Tst_{max}$ (0,6767; 16; 25) means the best value 0,6767 for the size grid 16 × 16 and for the number of output neurons 25.

| Level | min | max |
|-------|-----|-----|
| 1 – 0 | $Tst_{min}(0,0010; 26; 20)$ | $Tst_{max}(0,47671; 10; 25)$ |
|       | $Tr_{min}(0,0008; 24; 20)$ | $Tr_{max}(0,3709; 8; 5)$ |
| 2 – 0 | $Tst_{min}(0,0033; 28; 10)$ | $Tst_{max}(0,6767; 16; 25)$ |
|       | $Tr_{min}(0,0094; 30; 10)$ | $Tr_{max}(0,6002; 13; 25)$ |
| 2 – 1 | $Tst_{min}(0,0015; 29; 5)$ | $Tst_{max}(0,2869; 13; 15)$ |
|       | $Tr_{min}(0,0020; 29; 5)$ | $Tr_{max}(0,2178; 9; 25)$ |
| 3 – 0 | $Tst_{min}(0,0368; 22; 5)$ | $Tst_{max}(0,5277; 8; 15)$ |
|       | $Tr_{min}(0,0505; 16; 25)$ | $Tr_{max}(0,5384; 29; 25)$ |
| 3 – 1 | $Tst_{min}(0,1221; 16; 15)$ | $Tst_{max}(0,5077; 25; 10)$ |
|       | $Tr_{min}(0,1471; 21; 20)$ | $Tr_{max}(0,4700; 16; 10)$ |
| 3 – 2 | $Tst_{min}(0,0299; 30; 25)$ | $Tst_{max}(0,4396; 12; 5)$ |
|       | $Tr_{min}(0,0299; 26; 20)$ | $Tr_{max}(0,4045; 12; 5)$ |
| 4 – 0 | $Tst_{min}(0,000; 9; 20)$ | $Tst_{max}(0,7189; 21; 25)$ |
|       | $Tr_{min}(0,0000; 9; 20)$ | $Tr_{max}(0,6412; 21; 25)$ |
| 4 – 1 | $Tst_{min}(0,0016; 24; 10)$ | $Tst_{max}(0,3977; 28; 20)$ |
|       | $Tr_{min}(0,0058; 16; 10)$ | $Tr_{max}(0,4086; 16; 25)$ |
| 4 – 2 | $Tst_{min}(0,0043; 21; 15)$ | $Tst_{max}(0,5420; 12; 5)$ |
|       | $Tr_{min}(0,0069; 30; 25)$ | $Tr_{max}(0,4965; 12; 5)$ |
| 4 – 3 | $Tst_{min}(0,0022; 19; 5)$ | $Tst_{max}(0,2701; 9; 15)$ |
|       | $Tr_{min}(0,0040; 28; 10)$ | $Tr_{max}(0,2823; 9; 15)$ |

# 8    Conclusion

In this paper, we described the modified SOM neural networks which could be used in the application of problems working with graph data structures, for example in the application of an academic information system. The models of the proposed neural networks analyse possibilities to remember connections among elements of data structures. We evaluated the tests for starting data structures and results give quite a good starting point to the following work. The SLWL and SLWT networks should be models to be prepared for more experiments. The plan for the following work is to continue with the modification of the basic searching of winners and of a learning procedure for weights between an input and a grid.

**Acknowledgments.** This work was supported by Slovak grant agency VEGA, the project /0479/-12 Combinatorial Structures and Complexity of Algorithms and the project /0492/12 Computational models and analytical tools for spatial hearing research.

# References

1. Aiolli, F., Da San Martino, G., Hagenbuchner, M., Sperdutti, A.: Self-organizing Maps for Structured Domains: Theory, Models, and Learning of Kernels. In: Bianchini, M., Maggini, M., Scarselli, F., Jain, L.C. (eds.) Innovations in Neural Infor. Paradigms & Appli. 247, vol. SCI, pp. 9–42. Springer, Heidelberg (2009), http://www.springerlink.com
2. Andrejková, G., Oravec, J.: Application of modified SOM Neural Networks on Acyclic Data Structures. Acta Electrotechnica et Informatica 12(2), 3–8 (2012)
3. Andrejková, G., Oravec, J.: Combination SOM neural network model and a layer for a traning with supervisor. In: Tatras, H. (ed.) Proc. of Cognition and Artificial Life, pp. 1–10 (2013) (in Slovak)
4. Chappell, G.J., Taylor, J.G.: The temporal Kohonen map. Neural Networks 6, 441–445 (1993)
5. Frasconi, P.M., Gori, M., Sperduti, A.: A general framework of adaptive processing of data structures. IEEE-NN 9(5), 768–786 (1998)
6. Hagenbuchner, M., Sperduti, A., Tsoi, A.C.: A self-organizing map for adaptive processing of structured data. IEEE Transactions on Neural Networks 14(3), 491–505 (2003)
7. Hagenbuchner, M., Tsoi, A.C.: A supervised self-organizing map for structures. In: Proceedings IEEE International Joint Conference on Neural Networks, July 25-29, vol. 3, pp. 1923–1928 (2004)
8. Hagenbuchner, M., Tsoi, C.A., Sperduti, A.: A supervised self-organising map for structured data. In: Proc. WSOM 2001: Advances in Self-Organizing Maps, pp. 21–28. Springer (2001)
9. Hammer, B., Micheli, A., Sperduti, A., Strickert, M.: A general framework for unsupervised processing of structured data. Neurocomputing 57, 3–35 (2004)
10. Hammer, B., Micheli, A., Sperduti, A., Strickert, M.: Recursive self-organizing network models. Neural Networks 17(8-9), 1061–1085 (2004)
11. Haykin, S.: Neural networks: a comprehensive foundation, 2nd edn. Prentice-Hall, New Jersey (1999)
12. Kohonen, T.: Self-organized formation of topologically correct feature map. Biological Cybernetics, 43, 56–69
13. Kohonen, T.: Essentials of the self-organizing map. Neural Networks 37, 52–65 (2013)

14. Koskela, T., Varsta, M., Heikkonen, J., Kaski, K.: Temporal sequence processing using recurrent SOM. In: Proceedings of the 2nd International Conference on Knowledge-Based Intelligent Engineering Systems, pp. 290–297 (1998)
15. Sperduti, A.: Tutorial on neurocomputing of structures. In: Cloete, I., Zurada, J.M. (eds.) Knowledge-Based Neurocomputing, pp. 117–152. MIT Press, Cambridge (2000)
16. Sperduti, A.: Neural networks for adaptive processing of structured data. In: Dorffner, G., Bischof, H., Hornik, K. (eds.) ICANN 2001. LNCS, vol. 2130, pp. 5–12. Springer, Heidelberg (2001)
17. Scarselli, F., Gori, M., Tsoi, A.C., Hagenbuchner, H., Monfardini, G.: The Graph Neural Network Model. IEEE Transactions on Neural Networks 20(1), 61–80 (2009)
18. Strickert, M., Hammer, B.: Self-organizing context learning. In: European Symposium on Artificial Neural Networks (ESANN), pp. 39–44. D-side Publications (2004)
19. Vančo, P., Farkáš, I.: Experimental comparison of recursive self-organizing maps for processing tree-structured data. Neurocomputing 73(7-9), 1362–1375 (2010)
20. Voegtlin, T.: Recursive self-organizing maps. Neural Networks 15(8-9), 979–991 (2002)
21. Yin, H.: The Self-organizing Maps: Background, Theories, Extensions and Applications. In: Fulcher, J., Jain, L.C. (eds.) Computational Intelligence: A Compendium. SCI, vol. 115, pp. 715–762. Springer, Heidelberg (2008)

# Event-Based Dialogue Manager for Multimodal Systems

Stanislav Ondáš and Jozef Juhár

Department of Electronics and Multimedia Communications, FEI, Technical University of
Košice, Slovakia Park Komenského 13, 041 20 Košice, Slovak Republic
{stanislav.ondas,jozef.juhar}@tuke.sk

**Abstract.** The proposed paper introduces a newly-designed and developed
dialogue manager for multimodal dialogue systems and robotics applications. It is
the distributed event-driven manager which operates in common data space. For
sharing knowledge about users, tasks, triggers and data objects, the solution based
on the SQLite database system has been implemented. The proposed dialogue
manager enables user-initiative, system-initiative and mixed-initiative dialogues.
The transition network can be also built where nodes are representing the dialogue
tasks and each transition may contain the transition condition.

## 1 Introduction

The ability to be an active participant of the human-human interaction involves
complex sophisticated processes which include information about the real world as
well as personal experiences, social competences and information obtained from
several input channels received through sense organs. Human senses enable us to
interact with each other in a very sophisticated manner. Each of them separately and
also all together provide us the complex view of the real world and they make us
capable of describing entities and relationships of the surroundings and to share ideas
about these things to other people. Speech can be recognized as one of the most
important input/output modalities because it is an acoustic expression of the language
which has a close relation to the real world representation in the human mind and it is
capable to share that representation with other people. The importance of speech as a
modality also consists of a fact that the speech is capable to replace other modalities
when they are not available (e.g. in telephony interaction). Therefore, enabling the use
of speech in communication with machines was the most important milestone in
human-machine interfaces (HMI) history.

The human-like interaction can be enabled by the multimodal dialogue systems
(MDS). Bernsen [1] defines: "A multimodal interactive system is a system which uses
at least two different modalities for input and/or output." Following this, we can
define the multimodal dialogue system (MDS) as an interactive system enabling the
human-machine dialogue-based interaction to use at least two different modalities,
e.g. combination of speech, gestures, touches, etc. The multimodal dialogue system
can be seen as one of the possible realizations of HMI.

© Springer International Publishing Switzerland 2015                                         227
P. Sinčák et al. (eds.), *Emergent Trends in Robotics and Intelligent Systems*,
Advances in Intelligent Systems and Computing 316, DOI: 10.1007/978-3-319-10783-7_25

A group of "core" technologies can be identified that are an obvious part of MDSs - speech recognition, meaning extraction, interaction management or dialogue management, output planning and generation or text-to-speech technology. Whereas technologies for modalities recognition and generation are becoming more available, there are still some challenging tasks, such as multimodal fusion and fission, modality interpretation, producing multimodal behaviour or management of interaction. Management of interaction, including intent planning, behaviour planning and realization, needs a lot of previous research before being successfully usable. However, we are very successful in the task-oriented interaction. The natural "free" interaction is still a difficult task, due to the need to involve a large range of human capabilities. To be able to add such capabilities to a virtual human, knowledge from several research areas is necessary, e.g. informatics, communications, cognitive science, neurology, psychology, linguistics, etc.

A lot of work has been performed which can be adopted to help developing a multimodal interface with humanlike communication attributes (e.g. see [2], [3], [4], [5], [6]) and should be considered during the design and development process.

In our work we are focused on the design and development of the dialogue manager for such multimodal systems. There are a lot of approaches to the dialogue management (see e.g. [7]) but their implementations are often not publicly available or are unusable due to weak documentation. Therefore, we have started to design and develop the new dialogue manager following requirements defined after the analysis of the state of the art in this area.

The proposed paper is organized as follows. The next section introduces and describes the newly-developed dialogue manager including the background and motivation of the research, the concept of the manager and the approach to sharing knowledge which is necessary for the management of interaction. The last part of the paper concludes the results of the proposed work.

## 2     The Event-Based Distributed Dialogue Manager

### 2.1     Background

At the beginning, the analysis of requirements was performed. In the case of designing the dialogue manager, the requirements of the overall communication system need to be analyzed. Natural interaction between human and machine can be identified as the obvious requirement. Fulfilling of this requirement is not contributed only by an appropriate approach to interaction management but also by things like production of feedback, modality interpretation and fusion, and real world representation. On the other hand, as was concluded in the introduction, the complexity which is required for such natural interaction is relatively high. Therefore, undemanding dialogue design, maintainability and reliability need to be taken into consideration during the design process.

Approaches used in several popular dialogue systems, e.g. Jaspis, described by Turunen in [8], Olympus, introduced by Bohus in [9], or multilayer architecture, described by Raux and Eskenazi in [10], were studied and the following general requirements on the newly designed multimodal dialogue system and its dialogue manager were defined:

- The system should be able to perceive a user continuously, meaning it continuously receives input signals and processes them.
- The user may, at any time during the interaction, provide more information than the system expects in that part of interaction.
- The user should, at any time, be able to interrupt the system turn and to route interaction to another topic.
- The system should be able to control turn-taking independently. It means that the turn-taking should be independent of the dialogue management processes. Dialogue management processes should be asynchronous to interaction realization.
- The user may ask questions.

For managing interaction in MDS, several approaches were studied and considered. Our first idea was to take our previously developed dialogue manager based on VoiceXML 1.0 (described in [19]). We were supposed to use it for the realization of a set of small sub dialogues prepared in a more sophisticated management process, but the intended technique had proven to be cumbrous and unusable for several reasons. One of them is that the turn-taking is very limited and such approach enables only a "question-answer" interaction. Approaches described by Wu in [11], by Bohus and Rudnicky in [12], by Nestorovic in [13] or by Raux and Eskenazi in [10] were analysed. Accordingly, following criteria for a newly designed dialogue manager were established:

- Data, flow and presentation should be separated, as proposed by the Voice Browser Working Group of W3C Consortium.
- The creation of new applications for MDS should be effective and undemanding.
- The system should have the ability of personalization, meaning that the interaction should adapt to the user. Information about the user obtained by the system should be saved for being used in the next interaction.
- Interaction should be driven by events.
- Proposed solutions should enable managing the dialogue both in finite-state and a frame-based way.

## 2.2    The Concept of DM

According to appointed criteria, we started to work on the design of the dialogue manager. During the design process, the profile of the manager was formulated: The proposed solution is the distributed dialogue manager which consists of several agents. These agents cooperate over the common data space in a form of the SQL database (described later). The interaction is driven by events that are represented by triggers and data objects. These may be invoked by the system, user, or data. The triggers initiate small tasks. Each task has associated data objects and constructions for system prompts. Tasks can have different priorities. If the user is passive, the dialogue manager can invoke new tasks (system-initiative dialogue) by putting a new trigger into the trigger queue. The designed approach also enables it to build transition networks or state machines which enable it to create trees or networks, where tasks are being nodes of such structure. Each transition also holds a decision condition which must be true to use the particular transition from one task to another.

## 2.3    The Architecture of DM

Designed distributed dialogue manager (Fig. 1) consists of two standalone servers:

- Interaction Manager with incorporated User & History manager
- Task Manager

The particular managers are realized as Galaxy hub framework compatible servers which are written in C++ language. Each server has the SQL wrapper for the communication with the common data space in the form of SQL database.

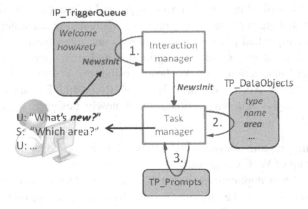

**Fig. 1.** Components of DM and the workflow

The interaction manager is the main server of DM. It is responsible for event loop mechanisms, initialization of the Interaction Pool (the space for interaction data), event-selection algorithm (the loop 1. in Fig. 1) and the destruction of the pool. During the initialization phase, the initial set of triggers which invoke a welcome task and the next tasks according to the user experience level are included to a trigger queue. The Interaction manager cooperates with the Users Pool, where data about users is stored. The event-selection algorithm checks the trigger queue and selects the next trigger which will be handled by the Task Manager. The Interaction manager has two algorithms for selecting triggers. The first one searches for the first unhandled trigger in the trigger queue. When the trigger handling is finished, the second algorithm passes the transition network and searches for the next node (task) which should be processed. The transition condition is checked to evaluate whether the transition is possible. The transition network enables to concatenate several tasks together, but the flow of the interaction is more flexible because the user can invoke arbitrary tasks by their input.

The Task Manager is responsible for performing particular tasks which were selected by the Interaction Manager. The task handling mechanism has three fundamental algorithms – data object values collection (Loop 2. in Fig. 1), cooperation with external data and output concepts generation (Loop 3. in Fig. 1). Each task can have a concept for a general prompt which introduces the particular task. Tasks can require filling zero or more data objects which are attribute-value pairs for holding information obtained from the user related to a particular task. When selected (or all) data objects have their values filled, the Task Manager may perform specific functions, e.g. writing data to the database, querying the database, performing transitions to another task or simply do nothing.

The User & History manager operates upon the Users pool, where information about the users and history of the interaction are stored. It cooperates with both Interaction and Task Manager to personalize interactions with the particular user.

## 2.4   The Common Data Space

Multimodal systems may use several approaches to knowledge sharing. One of the popular solutions of knowledge sharing in the Artificial Intelligence area is "blackboard" system. Blackboard approach was first proposed in the Hearsay-II speech understanding system [14] and after that it was used in several systems, e.g. in MIRAGE ECA [15] or in the embodied agent Greta proposed by Niewiadomski in [16] or in the solution introduced by Huang et al. in [17]. The blackboard works like a shared memory which enables to read and write information by several cooperating components. One of the best known frameworks for building a blackboard system is Psyclone. Another popular approach is to use database for sharing common knowledge.

Common data space in a form of embedded database was selected as an appropriate solution also for our MDS. The main reason was simple: a well-known data interface (also allowing access from external environments) through SQL language. The possibility to browse and edit the data by a lot of various editors was also taken into consideration.

The common data space is divided into four separate pools for saving interaction data, user's data, task data and domain data. Each pool consists of a few tables holding appropriate data. The database is stored in one local file. Each server includes the SQLite software library that implements a self-contained, serverless, zero-configuration and transactional SQL database engine.

The *Interaction Pool* (IP) contains data related to the actual interaction with the user. It serves mainly for Interaction Manager to control interaction flow by checking and picking up triggers to be processed. It consists of three tables:

- *IP_ActualInteraction* table stores unique user ID and interaction ID.
- *IP_TriggerQueue* table stores triggers which appear during interaction
- *IP_DataObjectQueue* table contains data objects and their values which are collected during the interaction.

The Users Pool (UP) stores information about users and interaction history to enable the personalization of the interaction. It consists of two tables:

- *UP_UserInfo* table which stores user name, gender, age, e-mail address and experience level and
- *UP_History* table which holds information about previous interactions with the user.

The Tasks Pool (TP) represents the dialog and domain models. It contains information related to the task. Tasks represent small segments of the interaction or the smallest dialogue units. For the dialogue management purposes, each interaction between the system and the user can be split into the sequence of tasks.

Here, each task is defined by its trigger, data objects and related prompts. Tasks are invoked by picking the trigger from the trigger queue by the Interaction Manager. When all required data objects are filled in, the task is completed. Appropriate prompts are selected according the user model, mainly taking in consideration the experience level of the user.

Tasks Pool contains four tables:

- *TP_Tasks* table contains basic information about the task – ID, name, and name of the trigger.
- TP_Prompts table stores information about system prompts including ID, type, text, URL of surrounding BML script, user experience level and task ID.
- *TP_DataObjects* table holds name, type, prompt type and task ID of particular data objects, and
- *TP_TNetwork* table enables building a finite state network of tasks. Task network defines transitions and condition of transitions between particular tasks to concatenate them.

The last pool of the knowledge database is the Data Pool (DP) which consists of two tables:

- DP_BACKDATA table is a space for data obtained from external sources (web pages, remote databases). It contains attribute-values pairs which are incorporated into the system output (prompts).
- DP_ITEMS table contains data for keyword spotting technique for analyzing input user's utterances.

## 2.5     Interaction Examples

The proposed dialogue manager was implemented into two systems – the multimodal dialogue system SIMONA [18] and the service robot speech interface.

The simple weather forecast service was written for SIMONA agent. It consists of four tasks – welcome, how help, weather and weather data. An example of the interaction with the experienced user looks like this:

**Example 1:** Interaction **with SIMONA agent; the Weather forecast service.**
S: (smile) Hi Stanislav!                                    (*welcome* task)
S: (hands forward) How may I help you?                (howhelp task)
U: Hi! What is the <u>weather</u> going to be like <u>tomorrow</u>?   (*weather* task)
S: In Košice?
U: Yes.
S: (head nod)
S: Tomorrow will be sunny, 27 degrees.              (*weatherdata* task)

The proposed dialogue manager was implemented also into the service robot speech interface [20]. The simple application was created to enable entering commands by voice. This application consists of four tasks and the transition network, as shown on Fig. 2.

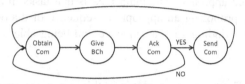

**Fig. 2.** The network of tasks for controlling the service robot by voice commands

The flow of the interaction has a loop character. It starts with the task for obtaining a command from the user (ObtainCom task), continues by giving some backchannel from the system (GiveBch) and confirming the task (AckCom). The confirmation of the command is done by pushing the "dead man" joystick button on the control panel robot interface. Subsequently, in the SendCom task, the recognized command is delivered to the robotic vehicle. The next example shows the user-initiative piece of interaction with such an application:

**Example 2: Service robot** speech **interface application.**
U: Robotic arm, base position.                      (*ObtainCom* task)
S: Arm, base position                              (GiveBCh task)
U: *(User pushes "dead man" button of the joystick in specified time interval)*
                                                  (*AckCom* task)
*System performs required task.*                  (*SendCom*)
U: Front camera                                   (*ObtainCom* task)
S: Rear camera                                    (*GiveBCh* task)
U: Front camera                                   (*ObtainCom* task)
S: Front camera                                   (*GiveBch* task)
U: *(User pushes "dead man" button of the joystick in specified time interval)*
                                                  (*AckCom* task)
System performs required task.                    (SendCom)

## 3     Conclusion

The newly-designed dialogue manager has been implemented for controlling interaction in the SIMONA – Slovak embodied conversational agent [18] and also into the service robot speech interface [20]. It is the event-driven manager with a knowledge sharing system that has a form of database with pools. Proposed approaches enable to control interaction in a simple, effective and variable manner. The interaction can be controlled solely by the user utterances (user-initiative interaction) from which triggers are extracted and conveyed into the trigger queue. Manager also enables a strict system-initiative dialog, where the dialog flow is determined by the transition network (system-initiative interaction). Mixed-initiative dialogues are allowed by combining the previous two scenarios.

Our future work will be focused on the incorporating of speech acts (or dialogue acts) into the proposed approach. The main idea is that tasks in dialogue interaction can be performed by proposing an appropriate sequence of dialogue acts. There are two main challenges – finding such a sequence and being able to recognize a user's dialogue acts.

**Acknowledgments.** The research presented in this paper was supported by Research & Development Operational Program funded by the ERDF through project with ITMS project code 26220220141 (90%) and by the Ministry of Education, Science, Research and Sport of the Slovak Republic under research project VEGA 1/0386/12 (10%).

## References

1. Bernsen, N.O.: Multimodality Theory. In: Tzovaras, D. (ed.) Multimodal User Interfaces. Signals and Communication Technology, pp. 5–29. Springer, Heidelberg (2008)
2. Esposito, A.: The amount of information on emotional states conveyed by the verbal and nonverbal channels: Some perceptual data. In: Stylianou, Y., Faundez-Zanuy, M., Esposito, A. (eds.) WNSP 2005. LNCS, vol. 4391, pp. 249–268. Springer, Heidelberg (2007)
3. Esposito, A.: COST 2102: Cross-modal analysis of verbal and nonverbal communication (CAVeNC). In: Esposito, A., Faundez-Zanuy, M., Keller, E., Marinaro, M. (eds.) Verbal and Nonverbal Commun. Behaviours. LNCS (LNAI), vol. 4775, pp. 1–10. Springer, Heidelberg (2007)
4. Vicsi, K., Sztaho, D.: Recognition of Emotions on the Basis of Different Levels of Speech Segments. Journal of Advanced Computational Intelligence and Intelligent Informatics 16(2), 335–340 (2012)
5. Jokinen, K.: Gaze and gesture activity in communication. In: Stephanidis, C. (ed.) Universal Access in HCI, HCII 2009, Part II. LNCS, vol. 5615, pp. 537–546. Springer, Heidelberg (2009)
6. Bevacqua, E., Pammi, S., Hyniewska, S.J., Schröder, M., Pelachaud, C.: Multimodal backchannels for embodied conversational agents. In: Safonova, A. (ed.) IVA 2010. LNCS (LNAI), vol. 6356, pp. 194–200. Springer, Heidelberg (2010)

7. McTear, M.: Spoken Dialogue Technology: Towards the Conversational User Interface. Springer (2004) ISBN 1852336722
8. Turunen, M., Hakulinen, J.: Jaspis - an architecture for supporting distributed spoken dialogues. In: Proceedings of Eurospeech 2003, pp. 1913–1916 (2003)
9. Bohus, D., Raux, A., Harris, T.K., Eskenazi, M., Rudnicky, A.I.: Olympus: an open-source framework for conversational spoken language interface research. In: Bridging the Gap: Academic and Industrial Research in Dialog Technology Workshop at HLT/NAACL 2007, pp. 32–39 (2007)
10. Raux, A., Eskenazi, M.: A Multi-layer Architecture for Semi-synchronous Event-driven Dialogue Management. In: Proc. of Automatic Speech Recognition & Understanding, ASRU 2007, Kyoto, pp. 514–519 (2007)
11. Wu, X., Zheng, F., Xu, M.: Topic Forest: A Plan-based Dialog Management Structure. In: Proc. of the IEEE International Conference on Acoustics, Speech, and Signal Processing (ICASSP 2001), Salt Lake City, Utah, vol. 1, pp. 617–620 (2001)
12. Bohus, D., Rudnicky, A.I.: The RavenClaw dialog management framework: Architecture and systems. Computer Speech and Language 23(3), 332–361 (2009)
13. Nestorovic, T.: General Agent-based Architecture for Collaborative Dialogue Management. In: 2nd International Conference on Software Technology and Engineering (ICSTE 2010), San Juan, USA, vol. 2, pp. 207–211 (2010)
14. Erman, L.D., Lesser, V.R.: The Hearsay-II speech-understanding system: Integrating knowledge to resolve uncertainty. Computing Surveys 12(2), 213–253 (1980)
15. Thórisson, K.R., Benko, H., Abramov, D., Arnold, A., Mas-key, S., Vaseekaran, A.: Constructionist Design Methodology for Interactive Intelligences. AI Magazine 25(4), 77–90 (2004)
16. Niewiadomski, R., Bevacqua, E., Mancini, M., Pelachaud, C.: Greta: an interactive expressive ECA system. In: AAMAS 2009 Proceedings of the 8th International Conference on Autonomous Agents and Multiagent Systems, Budapest, Hungary, vol. 2, pp. 1399–1400 (2009)
17. Huang, H.-H., Cerekovic, A., Tarasenko, K., Levacic, V., Zoric, G., Pandzic, I.S., Nakano, Y., Nishida, T.: Integrating embodied conversational agent components with a generic framework. In: Multiagent and Grid Systems - Innovations in Intelligent Agent Technology, vol. 4(4), pp. 371–386. IOS Press, Amsterdam (2008)
18. Ondas, S., Juhar, J.: Design and development of the Slovak multimodal dialogue system with the BML Realizer Elckerlyc. In: Cognitive Infocommunications (CogInfoCom), pp. 427–432 (2012)
19. Ondáš, S., Juhár, J.: Dialog manager based on the VoiceXML interpreter. In: Proceedings of the DSP-MCOM 2005, Košice, Slovak Republic, TU Košice, September 13-14, pp. 80–8073 (2005) ISBN 80-8073-313-9
20. Ondáš, S., Juhár, J., Pleva, M., Čižmár, A.: Roland Holcer, Service robot SCORPIO with robust speech interface. International Journal of Advanced Robotic Systems 10(3), 1–11 (2013)

# Distributed Multi-agent System for Area Coverage Tasks: Architecture and Development

Ivana Budinská, Tomáš Kasanický, and Ján Zelenka

Institute of Informatics Slovak Academy of Sciences, Bratislava, Slovakia

**Abstract.** The article aims to present a new concept in mobile robotics – swarm robotics. It gives an overview of related research in coordination of mobile robots in a group and presents recent research activities in swarm robotics at the Institute of Informatics SAS. The architecture and development of a distributed multi-agent system for area coverage tasks are introduced.

## 1 Introduction

Robotics became a very important part of our everyday life and its importance increases with new developments in many related areas such as informatics and information technologies, sensors, communication technologies and others. Mobile robots have been successfully utilized in many applications where they can substitute humans working in dangerous environments and doing monotonic and hard work operations, e.g., demining, fire rescue activities, seeking for survivors of natural disasters, monitoring and guarding of large and hardly accessible areas, etc.

There are many open research problems related to the control of mobile robots that can work autonomously in such applications. A mobile robot as a complex system consists of two basic control parts: actuations that are used for locomotion and sensors for guidance. A control system of each mobile robot has to handle basic tasks of localization, local navigation and obstacle avoidance. An objective of our research is to design and develop a swarm of autonomous mobile agents that cooperate while pursuing a common goal – exploring and monitoring of large areas. The aim is to contribute to a large group of coordinating methods, enhance some of the specific features, and eliminate drawbacks of existing methods. The focus is on a higher level of control within a group of mobile robots. It is assumed that all robots handle basic control tasks (stabilization, localization, collision avoidance) on a lower level – local control.

There are many possible applications for such multiple robot systems. They can be used for environmental monitoring, surveillance, intelligent data gathering, etc. One of the most prospective application domains for multiple robot systems can be found in the area of humanitarian demining. In this case, two main tasks are considered: mine detection and mine neutralization. Robotic systems have to be equipped with specific sensing systems, mine elimination systems and should be able to work in a dangerous environment. (Havlik, 2008a,b) Robotic systems employed in demining tasks are often damaged by explosions of mines. That is why a group of autonomous robots can accomplish the task regardless of how many members of the group remain in a good working condition (Havlik, 2012).

© Springer International Publishing Switzerland 2015

P. Sinčák et al. (eds.), *Emergent Trends in Robotics and Intelligent Systems*,

Advances in Intelligent Systems and Computing 316, DOI: 10.1007/978-3-319-10783-7_26

Coordination problem within a group of mobile robots refers to methods and algorithms that belong to a new concept of robotics called swarm "robotics". The term "swarm robotics" was first used by Gerardo Beni, a professor at UC California and Jing Wang in 1989, in order to impact a nation of swarm intelligence to cellular robotic systems.

Swarm robotics (Budinska, 2012) is based on the idea that a group of relatively simple autonomous robots working on the same task can be a more effective and sometimes also cheaper solution than one very complex robotic system. Swarm robotics is a new approach in decentralized coordination of the behaviours of a large number of autonomous robots that can be represented as autonomous agents for simulation and coordination purposes. A robot in a group/swarm has only simple behaviour, local perceptions and limited communications abilities.

Swarm robotics takes inspiration from the nature. Biological swarm features a great behaviour that can be applied in coordination behaviour of a large number of robots in a decentralized manner. Examples are flocks of birds, shoals of fish, ant colonies, bee swarms and also, e.g., crowds of people. Besides the application of simple rules to control various biological societies, dynamics of ecosystems can also be applied in multi robot systems in order to simulate emergent cooperation as a result of selfish behaviour. In the sense of the above mentioned, the motivation for the research of the coordination of a group of mobile robots is: robustness, scalability, stability, flexibility/versatility, super linearity, and maybe low costs.

In most of the application cases, mobile robots pursue their individual plans while trying to accomplish the overall task for the whole multi-robot system.

## 2    Formulation of a Problem

The goal is to develop a multi-agent system intended for exploring and monitoring large areas.

Agents in this research are considered as autonomous mobile entities that can perceive their environments through sensors and act in that environment through actuators. A multi-agent system for area coverage tasks is a system of autonomous mobile agents (e.g. automated guided vehicles AGVs, unmanned aerial vehicles UAVs, etc.) that communicate and interact with each other and operate in certain area. Each agent can only acquire information from its nearby area and agents are continuously building a common knowledge base for sharing information about the area. The main issue is to build a multi-agent system that effectively coordinate and cooperate all members' activities to complete a global goal – wide area searching, surveillance, rescue missions, etc. The coordination algorithms have to be distributed, robust, flexible, platform independent and adaptive. An area coverage tasks were defined in (Dasgupta, 2009) as follows:

There are two areas defined for each agent: a working area that is defined by a range of a robot´s detection system and an operating area that is an area where the agent r can move from its local position l doing action a in the next step in time t. A coverage function transforms an action of the robot r in the next time period into the unit of area c: $f : a_r^t \times l_r^t \rightarrow c_r^t$

The problem is to find a set of actions $a$ for each robot $r$ that the union of areas $c_r$ for all robots and time intervals are equal to the monitored area.

Depending on a specific application, there are more criteria given to the problem definition, e.g. non-overlap criterion that optimise the searching task in such a way, that the minimum of areas are repeatedly explored by robots.

## 3    Related Works

Bio-inspired search strategies have been recently extensively researched. Hereford and Siebold (2010) present an algorithm inspired by a swarm behaviour to control a group of robots that are searching a target in an environment. Another method for space exploration tasks was introduced by Masar and Zelenka (2012). Both methods apply to the idea taken from the Particle Swarm Optimisation (PSO) method. The PSO was originally designed to find an optimal solution and it has a strong tendency to converge early to suboptimal solutions. In order to overcome this drawback, a combination of PSO and Ant Colony Optimization methods was suggested in (Masar, 2013a,b). An Ant Colony Optimization method (ACO), proposed by Dorigo and Stützle (2004), employs virtual pheromone marks that help finding an optimal path through a graph. A combination of PSO with virtual pheromones for constrained optimization problems was described (Kalivarapu, 2008). Shang et al. (2009) applies virtual pheromones on communication mechanisms in order to decrease communication costs in a map coverage task. A multi-agent system introduced in this paper employs also simple rules adopted from an artificial flock of birds, created by Reynolds (1987). The agents use a model of an unknown environment created with the help of a sensing system that is used also for navigation tasks (Hanzel, 2012).

Some excellent results in the area of developing distributed algorithms for multi-robot systems with the aim to operate large scale multi-robot systems with desired behaviour were achieved by James McLurkin[1]. Also Marco Dorigo and his group proved an extraordinary achievement in the area of multi-robot coordination and cooperation, as it can be seen in Fig. 1.

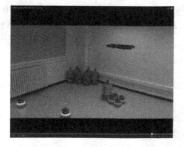 

**Fig. 1.** "Spatially Targeted Communication and Self-Assembly" by Nithin Mathews, Anders Lyhne Christensen, Rehan O'Grady and Marco Dorigo from Universite Libre de Bruxelles and Instituto Universitario de Lisboa, was presented at IROS 2012 in Vilamoura, Portugal.[2]

---

[1] http://www.cs.rice.edu/~jm23/

[2] http://www.youtube.com/watch?v=i3ernrkZ91E

# 4    Coordination Methods and Algorithms

A new method for coordination of a group of autonomous mobile agents was developed. The method is inspired by the Particle Swarm Optimisation (PSO) (Kennedy, 1995) and Ant Colony Optimisation (ACO) (Dorigo, 2004) methods. The virtual model of the area is divided into squares with the range that corresponds to the range of working areas of robots. Each square has two basic states: revealed and unrevealed. The idea behind the method is to distribute agents over the monitored area so that all parts of the area are covered by sensing of at least one agent in a defined time. Agents evaluate their nearby areas and they are forced to move in the direction of the most unrevealed areas. Agents were added the ability to deposit pheromone marks in the space they covered by their sensors. The value of virtual pheromone marks decrease in time. Recently visited areas have the highest values of pheromones and unrevealed areas have zero value of pheromones. On the contrary to the original ACO method, agents are attracted by the lowest value of pheromones.

The next position of an agent is then computed by the following equation (Masar, 2013b):

$$x_{id}(t+1) = x_{id}(t) + v_{id}(t),$$

$$v_{id}(t+1) = c_1(p_{ap}(t) - x_{id}(t)) + c_2(p_{c1}(t) - x_{id}(t)) + c_3(p_{c2}(t) - x_{id}(t)),$$

where

- $x_{id}(t)$ denotes a current position of the agent and c1, c2, c3 are weight constants,
- vector $p_{c1}$ denotes geometrical gravity of positions of all robots that are in the zone of attraction,
- vector $p_{c2}$; opposite vector to the vector denoting a geometrical centroid of positions of all robots that are in the zone of repulsion,
- vector $p_{ap}$ denotes position of the lowest pheromone value found in the defined area around the agent.

There are positions computed independently for each of the agents. In such way the algorithms are distributed among agents. All agents perform the following steps (Masar, 2013a, b):

*LOOP*
*{find the lowest pheromone value in predefined range if (there are any agents in the zone of attraction)*
*{ find a vector denoting a geometrical centroid of positions of these agents}*
*if (there are any agents in the zone of repulsion)*
*{ find an opposite vector to the vector denoting the geom. centroid of positions of these agents }*
*compute the next waypoint using Eq. 3*
*if (the distance to the waypoint is outside the predefined range)*
*{ transform this waypoint to fit the predefined range}}*

The computed positions are sent to real mobile agents-robots.

# 5    A Multi-agent System Architecture

A multi-agent system consists of a group of mobile autonomous robots (UAV, AGV, etc.) with a certain unit of intelligence. All real robots are presented as virtual agents that cooperate on the basis of common knowledge of the environment. There is a limited communication between each of the agents and a central computer - only robots´ positions and a common map are exchanged for coordination purposes. A three-layered architecture is suggested.

The general architecture scheme is depicted in Fig. 2

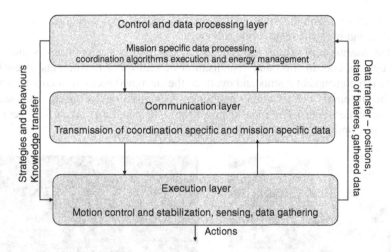

**Fig. 2.** General architecture of the system

The execution layer (Budinska, 2013) is built from any number of mobile robotic systems. The maximum number of the robots in the system is limited by the throughput of the communication system. We consider two types of mobile robots - UAVs (unmanned aerial vehicles) and AGVs (automated guided vehicles). There are many different implementations of UAVs. Chovancova et al. (2012) suggests micro aerial mobile robots. Our solution has been testedon Lego Mindstorm robots and arducopters as well. Mobile robots are equipped with sensing systems, a camera and microcontrollers. An inertial navigation system integrated with global navigation satellite systems is used in outdoor testing. Local navigation and control are managed by Ardupilot[3] and Mission planner[4].

The communication layer supports communication in two ways: among mobile agents, and between a mobile agent and a base computer, either for coordination data or mission specific data. The communication network architecture has to be scalable and distributed with capabilities to communicate with a range of different data types depending on a specific application (Zolotova, 2013).  The communication layer

---

[3] http://ardupilot.com/
[4] http://code.google.com/p/ardupilot-mega/wiki/Mission

consists of transmitters and receivers on both sides. Specific channels are suggested for coordination data and for mission specific or payload data.

The control and data processing layer consists of two computers. The first one is responsible for agent´s movement coordination within the system. It collects localisation data and computes new positions for all robots in separate threads. The second computer is intended to process mission specific (or payload) data. The computers are interconnected in order to enable modification of agents' collaboration according to the mission goals.

# 6     Experiments and Testing Results

The concept of a robotic swarm for area coverage tasks was tested by computer simulation and by using real mobile robots.

A visualisation tool VERA (Masar, 2012) has two basic functions: simulation of virtual agent's coordination and visualisation of real robot movements. It enables to simulate large groups of agents and evaluate the designed method in comparison with other bio-inspired coordination methods. Visualisation is used for semi-real simulation experiments. The input data for the visualisation is taken from real robots (from Lego control boxes) as it is depicted in Fig. 3.

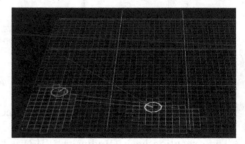

**Fig. 3.** Simulation experiments using the VERA tool and Lego Mindstorm control boxes

Some indoor experiments took place using a robotic platform Lego Mindstorms programmed in NXC. There are communication restrictions in the Lego Mindstorms platform - only a group of 4 robots can be tested directly. The test field with robots is depicted in Fig. 4.

**Fig. 4.** Indoor simulation environment

Outdoor experiments have been conducted using arducopters (Fig. 5). The computation of new waypoints is realized on a control ground computer in order to alleviate the on-board batteries consumption. There is a separate thread for each real mobile agent in the computer. We have developed a terminal that interconnects virtual agents with the Mission planner. The computed waypoints in Cartesian coordinates are transformed into GPS coordinates and are sent through the terminal into the Mission planner. At the same time, robots are sending their actual positions in GPS coordinates through the terminal. The GPS coordinates are transformed into Cartesian coordinates and they are considered as inputs for new waypoint computation. By now it is not possible to dynamically change waypoints while UAVs are completing an autonomous flight. The set of computed waypoints are sent to each UAV and after completing the fly, it waits for new waypoints. In the future the drawback will be eliminated.

**Fig. 5.** Indoor experiments using arducopters and Mission planner

# 7   Conclusion and Future Work

A multi agent system for long term observation is presented in the paper. The agents in the system cooperate to accomplish a common task – the monitoring of an area. The presented distributed control algorithm was developed with the consideration of limited computational capacity and complexity of on-board computers. Coordination algorithms can be implemented on many robotic platforms. They do not depend on the number of robots in a group. According to the specific robotic platform they can be adapted to various environment types. Behaviour of the system can be easily changed from exploring to monitoring through parameter settings.

Experimental tests of the presented approach were successfully conducted in interior with Lego Mindstorm robots. Ongoing experiments with arducopters show the possibility for larger exploitation of the approach. The basic coordination algorithm will be enhanced with new features related to the specific application requirements, e.g. energy management for long term observations.

**Acknowledgments.** The research presented in this paper is partially supported by the national grant agency SRDA under the grant No. 0261-10 and VEGA 2/0194/13.

# References

1. Budinská, I.: Multi-agents coalitions for multi robot coordination. In: Kybernetika a Informatika 2012, 4p. Vydavateľstvo STU, Bratislava (2012) ISBN 978-80-227-3642-8
2. Chovancova, A., Rodina, J., Hubinský, P.: Modelling of the VTOL micro aerial mobile robot. ATP Journal Plus 1, 104–107 (2012)
3. Dasgupta, P., Cheng, K., Fan, L.: Flocking-based Distributed Terrain Coverage with Mobile Mini-robots. In: Proc. Swarm Intelligence Symposium, Nashville, TN, pp. 96–103 (2009)
4. Dorigo, M., Stutzle, T.: Ant Colony Optimization. MIT Press (2004) ISBN 0-262-04219-3
5. Hanzel, J., Kľúčik, M., Jurišica, L., Vitko, A.: Range finder models for mobile robots. In: Proceedings of MMaMS 2013, Procedia Engineering, vol. 48, pp. 189–198, 1877–7058. © 2012 Published by Elsevier Ltd. (2012)
6. Havlik, S.: Some Robotic Approaches and Technologies for Humanitarian Demining. In: Habib, M.K. (ed.) Humanitarian Demining Innovative Solutions and the Challenges of Technology, pp. 289–314. I-Tech Education and Publishing (2008a)
7. Havlik, S.: Land Robotic Vehicles for Demining. In: Habib, M.K. (ed.) Humanitarian Demining Innovative Solutions and the Challenges of Technology, pp. 315–326. I-Tech Education and Publishing (2008b)
8. Havlík, Š., Budinská, I., MASÁR, M.: Multi-robot coordination in performing hazardous operations. In: RAAD 2012: 21th International Workshop on Robotics in Alpe-Adria-Danube Region, pp. 169–176. Edizioni Scientifiche e Artistiche, Naples (2012) ISBN 978-88-95430-45-4
9. Hereford, J.M., Siebold, M.A.: Bio-inspired search strategies for robot swarms. In: Martin, E.M. (ed.) Swarm Robotics From Biology to Robotics. InTech (2010) ISBN: 978-953-307-075-9
10. Kalivarapu, V., Winer, E.: Implementation of Digital Pheromones in Particle Swarm Optimization for Constrained Optimization Problems. In: 49th AIAA/ASME/ASCE/AHS/ASC Structures, Structural Dynamics, and Materials Conference, Schaumburg, IL (2008)
11. Kennedy, J., Eberhart, R.: A new optimizer using particle swarm theory. In: Proceedings of the IEEE Sixth International Symposium on Micro Machine and Human Science, pp. 39–43 (1995) ISBN: 0-7803-2676-8
12. Masár, M.: Doctoral thesis: Bio-inspired multi robot coordination method (2013a) (in Slovak)
13. Masár, M., Budinská, I.: Robot coordination based on biologically inspired methods. In: Advanced Materials Research, vol. 664, pp. 1022–6680 (2013b) ISSN 1022-6680

14. Masár, M., Zelenka, J.: Modification of PSO algorithm for the purpose of space exploration. In: INES 2012: IEEE 16th International Conference on Intelligent Engineering Systems 2012, pp. 51–54. IEEE Operations Centre, Piscataway (2012) ISBN 978-1-4673-2692-6
15. Reynolds, C.W.: Flocks, Herds, and Schools: A Distributed Behavioural Model. Computer Graphics 21(4), 25–34 (1987) ISBN: 0-89791-227-6
16. Shang, L., Kai, C., Guan, H., Liang, A.: A Map-Coverage Algorithm Basing on Particle Swarm Optimization, Scalcom-embeddedcom. In: 2009 International Conference on Scalable Computing and Communications; Eighth International Conference on Embedded Computing, pp. 87–91 (2009) ISBN 978-0-7695-3825-9
17. Zolotová, I., Laciňák, L., Lojka, T.: Architecture for a universal mobile communication. In: SAMI 2013: IEEE 11th International Symposium on Applied Machine Intelligence and Informatics: Proceedings, Herľany, Slovakia, January 31-February 2, pp. S. 61–S. 64. IEEE, Budapest (2013) ISBN 978-1-4673-5926-9

# Vowel Recognition Supported by Ordered Weighted Average

Martin Klimo, Ondrej Škvarek, Juraj Smieško, Stanislav Foltán, and Ondrej Šuch

University of Zilina, Slovakia
Martin.Klimo@uniza.sk

**Abstract.** Ordered Weighted Average is a class of aggregators that generalize the concepts of median, minimum and maximum. An important feature of these aggregators is that they can be readily implemented in hardware. In our work we study whether these aggregators can be advantageously used to aggregate decisions by the group of experts. The experts in question are fuzzy logic recognition structures, a kind of a neural network. This paper compares the efficiency of group decision against the best trained system (within a class of fuzzy logic functions) on instances of two vowel discrimination problem.

## 1 Introduction

The development of robust speech recognition remains an important challenge that needs to be solved in order to allow humans to naturally interact with robots and industrial machinery. Various statistical methods have been successfully applied, for instance Hidden Markov Models, Neural Networks, Bayesian Networks, min-max circuits [1-7]. Discriminative techniques have been able to consistently outperform maximum likelihood ones in automatic speech recognition [8]. In this work we consider the question whether a discriminant combining knowledge of experts performs significantly better than the best performing expert.

R. Yager proposed in 1988 [9] a new type of aggregation operator called Ordered Weighted Average (OWA). Our idea is to use this aggregation for multi-criteria speech recognition in which the decision is taken by a set of recognition structures, not only by the best one. To demonstrate the impact of the OWA approach, we have chosen a simple case – vowel recognition

Usually, short term power spectrum features are used as the features of a discriminant function, such as Mel frequency cepstral coefficients or perceptual linear prediction. In this work we use for features (experts) the outputs of recognition structures, similarly to the tandem approach which used neural networks [10]. Our motivation for using these structures is that they can be implemented solely with a new processing element, the memristor [14], leading to a very low power recognition module. The recognition structures can be described as being fuzzy logic functions composed from minimum, maximum and strong negation elementary operations [14]. Therefore, the whole recognition system can be implemented in analogue hardware (in voltage mode) and it can perform recognition in real time. The best experts (neural networks) were obtained by evolutionary heuristics [11].

© Springer International Publishing Switzerland 2015

P. Sinčák et al. (eds.), *Emergent Trends in Robotics and Intelligent Systems*,

Advances in Intelligent Systems and Computing 316, DOI: 10.1007/978-3-319-10783-7_27

## 2    Methodology

The main goal of the paper is to study the question whether a collective decision of a population of recognition structures aggregated by OWA is better than the decision given by a single best structure. A block diagram of the proposed method is in Fig. 1. Speech pre-processing consists of speech windowing in time, spectral transform by Fast Fourier Transform and log-power spectrum calculation. Three vowels are taken from TIMIT Speech Corpus [12] and they are labelled according to the Corpus as "IY" as the vowel in "she", "AA" as the vowel in "dark", and "AO" as the first vowel in "water". The reason was to include a pair of dissimilar vowels (AA, IY) and a pair of vowels that are quite similar (AA, AO).

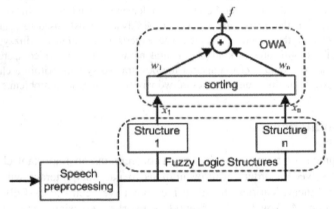

**Fig. 1.** Block diagram of the vowel recognition

Spectra of the vowels are fed to the fuzzy logic functions (structures) that are trained to maximise average distance for outputs of two selected vowels, e.g., AA vs. IY. The maximisation is based on the evolutionary heuristic [10]. From the last population, 8 structures were chosen with the best and distinct results, and their outputs provide arguments for the OWA method. Weighting coefficients were optimised by a gradient method and by the Nelder–Mead method. As a discriminant was used the difference of average outputs $\delta = |E(f(AA)) - E(f(AO))|$, the square difference of average outputs $\Delta = [E(f(AA)) - E(f(AO))]2$ as well as Fisher linear discriminant.

Correlation based recognition methods are compute intensive, therefore smaller models are preferred. On the other hand, pattern matching based recognition methods need large models and, therefore, prefer similarity metrics that can be computed very quickly [13]. We also believe that further progress can be achieved by large structures that can be natively parallelized. Therefore, we are concentrating on fuzzy logic functions that can be implemented by memristors [14]. This is the reason why, within this paper, we are focused on fuzzy logic functions with min, max and negation elements, although we know that functions based on Łukasiewicz implication give a better result [11].

An OWA aggregation operator of dimension n is a mapping $F: R^n \rightarrow R$ that has an associated weighting vector $\boldsymbol{w} = (w_1, w_2, \ldots, w_n)$ of dimension n having the properties $\sum_{j=1}^{n} w_j = 1$ and such that $F(a_1, a_2, \ldots, a_n) = \sum_{j=1}^{n} b_j w_j = \boldsymbol{b} \cdot \boldsymbol{w}^T$ where $b_j$ is the j-th largest value of $a_i$ and $\boldsymbol{b} = (b_1, b_2, \ldots, b_n)$.

We take AA vs. IY as an example to explain OWA coefficient optimization. There are 400 samples of AA and 400 samples of IY that are the outputs from 8 fuzzy logic systems. The samples form the matrix $A = \{a_{k,i}\}$ with size (400+400)x8. Matrix rows are inputs into OWA aggregation function and the output of OWA we denote $d_k$. The optimal output value is $d_k = 1$ for AA and $d_k = 0$ for IY. Then the maximization of discriminant $\Delta = [E(f(AA)) - E(f(AO))]2$ can be replaced by minimization of E(Err)= [1-E(d(AA)) + E(d(IY))]2. The problem of learning the optimal weights $\boldsymbol{w}$ can be simplified by taking advantage of the linearity of the OWA aggregation with respect to the ordered arguments. We denote the reordered object of the k-th sample by $\boldsymbol{b}_k = (b_{k,1}, b_{k,2}, \ldots, b_{k,n})$ where $b_{kj}$ is the j-th largest element of the arguments $a_{k,i}$. The problem of modelling the aggregation process is to find the vector of weights $\boldsymbol{w}$ minimizing

$$\tfrac{1}{2}||\boldsymbol{B} \cdot \boldsymbol{w}^T - \boldsymbol{d}||^2, \text{ where } \boldsymbol{B} = \{b_{k,i}\} = (\boldsymbol{b}_1, \boldsymbol{b}_2, \ldots, \boldsymbol{b}_m)^T \text{ and } \boldsymbol{d} = (d_1, \ldots, d_m)$$

and m is the number of all given samples of AA and IY. According to [15], the constraints on $\boldsymbol{w}$ can be eliminated by the following reparametrisation:

$$w_i = \frac{e^{\lambda_i}}{\sum_{j=1}^{n} e^{\lambda_j}}. \tag{1}$$

It becomes clear that for any value of the parameters $\lambda_i$ the weights $w_i$ will be within the unit interval and their sum is 1.

We have applied the gradient descent techniques (see [15]) with some modifications. Let $\lambda_i(l)$ indicate the estimate of $\lambda_i$ after the l-th iteration and so $w_i(l)$ is current estimate of the weights. For example, if at the beginning, i.e. $l = 0$, we put $\lambda_i(0) = 0$ for all i, then the starting weights have the uniform distribution $w_i(0) = \frac{1}{n}$.

We have proposed a three step algorithm to optimize OWA parameters:

Step 1 We use the $\lambda_i(l)$ to estimate weights

$$w_i(l) = \frac{e^{\lambda_i(l)}}{\sum_{j=1}^{n} e^{\lambda_j(l)}}. \tag{2}$$

Step 2 We use the estimated weights to get a calculated estimation aggregated values

$$\hat{\boldsymbol{d}} = \boldsymbol{B} \cdot \boldsymbol{w}(l)^T. \tag{3}$$

Step 3 We now update our estimates of the $\lambda_i$:

$$\lambda_i(l+1) = (1 - \beta)\lambda_i(l) - \beta w_i(l) \cdot \overline{Err}_i, \tag{4}$$

where

$$\overline{Err_i} = \sum_{j=1}^{m}(B_{j,i} - \hat{d_j})(\hat{d_j} - d_j)/m. \tag{5}$$

## 3    Results

To have an idea about chosen vowels similarity, Fig. 2. shows vowels' separation by coordinates of the their first three PCA components. As one can expect, dissimilar

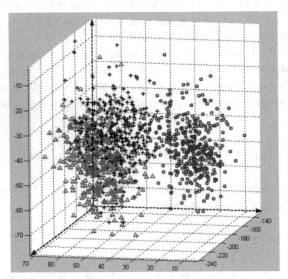

vowels (AA vs. IY) can be recognized very well. Seven of eight structures give very low error probability p, and structure number 3 gives all decision correctly (see Table 1). OWA then cannot give any improvement compared with the best individual structure S3. As we can see in more details in Fig. 3 which shows outputs histograms, results for the best structure S3 and for OWA outputs are very similar and both of them give zero error probability.

**Fig. 2.** First three PCA components of chosen vowels (AA (triangles), AO (pluses), IY (circles))

**Table 1.** Error probability p for single structure S and OWA weights (Nelder–Mead) AA-IY

|   | $S_1$ | $S_2$ | $S_3$ | $S_4$ | $S_5$ | $S_6$ | $S_7$ | $S_8$ |
|---|-------|-------|-------|-------|-------|-------|-------|-------|
| p | 0,036 | 0,025 | 0,000 | 0,004 | 0,003 | 0,003 | 0,009 | 0,011 |
| w | 0,000 | 0,000 | 0,194 | 0,469 | 0,108 | 0,228 | 0,000 | 0,000 |

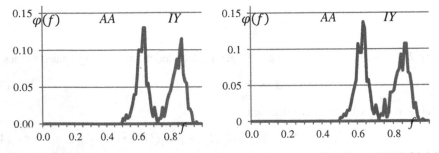

**Fig. 3.** Output value histograms for AA – IY on single structure 3 (left) and OWA (right)

Recognition is much harder for similar vowels AA, AO (see Fig. 2). The best structures S6, S7 return the error probabilities: p(S6)= p(S7)=0.203 that were improved by OWA with quadratic discriminant Δ that gave the value p(OWA) = 0.198. Because of the same number AA, AO, this is an average value of true negative probability p(OWAt-) = 0.133 and false positive probability p(OWAf+) = 0.267. Also the output histograms are slightly different (see Fig. 4.). This OWA result was obtained by the gradient method, while the Nelder–Mead numerical optimization algorithm did not find the optimum weighting because of p(S7) < p(OWANM) = 0.213. Table 2 gives error probabilities of the individual structures and the optimum weights obtained in OWA calculated by gradient method.

**Table 2.** Error probability $p$ and OWA weights (gradient) AA-AO

|   | $S_1$ | $S_2$ | $S_3$ | $S_4$ | $S_5$ | $S_6$ | $S_7$ | $S_8$ |
|---|-------|-------|-------|-------|-------|-------|-------|-------|
| p | 0,440 | 0,258 | 0,254 | 0,204 | 0,204 | 0,203 | 0,203 | 0,291 |
| w | 0,126 | 0,126 | 0,126 | 0,125 | 0,125 | 0,125 | 0,125 | 0,123 |

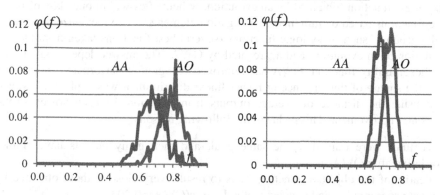

**Fig. 4.** Output histograms for AA – AO on structure 7 (left) and OWA (right)

We can conclude that collective OWA based decision with $\Delta = [E(f(AA)) - E(f(AO))]^2$ discriminant does over-perform the best fuzzy logic individual structure. To decide whether this result is caused by OWA aggregation method or by discriminant, we also optimized OWA coefficients using the difference of average outputs δ and the Fisher linear discriminant [2] as utility functions. Fig. 5 shows the output histograms.

Even though the discriminants are not directly comparable, they can be compared by error probability reached in recognition. Recognition error probability obtained with the difference of average outputs δ as the utility function gives p(OWA)=0.213, while with Fisher discriminant p(OWA)=0.196, that is better than p(S7)=0.203 obtained by the best individual structure and very similar to p(OWA)=0.197 obtained by the maximization of square average output difference. It should be remarked that we did not calculate the gradient for these utility functions (except square average) and the Nelder–Mead method was applied.

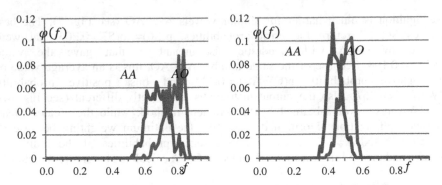

**Fig. 5.** Output histograms for AA – AO on OWA with Δ (left) and Fisher discriminants (right)

## 4    Conclusions

We have studied the question whether it is better to recognize two vowels by the best fuzzy logic function (obtained by an evolutionary heuristics within one class of fuzzy logic functions based on min, max and negation elements), or to recognize the vowels by the group of such fuzzy logic functions (several best functions taken from the last population of an evolution) and aggregated by OWA. The answer depends greatly on the chosen utility function. OWA has improved recognition performance noticeably when the square of this distance or Fisher linear discriminant was applied. However, when using the distance of average outputs from function for each vowel, OWA brings no improvement. This is due to the following reasons:

- structures were trained to the same goal, and then many outputs have similar ordering after OWA sorting.
- linearity of the objective function leads to results very close to those obtained by linear programming and simplex method, i.e. $p(OWA)=0.203$.

Our recommendation to obtain better vowel recognition results by OWA with linear utility function is to train structures to the specific goals to obtain their higher specialization and higher heterogeneity of their outputs.

**Acknowledgments.** This contribution is the result of the project implementation: Centre of excellence for systems and services of intelligent transport II., ITMS 26220120050 supported by the Research & Development Operational Programme funded by the ERDF. Partially supported by VEGA 2/0112/11 grant.

## References

1. Bishop, C.M.: Pattern Recognition and Machine Learning. Springer (2006) ISBN 0-387-31073-8
2. Flanagan, J.L., Allen, J.B., Hasegawa-Johnson, M.A.: Speech Analysis Synthesis and Perception, 3rd edn. (2008)

3. Mariani, J.: Spoken language processing. Wiley (2009) ISBN 978-1-84821-031-8
4. Huang, X.D., Ariki, Y., Jack, M.A.: Hidden Markov Models for Speech Recognition. Edinburgh University Press (1990)
5. Katagiri, S. (ed.): Handbook of Neural Networks for Speech Processing. Artech House (2000) ISBN 0-89006-954-9
6. Wiggers, P., Rothkrantz, L.J.M., van de Lisdonk, R.: Design and Implementation of a Bayesian Network Speech Recognizer. In: Sojka, P., Horák, A., Kopeček, I., Pala, K. (eds.) TSD 2010. LNCS (LNAI), vol. 6231, pp. 447–454. Springer, Heidelberg (2010)
7. Šuch, O.: Using min-max Circuits for Speech Recognition. Habilitation thesis, University of Žilina (2013), http://www.savbb.sk/~ondrejs/Phoneme/hab.pdf
8. Heigold, G., Ney, H., Schlüter, R., Wiesler, S.: Discriminative Training for Automatic Speech Recognition. IEEE Signal Processing Magazine 29(6), 58–69 (2012)
9. Yager, R.R.: On Ordered Weighted Averaging Aggregation Operators in Multicriteria Decision-making. IEEE Transactions on Systems, Man, and Cybernetics 18(1) (January/February 1988)
10. Hermansky, H., Ellis, D., Sharma, S.: Tandem connectionist feature extraction for conventional HMM systems. In: Proc. IEEE Int. Conf. Acoustics, Speech, and Signal Processing (ICASSP), Istanbul, Turkey, pp. 1635–1638 (June 2000)
11. Foltán, S.: Speech recognition by means of *fuzzy* logical circuits. In: 18th International Conference on Soft Computing, Brno (2012) ISBN 9788021445406
12. Hastie, T., Tibshirani, R., Friedman, J.: Elements of Statistical Learning. Springer, dataset, http://cran.r-project.org/package=ElemStatLearn
13. Dean, T., Ruzon, M.A., Segal, M., Shlens, J., Vijayanarasimhan, S., Yagnik, J.: Fast, Accurate Detection of 100,000 Object Classes on a Single Machine. In: Proceedings of IEEE Conference on Computer Vision and Pattern Recognition. IEEE Computer Society, Washington (2013)
14. Klimo, M., Šuch, O.: Memristors can implement fuzzy logic. Cornell University, Ithaka (2011), http://arxiv.org/pdf/1110.2074.pdf
15. Yager, R.R., Filev, D.P.: Induced Ordered Weighted Averaging Operators. IEEE Transactions on Systems, Man, and Cybernetics, Part B: Cybernetics 29(2) (April 1999)

# Advanced Generalized Modelling of Classical Inverted Pendulum Systems

Slávka Jadlovská, Ján Sarnovský, Jaroslav Vojtek, and Dominik Vošček

Department of Cybernetics and Artificial Intelligence,
Faculty of Electrical Engineering and Informatics, Technical University of Košice, Letná 9,
042 00 Košice, Slovak Republic
{slavka.jadlovska,jan.sarnovsky}@tuke.sk,
{jaroslav.vojtek,dominik.voscek}@student.tuke.sk

**Abstract.** The purpose of this paper is to present the design and program implementation of expansions made to the existing general algorithmic procedure which yields the mathematical model for a classical inverted pendulum system with an arbitrary number of pendulum links. The expansions include the option to define the reference position of the pendulum in a planar coordinate system, to choose the reference direction of pendulum rotation and to select the shape of a weight attached to the last pendulum link. The underlying physical formulae based on the generalized inverted pendulum concept are implemented in the form of a symbolic MATLAB function and a MATLAB GUI application. The validity and accuracy of motion equations generated by the application are demonstrated by evaluating the open-loop responses of simulation models of the classical single and double inverted pendulum system using the newly-developed MATLAB blocks and applications.

**Keywords:** classical inverted pendulum system, attached weight, reference pendulum position, automatic model generation, symbolic MATLAB function.

## 1 Introduction

Stabilization of a physical pendulum or a system of interconnected pendulum links in the upright unstable position is a benchmark problem in nonlinear control theory 1: in recent years, several types of stabilizing mechanisms such as a cart moving on a rail 2, rotary arm 3 or vertical oscillating base have been introduced. Inverted pendulum systems (IPSs) are therefore regularly employed as typical examples of unstable nonlinear underactuated systems in the process of verification of linear/nonlinear control strategies in corresponding control structures. Direct practical applications include walking humanoid robots, launching rockets, earthquake-struck buildings and two-wheel vehicles, such as Segway PT. The principles of modelling and control of IPSs can further be considered as the basic starting point for the research of advanced underactuated systems such as mobile robots and manipulators 4 as well as aircraft and watercraft vehicles 5.

© Springer International Publishing Switzerland 2015
P. Sinčák et al. (eds.), *Emergent Trends in Robotics and Intelligent Systems*,
Advances in Intelligent Systems and Computing 316, DOI: 10.1007/978-3-319-10783-7_28

We focused our research on the mutual analogy among mathematical models of IPSs with a varying number of pendulum links. Consequently, we introduced the concept of a generalized n-link inverted pendulum system with n+1 degrees of freedom (DOFs) and a single actuator which allows to treat an arbitrary IPS as a particular instance of the system of n pendula attached to a given stabilizing base. General procedures which determine the Euler-Lagrange equations of motion for a user-specified instance of a generalized classical (i.e. on a cart) and rotary IPS were developed and implemented via MATLAB's Symbolic Math Toolbox 1.

The design and implementation of the general procedures has so far been based on the assumption of interconnected pendulum rods which angles are determined relative to a fixed planar/spatial coordinate system and their centres of gravity (CoG) are identical to their geometric centre. The goal of this paper is to expand the existing procedures for automatic model generation of classical IPSs with further, practically motivated generalizations. The paper is organized as follows. Firstly, the generalized classical IPS, which has been studied so far, is presented, this time as the basic starting point for further research. Next, two categories of expansions are described in terms of their impact on general mathematical model derivation: change in CoG position of a pendulum caused by a weight attached to its end and the option to specify the orientation of the pendulum reference position together with the reference direction of pendulum rotation. The next section details the implementation of an expanded general procedure which aim is to ultimately cover all possible forms of models found in relevant literature by including all possible combinations of underlying assumptions for pendulum reference position/direction and the existence of attached weights. Finally, the validity and accuracy of the procedure is verified using the classical single and double IPS, both represented by pre-prepared Simulink blocks encompassing all investigated features, and the paper is concluded with an evaluation of achieved results.

## 2    Expanded Generalized Classical Inverted Pendulum System

The generalized system of classical inverted pendula was introduced in 1 as a set of $n \geq 1$ rigid, homogenous, isotropic rods (pendulum links) which are interconnected in joints and attached to a stable cart which enables movement along a single axis. A multi-body mechanical system defined this way is underactuated since it has fewer actuators than DOFs 4: the only input (force F(t) acting upon the cart actuates the cart position [m] as well as the n pendulum angles [rad]. Through the mathematical model derivation process in 1, it was assumed that all motion was bound to a planar coordinate system with the cart moving along the x-axis, which was simultaneously identified with the projection of the zero potential energy level into the xy-plane. The value of every pendulum angle was determined clockwise with respect to the vertical upright position of the pendulum, which was defined as parallel to the y-axis (Fig. 1).

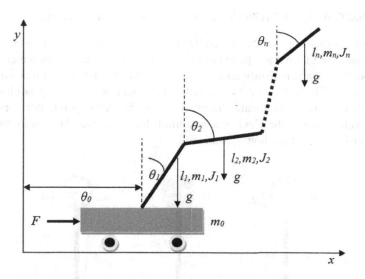

**Fig. 1.** Generalized system of classical inverted pendula – scheme and basic nomenclature

According to the Lagrangian formulation of classical mechanics, every possible configuration of a multi-body system can be uniquely defined by a vector of generalized coordinates equivalent to the system's DoFs which are, in the case of a generalized classical IPS, identified as cart position and pendulum angles:

$$\boldsymbol{\theta}(t) = \begin{pmatrix} \theta_0(t) & \theta_1(t) & \dots & \theta_n(t) \end{pmatrix}^T \tag{1}$$

For every generalized coordinate, a nonlinear second-order differential motion equation is specified by employing Euler-Lagrange equations defined in the form:

$$\frac{d}{dt}\left(\frac{\partial L(t)}{\partial \dot{\boldsymbol{\theta}}(t)}\right) - \frac{\partial L(t)}{\partial \boldsymbol{\theta}(t)} + \frac{\partial D(t)}{\partial \dot{\boldsymbol{\theta}}(t)} = \boldsymbol{Q}^*(t) \tag{2}$$

where L(t) is the difference between the multi-body system's kinetic and potential energy, D(t) stands for the dissipation properties and Q*(t) is the vector of generalized external inputs 6. The process of mathematical model derivation for a selected IPS hence transforms into a procedure to determine its kinetic, potential and dissipation energy, each defined as a sum of energies of the multi-body system's individual bodies, i.e. the cart (i=0) and all pendulum links (i=1,... n):

$$E_K(t) = \sum_{i=0}^{n} E_{Ki}(t) \ , E_P(t) = \sum_{i=0}^{n} E_{Pi}(t), \ D(t) = \sum_{i=0}^{n} D_i(t) \tag{3}$$

While the potential energy of the i-th body depends on CoG position coordinates, CoG velocity components need to be obtained to specify kinetic energies or dissipative properties 7. The actual physical formulae which form the core of the procedure for a generalized IPS were derived in 1 and presented together with generated motion equations of example IPSs and the verification of their validity.

258     S. Jadlovská et al.

## 2.1     Modified Mass Distribution as a Result of Attached Weight

As the first expansion to the generalized IPS, we considered a system with an arbitrary number of inverted pendulum links mounted on a cart, in which a weight with a specified shape is firmly attached to the end of the pendulum link furthermost from the cart. Therefore, for i=1,..., n-1, the whole mass of a pendulum rod is concentrated to the CoG located midway from the pivot point, but for i=n, the attached weight causes the CoG of a pendulum link with weight to shift away from the geometric centre of the homogenous rod.

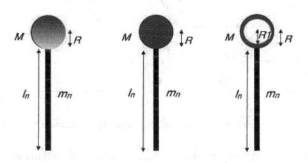

**Fig. 2.** Considered weight shapes: sphere, cylinder, ring

By computing the distance between the CoG of the weighted pendulum link and the pivot point, CoG position and velocity coordinates followed by related potential and kinetic energies were derived and can be found in 8. To fully express the kinetic energy, the moment of inertia of the weighted pendulum was calculated as the sum of moments of inertia of the pendulum rod and the weight itself. Three shapes of weight – sphere, cylinder and ring (Fig. 2) were considered, however the algorithm can be applied to all symmetric isotropic bodies with known moments of inertia, since the resulting equations of motion only differ by the moment of inertia of the attached weight.

## 2.2     Modified Pendulum Reference Position

Most differences between the correctly derived inverted pendulum models found in various sources can be attributed to the initial choice of reference position for all pendulum links and the reference direction of pendulum rotation, both of which determine the numeric value of the pendulum angle at every time instant. Our next goal was, therefore, to expand the procedure of mathematical model derivation for generalized IPS so that all feasible combinations of initial assumptions would be covered. Eight possible combinations of reference pendulum positions with respect to a planar coordinate system (top, bottom, right, left) and pendulum movement directions (clockwise, counterclockwise) were considered (see Fig. 3 for examples).

During the derivation process which was recorded step-by-step in 9 the selected combination of initial assumptions has shown to have direct influence on the coordinates of the CoG position of each pendulum link and, subsequently, on the related CoG velocity components, on the expressions for kinetic and potential energy and, finally, on the motion equations in their final form.

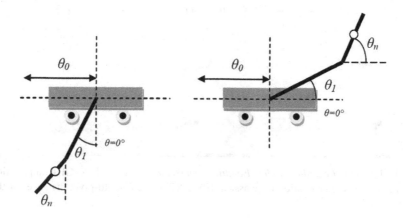

**Fig. 3.** Examples of reference frame definitions – a) bottom clockwise b) right counterclockwise

# 3    Expanded General Procedure for Mathematical Model Derivation – Program Implementation and Application

Using the familiar theoretical background as well as the newly-derived physical formulae, we completely reworked the earlier general algorithmic procedure so that it would yield a mathematical model of an arbitrary classical IPS with respect to user-selected criteria for attached weights and pendulum reference position. The procedure was once again implemented as a MATLAB function which generates the nonlinear equations of motion via Symbolic Math Toolbox in the simplified and rearranged form, equivalent to the most likely form obtained by manual derivation. An application with a graphical user interface, Inverted Pendula Model Equation Derivator_v2, was also developed to provide a user-friendly access to the function. Compared to the earlier version of the Derivator where the user could only select the number of pendulum links 10, four options for weight type (including none) and eight for reference position/direction are now provided. As a further improvement, the equations can now be displayed in form of LaTeX expressions in addition to the original MATLAB representation. Fig. 4 shows an example preview of the Derivator_v2 window which contains the generated model equations for the classical double IPS with an attached cylinder-shaped weight.

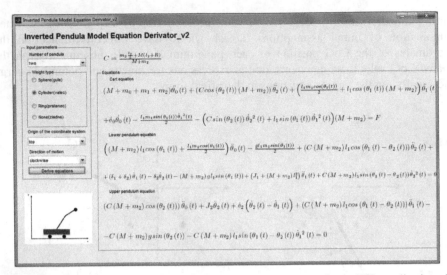

**Fig. 4.** Inverted *Pendula Model Equation Derivator – ver. 2* – GUI application for mathematical model generation for classical IPSs, "C" is the *CoG*-to-pivot distance for the *n*-th pendulum link

By evaluating the results of the implemented expanded general procedure, we concluded that the mathematical model of the original generalized classical IPS (i.e. without attached weight) represents a special case of a model of the weighted IPS in case the weight mass is set to zero, which serves as a confirmation of the procedure's accuracy. Analogically, the original specification of a generalized IPS, characterized by a particular combination of initial assumptions about the reference position and rotation of the pendulum, now becomes a representative of a family of models related to the same physical system with confirmed validity.

## 3.1    Verification of Generated Mathematical Models of Classical Inverted Pendulum Systems

A structured Simulink block library, Inverted Pendula Modelling and Control (IPMaC), has been developed since 2009 as a comprehensive software framework for the analysis and control of IPS in the simulation environment. To reflect the expansions outlined in previous sections, library blocks which implement the motion equations of the classical single and classical double IPS were equipped with newly-implemented properties. As a result, the subsystem mask of both blocks now allows the user to dynamically change the parameters of the simulation model, specify the number of input and output ports, the shape of the attached weight, reference pendulum angle value and reference direction of pendulum rotation. The possibility to switch to a simplified model which neglects friction and omits the backward impact of the pendulum links on the cart was also implemented. The necessity to create a separate block for each set of equations was eliminated by callbacks which ensure the dynamic adjustment of the block structure so that it will always correspond to a specific set of motion equations. A GUI application, Analysis of Inverted Pendulum Systems (AoIPS), was developed in MATLAB as a graphical tool to monitor, analyse and evaluate the open-loop dynamics of a selected classical IPS in a single window. For every simulation experiment, a scheme containing a suitable Simulink block is run in the background, block parameters are set to values specified in the GUI, and simulation results are exported into separate figures for further investigation.

Next, it will be assessed whether the generated mathematical models of classical single and double IPSs can be considered as valid and accurate for the control design purposes. Using the AoIPS tool, open-loop responses to an impulse signal constrained in time/amplitude were obtained for both simulated models, starting from the initial upright equilibrium. Numeric parameters were specified in 89.

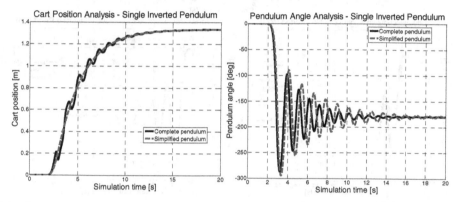

**Fig. 5.** Classical single inverted pendulum system – cart position and pendulum angle in a complete vs. simplified model

Simulation experiments which illustrate the dynamics of simplified models of IPSs or analyse the influence of the pendulum's modified mass distribution on the system dynamics were evaluated first. The comparison of time behaviour of the cart position and pendulum angle for the complete and simplified model of a classical single IPS is depicted in Fig. 5, while the dynamics of the classical double IPS with different attached weights is evaluated in Fig. 6. In both cases, all pendulum links fall from the upright into the downward equilibrium through a damped oscillatory transient state and stabilize there, in compliance with the empirical observations of pendula behaviour. In the simplified model, the pendulum makes a very long transition to the steady state as a result of neglected friction, and the cart trajectory correctly shows no signs of the backward impact caused by the pendulum movement. If weight is attached to the upper pendulum, the pronounced "jerky" cart movement is caused by the inertia of the heavier pendulum link. Total damping of a weighted system is much lower than that of the system with no weight load, which is reflected on larger oscillations of pendulum links and their prolonged settling time. The differences between the dynamics of systems with different types of weights are minimal.

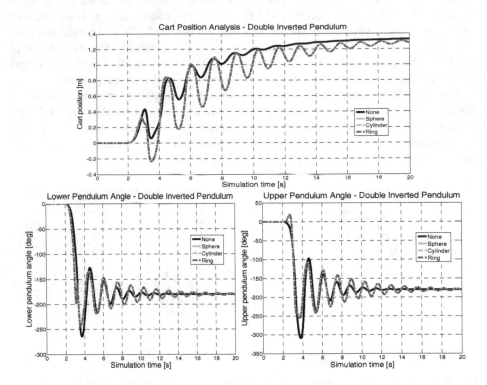

**Fig. 6.** Classical double inverted pendulum – cart position and pendula angles for different shapes of weights, including no weight

The following simulation experiments evaluate the influence of combinations of initial assumptions on the response of IPSs. Fig. 7 depicts the dynamical behaviour of a classical single IPS after selecting four starting positions which vary by 90° (top, left, down, right), while the pendulum angle is determined clockwise in all cases. The effect of the direction in which the pendulum angle is determined (clockwise / counterclockwise) is shown on a classical double IPS in Fig. 8. It has been proven that the changes in initial assumptions have no effect on the dynamics of either the cart or the pendulum links, and only the graphical representation of pendulum behaviour is subject to change. Choice of the reference value of the pendulum angle determines the numeric value corresponding to the upright/downward position of the pendulum, and the selected reference direction defines whether the pendulum angle will increase or decrease during simulation, as it is clear from the „mirror image" depicting the pendula behaviour in Fig. 8.

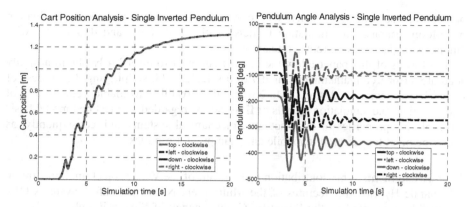

**Fig. 7.** Classical single inverted pendulum system – cart position and pendulum angle – effect of the changing reference position of the pendulum – top, left, down, right

Reasonable behaviour of the open-loop responses of both simulation models means that under, all criteria, systems described by the generated motion equations can be considered accurate enough to serve as a reliable test bed for the verification of linear and nonlinear control algorithms.

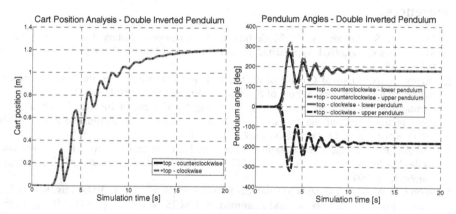

**Fig. 8.** Classical double inverted pendulum system – cart position and pendulum angles determined in a clockwise / counterclockwise reference direction

## 4    Conclusion

The purpose of this paper was to expand and further generalize the existing algorithmic procedure for obtaining the equations of motion of classical inverted pendulum systems (IPSs) with an arbitrary number of pendulum links. The expanded general procedure covers all feasible combinations of initial assumptions for the pendulum reference position and direction of rotation and considers various shapes of weight load attached to the last pendulum link. A GUI application was developed to provide an intuitive interface to the MATLAB function which implements the procedure. The validity of generated motion equations was confirmed by evaluating open-loop responses of simulation models of classical single and double IPS with emphasis on the newly-introduced features.

The results of this paper allow the control engineer to effortlessly obtain a highly accurate, error-free mathematical model of a selected IPS, simplifying the process of model-based control design. Moreover, the readily available collection of mathematical and simulation models of IPSs can be regarded as a test bed model basis for exploring properties of underactuated mechanical systems and, consequently, as a starting point for research in mobile and manipulator robotics.

**Acknowledgments.** This contribution is the result of the Vega project implementation – Dynamic Hybrid Architectures of the Multi-agent Network Control Systems (No. 1/0286/11), supported by the Scientific Grant Agency of Slovak Republic – 50% – and of the KEGA project implementation – CyberLabTrainSystem – Demonstrator and Trainer of Information - Control Systems (No. 021TUKE-4/2012) – 20%. It was also supported by the Research and Development Operational Program for the "University Science Park Technicom for innovative applications with knowledge technology support" project (ITMS code 26220220182), which was co-financed by the ERDF – 30%.

# References

1. Jadlovská, S., Sarnovský, J.: Modelling of Classical and Rotary Inverted Pendulum Systems – a Generalized Approach. Journal of Electrical Engineering 64(1), 12–19 (2013) ISSN 1335-3632
2. Bogdanov, A.: Optimal Control of a Double Inverted Pendulum on the Cart. Technical Report CSE-04-006, OGI School of Science and Engineering, OHSU (2004)
3. Furuta, K., Yamakita, M., Kobayashi, S.: Swing Up Control of Inverted Pendulum. In: Proc. of the Int. Conf. on Industrial Electronics, Control and Instrumentation (IECON 1991), Kobe, Japan, October 28-November 1 (1991)
4. Tedrake, R.: Underactuated Robotics: Learning, Planning and Control for Efficient and Agile Machines. Course Notes for MIT 6.8. Massachusetts Institute of Technology, Cambridge (2009)
5. Spong, M.W.: Underactuated Mechanical Systems. In: Siciliano, B., Valavanis, K.P. (eds.) Control Problems in Robotics and Automation. LNCIS, vol. 230, pp. 135–150. Springer, Heidelberg (1998)
6. Goldstein, H., Poole, C., Safko, J.: Classical Mechanics, 3rd edn., p. 680. Addison-Wesley (2001)
7. Halliday, D., Resnick, R., Walker, R.J.: Fundamentals of Physics, Ext. 7th edn. Wiley (2004)
8. Vošček, D.: Modelling and Control of Inverted Pendulum Systems II (Modelovanie a riadenie systémov inverzných kyvadiel II.). Bachelor Thesis. Supervisor: prof. Ing. J. Sarnovský, CSc, consultant: Ing. S. Jadlovská, FEEI-TU (2013)
9. Vojtek, J.: Modelling and Control of Inverted Pendulum Systems I (Modelovanie a riadenie systémov inverzných kyvadiel I.). Bachelor Thesis. Supervisor: prof. Ing. J. Sarnovský, CSc, consultant: Ing. S. Jadlovská, FEEI-TU (2013)
10. Jadlovská, S., Sarnovský, J.: An extended Simulink library for modelling and control of inverted pendula systems. In: Proc. of the Int. Conf. Technical Computing Prague 2011, Prague, Czech Republic (November 8, 2011) ISBN 978-80-7080-794-1

# Towards Acceptable Products: Grasping User's Mind by the Means of Cognitive Science and Rational Choice Theory

David Zejda

University of Hradec Králové, Faculty of Informatics and Management, Czech Republic
david.zejda@uhk.cz

**Abstract.** Intelligent assistive robots as well as other products of applied science have great potential to improve different aspects of our lives. In the past, some promising products and whole technologies failed to meet goals set for them by the scientists, inventors and manufacturers. Some of them have been set on the side line, the rest were left in oblivion. Our study of both cognitive science and rational choice theory is being motivated by the goal to help all the parties to avoid or prevent such fate for the results of their work. Psychological and social aspects of the products and their properties have to be considered carefully in order to result in technologies resonating with user's real needs. On the following lines, we present two different conceptualizations of human mind - cognitive science and rational choice theory - and compare the main focus of each of them in regards to the means to capture mental processes which likely take place in the mind of the person who is appropriating a technology or a product. Two models, one based on rational choice theory and one based on cognitive science, are introduced briefly. In the concluding section we bring some implications for the process of developing, designing and presenting hi-tech products, including intelligent assistive robots.

**Keywords:** appropriation, rational choice theory, cognitive science, technology evaluation.

## 1 Introduction

Products and technologies are gaining more features, more complexity or even intelligence. The cost of the development matches the complexity. Developing a product which won't be accepted by users should be prevented as much as we can. There are more related questions. How to find the best target group? How to design the user interface with acceptability on mind? How to present and introduce the product? Mario Tokoro in his proposal of "open systems science" emphasizes the interdisciplinarity as a key aspect of science of the future. [1]

To find the answers to the questions mentioned above, we have to transcend the field of the technology itself. We have to find out how users think, how they evaluate, handle, apply, which aspects do they consider and which other circumstances affect these processes. Both rational choice theory, the subfield of economics, and cognitive

© Springer International Publishing Switzerland 2015
P. Sinčák et al. (eds.), *Emergent Trends in Robotics and Intelligent Systems*,
Advances in Intelligent Systems and Computing 316, DOI: 10.1007/978-3-319-10783-7_29

science, the current leading subfield of psychology aiming to describe our minds, seem to provide certain modelling tools and methods. In this paper we focus to present surfaces of contact between these fields and their applicability for our purpose.

In general, rational choice theory brings precise tools of economy to describe our reasoning in a great variety of life situations. But, we as human beings are not rationally reasoning machines, we do not think, recollect, decide or behave precisely, logically and deterministically. Actually, our mind brings more biases and less rationality than what we would like to believe. Rational choice theory brings good answers to the question how "we should think and behave" (its nature is naturally normative), whereas cognitive science is much more descriptive, aiming to capture how we indeed think and behave, thus reflects many of the inconsistencies, biases and flaws of the reasoning machines in our heads. Both rational choice theory and cognitive science research already concluded various useful conceptual models. Because processes in the human mind are highly complex, every model attempting to describe them has to apply various simplifications, which also means that the approximate results from all the models have to be applied with caution.

## 2    Model of Initial Evaluation Based on Rational Choice Theory

Rational choice theory, also known as choice theory or rational action theory, a descendant or a subfield of economy, comes with a palette of various formal models aimed to provide an insight into human social and economic behaviour. Gary Becker widened the scope of the theory to describe many aspects of human behaviour [2], including drug addiction [3], beggary and compassion [4], crime and punishment [5], human capital [6], love, marriage and family [7]. Rationality involves balancing costs against benefits in order to maximize advantage. Advantage may be defined in terms of money and alternative definitions are also available [8] (e.g. Bentham [9] or Marshall [10]).

On the foundations of rational theory we have built and published a simple model of technology evaluation with a specific focus on ambient intelligence for ageing users. [11] The model describes the first stage of technology evaluation, when a prospective user is thinking whether to try a certain product or not. We don't use money to define utility (or advantage). Utility in our model reflects specific subtle needs of the considered user group of the elderly, such as perceived self-worthiness and social relations. Personal utility function in the model assigns utility value to each combination of comfort sources. Comfort sources are fully determined by the complete set of influence vectors, where each vector is assigned to a certain aspect of life. Instead of the monetary expressed cost, the model comprises effort, time and external support as resources. The exert function defines how much resources are necessary to adopt an aspect of life.

The model describes initial evaluation as a sequence of three successive steps, leading to either acceptance or one of three possible kinds of refusal. If an aspect of life (e.g. a product) is perceived as 1. to be beneficial, 2. to be reachable and 3. the best choice from all available options (all three conditions are satisfied), it passes the evaluation stage and may proceed to the appropriation phase. The simple evaluation model may be solved as an optimization problem of linear programming. [12]

# 3    Cognitive Science to Understand Appropriation

Our further goal was to design a model of the appropriation phase which follows for the product which passed the initial evaluation described by the model mentioned above. During the appropriation stage, the user struggles to learn how to operate the product, how to use it effectively and how to harmonize it with his daily routine. The user is satisfied or dissatisfied, positively surprised or disappointed. The process of appropriation typically lasts for several weeks. Compared with appropriation, initial evaluation is a quick and almost immediate decision with short process under stable conditions. The user knowingly intends to make a rational decision. Appropriation is, on the other hand, more complex, longer, less conscious and, though primarily driven by rational reasoning, potentially less rational. Further, we wished to come with a descriptive, not normative model – our goal was not to explain how users should think rationally, but rather how they indeed think. These requirements lead us to put rational choice theory aside and look for different foundations for the model.

Psychology developed greatly during the last century. Where former theories such as Freudian psychoanalysis or behaviourism failed to explain significant phenomena in our reasoning, cognitive science, based on precise scientific methods including rigorous statistical examinations in reproducible scenarios, markedly deepened our understanding of ourselves. [13] Cognitive science concluded that there are flaws in our reasoning even in simple isolated tasks, as described further. Many more of such flaws, biases and irrationalities influence our reasoning in the tasks which are complex, difficult and scattered along a longer period of time. When we published our model of appropriation [14], we also listed relevant "principles of mind", revealed by the cognitive science. Following paragraphs describe them briefly.

## 3.1    Substages of Evaluation

A lot of current scientific understanding of human reasoning is based on studies of behaviour in highly controlled simple tasks, where participants decide between well-defined options, such as single gambles. Real world situations are both not so much controllable and also more complex, requiring sequential evaluation. Appropriation is an example of such a complex and dynamic task. Efforts to explain chains of successive related decisions may be traced, e.g., back to the study of Damasio et al. on performance of brain-injured patients. [15] Other researchers followed the path, e.g. Lejuez et al. [16] examined how participants think and behave in a sequential balloon inflating task.

As a conclusion, it is quite natural for the human mind to break a complex problem into distinctive steps or episodes, framed, due to the character of the task, such as single blows in the balloon experiment, distinct days or weeks, in evaluations lasting longer. [17]

## 3.2    Anchors or Reference Points

Another concept very well described in many studies is the tendency of our mind to focus on a certain fact for further reasoning. Kahneman and Tversky [18] (among others) revealed the phenomenon of anchor (or reference point). Instead of reasoning in absolute quantities and final outcomes, our mind tends to think in relative comparisons and in shorter time frames. Both positive and negative outcomes of our decisions have returns diminishing proportionally to the distance from the reference point.

Though initially introduced by an economist for money-related decisions, the concept of labile, vague and adaptive reference points which are not always rational, has never been adopted by the mainstream economy, following its primary focus on normative description of human reasoning. A relative approach of our minds, though tenable in certain situations when applied in tasks with utility, contradicts a traditional diminishing marginal utility law known in economy as well as in rational choice theory. With these findings Kahneman and Tversky refined the theory of utility and called it the prospect theory. As subsequent research revealed, palette of various reference anchors exist, either related to the problem, such as status quo or aspiration level, or totally irrelevant, as proven, e.g., in [19]. Some of them are more prevalent and more influential, such as status quo. Lopes and Oden [20] concluded that at least three reference points play a significant role in our evaluations under uncertainty. They act in parallel: main reference point (usually status quo), aspiration level and security level (danger of loss). According to Hastie and Dawes [17], the model built around the concept of reference points may compete with prospect theory with its ability to describe our reasoning.

## 3.3    Gradual Adaptation

The concept of gradual adjustment is nothing more than an application of the concept of mental anchors in complex problems with successive decision chains. Specific examples of this phenomenon are mentioned, e.g., in [17]. When we respond to stimuli such as loudness or temperature, the past and present context of experience defines an adaptation level or reference point, and stimuli are perceived in relation to this point. Cyert and March revealed that we tend to search for alternatives in the neighbourhood of our previous try. [21]

In complex schemes with successive steps, we tend to follow the anchor-and-adjust strategy which leads to successive adjustment on-the-fly. E.g. Slovic et al. confirmed the effect in pricing and choice in successive virtual gambles. [22] Kahneman and Tversky pointed out that justification for reasoning on consequences with status quo on mind can be found in the general principle of adaptation – the stepwise adjustment of the mind anchor allows it to adopt our mind to constantly changing environment. On the other hand, it may easily lead to illogical flaws in our reasoning, such as "money pump" described in [17] which contradicts rational (economic) choices.

### 3.4    Two Chains of Reasoning

Neuroscience (also called neuroeconomics) examines neural substrates of our judgment and behaviour. [17] One of the conclusions made by neuroscience is the fact that our reasoning runs over internally in two trails (chains, circuits). While the dopamine-mediated system is responsible for assessing positivity, acetylcholine-mediated circuit ensures negativity. From the relevant works we may mention, e.g., [23].

What is relevant to our model, any evaluation runs over in parallel. User at the same time evaluates benefits, utility, rejoice, pleasurable surprise, etc., on one hand, and, on the other hand, negative aspects such as costs, pain, anger and disappointment.

### 3.5    Linear Model

According to many studies, if the task is to make a decision based on a set of cues, we could hardly name anything better to describe our reasoning than a very simple linear model. Actually, even improper linear models with weights not based on statistical techniques (e.g. with random weights, where only the direction of relation is explicitly assigned) outperform experts in many expert tasks [24].

Simple linear models executed by a dumb computer program achieves the same or better results in medical diagnosis tasks [25], in prediction of bankruptcy [26], in assessments of applicants [27], in estimations of real estate values, in stock investments and many other areas. Since the first notable book on the topic had been published more than 50 years ago [28], huge amount of studies concluded again the same. Nowadays, there is no controversy on this general inability in reasoning – any "expert insight" is more than outweighed with inconsistency, incomplete memory and plenty of other flaws. According to March [17], the fact that we still rely on experts despite their real incompetence may serve a purely social function. As a conclusion, if we wish to formally describe our reasoning in situations based on a set of cues, there is no need to seek for anything more sophisticated than a simple linear equation. Most likely, our mind won't do better.

## 4    Implications from the Model of Appropriation

The discussed principles served as a foundation to develop a model of appropriation. [14] As in other models, initial simplifications have been made. For example, our model can't be used to directly compare alternatives, such as different products or activities where a user might put his effort and time. Our goal was to reach conclusions potentially applicable in the design and development of advanced technologies. Appropriation may lead either to final acceptance by the evaluating user, or its rejection. In terms of the model, a product is rejected whenever a user runs out of patience before a level of stability is being reached. The model suggests various causes of rejection – if a user lacks initial enthusiasm, if it becomes too difficult or boring to master the product or its interface, if it does not satisfy user's aspiration level reflected in expectations (disappointment).

Certain variables defined in the model are out of reach of the designers because they reflect user's personal psychological characteristics, e.g. significance of underestimation. Other variables are partially given (their subjective component is significant), such as bore induced by every step of evaluation, enthusiasm, expectations and levels of sustainability and stability. Bore may be reduced, e.g., by lowering demands on effort or making the effort more pleasurable. Certain variables may be influenced, e.g., through both rationally relevant and irrelevant aspects of the situation in which the product is being presented to the user. Some variables are almost in the hands of designers, such as the real beneficial potential of the product and objective component of an effort necessary for the appropriation.

Our examination based on the model led us to an interesting conclusion regarding the role of enthusiasm. Enthusiasm, on one hand, induces a higher level of patience, which is good, because sufficient "supply of patience" induced primarily from enthusiasm is necessary to overcome bore gradually induced in each step. But it is primarily derived from expectations and with higher expectations there is a higher chance of disappointment, on equal terms, which is bad. As a consequence, higher initial enthusiasm does not necessarily mean a higher success rate (or lower probability of rejection, which is complementary) in appropriation. Though explanations of reasons behind this effect differ, there is consensus among economists and cognitive scientists that loss is more painful than gain. If our goal has a higher chance of successful appropriation, disappointment should be avoided. At first, it seemed that the designer's or merchant's goal should be to induce the highest level of enthusiasm which will (hopefully) not lead to disappointment.

But, as both the theory and the model suggest, this maximal level of enthusiasm not inducing disappointment may be still too high. Mellers et al. arranged a specific gambling task to capture regret and rejoice reactions and concluded that experienced utility is intensified if it produces regret or rejoicing, in particular if it is a surprise. [29] The exact effects of surprise on our perception of utility are still not fully examined. According to [17], it is still not clear under which conditions people actually infer and consider counterfactual at the time they make the decision, that is, under which conditions the regret and rejoicing effects affect perception of utility and the process itself. But it seems that, in compliance with Mellers' conclusions, it may be better to slightly undervalue the benefits of the introduced product. Lower levels of enthusiasm (inducing patience) may be more than compensated with the intensified effects of positive surprise leading to higher probability of successful appropriation. Such a positive surprise is expressed as a disappointment with negative value in our model.

One more reason to avoid inducing too high enthusiasm is our general tendency to overvalue impacts of our decisions on our well-being or happiness, both positive and negative. [30], [17] It is likely, though not fully examined yet, that this bias is slightly compensated with a past-adjusting defensive strategy according to which we reshape our memories to fit well to our perception of presence. Fischhoff demonstrated that people who know present events falsely overestimate their accuracy in which they would have predicted them. [31] Though hindsight contributes to diminish impact of expectations which fell short, it does not change the fact that the optimal level of enthusiasm is probably somewhere below the level and not inducing disappointment. Subsequent research in real world scenarios will be necessary to evaluate conclusions inferred from the model.

# 5    Conclusions

Both rational choice theory and cognitive science bring valuable insights into our cognition and behaviour. Rational choice theory brings formal tools and a great toolbox of concepts from economy, fits in scenarios where decision making is driven consciously by mind and answers primarily normative questions. Cognitive science provides a primarily descriptive view of our reasoning, including our irrationality. Thanks to recent research, it's gradually suitable to explain even decision making in quite complex scenarios.

With the aim to assist developers, designers and merchants of smart technologies such as intelligent robots, we created two simple models of mental processes which, being used together in a chain, cover the reasoning from the initial decision to the final appropriation. The model of initial evaluation draws from rational choice theory, whereas the model of appropriation is based on principles of the mind revealed by cognitive science. Despite simplifications and their generic nature, the models provide a structured view on the process. Each model defined several potential reasons for rejection. Some of them were already examined with implications for developers, designers and merchants, such as a situation when high enthusiasm induced in the initial evaluation stage does not necessarily lead to a higher chance of success in the following appropriation stage, rather opposite. The analysis of other reasons for rejection as well as other aspects of both models is still in progress. Though conclusions inferred from the model have support in the available theory, further research is necessary to evaluate them in real world scenarios.

This work was supported by the project No. CZ.1.07/2.2.00/28.0327 Innovation and support of doctoral study program (INDOP), financed from EU and Czech Republic funds.

# References

1. Tokoro, M.: Open Systems Science: Solving Problems of Complex and Time-Varying Systems. In: 8th International Conference on Practical Applications of Agents and Multi-Agent Systems (PAAMS 2010) (2010)
2. Becker, G.S.: The economic approach to human behaviour. University of Chicago Press (1976)
3. Becker, G.S., Murphy, K.M.: A theory of rational addiction. The Journal of Political Economy 96(4) (1988)
4. Becker, G.S.: Accounting for tastes. Harvard Univ. Pr. (1998)
5. Becker, G.S.: Crime and punishment: An economic approach. Journal of Political Economy 76(2) (1968)
6. Becker, G.S., et al.: Human capital. National Bureau of Economic Research, New York (1975)
7. Becker, G.S.: A Treatise on the Family. Harvard Univ. Pr. (1991)
8. Friedman, M.: Essays in positive economics. University of Chicago Press (1953)
9. Bentham, J.: Theory of legislation. Adamant Media Corporation (1876)
10. Marshall, A.: Principles of economics: an introductory volume (1920)

11. Zejda, D.: Deep Design for Ambient Intelligence: Toward Acceptable Appliances for Higher Quality of Life of the Elderly. In: 2010 Sixth International Conference on Intelligent Environments, Kuala Lumpur, Malaysia, pp. 277–282 (2010)

12. Zejda, D.: Ambient intelligence acceptable by the elderly. In: Molina, J.M., Corredera, J.R.C., Pérez, M.F.C., Ortega-García, J., Barbolla, A.M.B. (eds.) User-Centric Technologies and Applications. AISC, vol. 94, pp. 103–110. Springer, Heidelberg (2011)

13. Baars, B.J.: The Cognitive Revolution in Psychology. Guilford Press (1986)

14. Zejda, D.: The Model of Appropriation: Contribution of Rational Choice Theory and Cognitive Science to a Better Technology. In: 2011 7th International Conference on Intelligent Environments (IE), pp. 262–269 (2011)

15. Bechara, A., Damasio, H., Tranel, D., Damasio, A.R.: Deciding Advantageously Before Knowing the Advantageous Strategy. Science 275(5304), 1293–1295 (1997)

16. Lejuez, C.W., Aklin, W.M., Zvolensky, M.J., Pedulla, C.M.: Evaluation of the Balloon Analogue Risk Task (BART) as a predictor of adolescent real-world risk-taking behaviours. Journal of Adolescence 26(4), 475–479 (2003)

17. Hastie, R., Dawes, R.M.: Rational Choice in an Uncertain World: The Psychology of Judgment and Decision Making. SAGE (2009)

18. Kahneman, D., Tversky, A.: Prospect theory: An analysis of decision under risk. Econometrica: Journal of the Econometric Society, 263–291 (1979)

19. Tversky, A., Kahneman, D.: Judgment under uncertainty: Heuristics and biases. In: Judgment and Decision Making: An Interdisciplinary Reader, p. 35 (2000)

20. Lopes, L.L., Oden, G.C.: The role of aspiration level in risky choice: A comparison of cumulative prospect theory and SP/A theory. Journal of Mathematical Psychology 43(2), 286–313 (1999)

21. Cyert, R.M., March, J.G.: A behavioural theory of the firm. Blackwell (2005)

22. Slovic, P., Fischhoff, B., Lichtenstein, S.: Why study risk perception? Risk Analysis 2(2), 83–93 (1982)

23. Damasio, A.R.: Descartes' error: Emotion, reason, and the human brain. Quill, New York (2000)

24. Dawes, R.M.: The robust beauty of improper linear models in decision making. American Psychologist 34(7), 571–582 (1979)

25. Einhorn, H.J.: Expert measurement and mechanical combination* 1. Organizational Behavior and Human Performance 7(1), 86–106 (1972)

26. Libby, R.: Man versus model of man: some conflicting evidence* 1. Organizational Behavior and Human Performance 16(1), 1–12 (1976)

27. Wiesner, W.H., Cronshaw, S.F.: A meta-analytic investigation of the impact of interview format and degree of structure on the validity of the employment interview. J. Occup. Psychol. 61, 275–290 (1988)

28. Meehl, P.E.: Clinical versus statistical prediction: A theoretical analysis and a review of the evidence. University of Minnesota Press, Minneapolis

29. Mellers, B., Schwartz, A., Ritov, I.: Emotion-based choice. Journal of Experimental Psychology General 128, 332–345 (1999)

30. Harrison, J.R., March, J.G.: Decision Making and Postdecision Surprises. Administrative Science Quarterly 29(1), 26–42 (1984)

31. Fischhoff, B.: Hindsight!= foresight: the effect of outcome knowledge on judgment under uncertainty. British Medical Journal 12(4), 304 (2003)

# Using of Low-Cost 3D Cameras to Control Interactive Exhibits in Science Centre

Zoltan Tomori[1], Peter Vanko[2], and Boris Vaitovic[3]

[1] Institute of Experimental Physics, Slovak Academy of Sciences, Kosice, Slovakia
[2] Institute of Computer Science, Faculty of Science, P.J. Safarik University, Kosice, Slovakia
[3] Faculty of Arts, Technical University Kosice, Kosice, Slovakia

**Abstract.** 3D cameras (Kinect, Creative gesture camera) acquiring both RGB image and depth map are the objects of growing interests in robotics. Their compatibility with popular open source libraries (OpenNI, OpenCV, Intel PCSDK) allows creation of a robotic visual system which is cheap but powerful enough to solve many challenging computer vision problems. We created a sitting robot (interactive statue) with such a visual system for the entertainment of visitors in the Steel Park science centre in Kosice. Interactive statue tracks the face of the closest visitor and analyses his or her emotions (smile, closed eyes). Reaction to the detected emotion is the movement of the head and the arms by servo motors. In other exposition we used a Kinect camera which recognizes the colour of the visitor's clothes and evaluates its similarity with the reference colour of a safety cloak. Only visitors wearing the safety cloak are displayed on the background picture of the factory, the others are hidden. Calibrated Kinect/Projector system is used for the augmented reality sandbox which is lit by the projector by specific height-dependent colour, following the change of terrain altitude in the real time.

**Keywords:** Microsoft Kinect, Creative gesture camera, interactive statue, emotions detection, clothes recognition, augmented reality sandbox.

## 1 Introduction

Interactive humanoid robots are attractive exhibits in a modern type of museums called "science centres". Their interactivity is very important for both keeping the attention of visitors active and teaching them in the most natural way. The interaction with such robots is based on the Natural User Interface (NUI) offering a human-friendly communication using gestures, head and body pose, voice commands, eye gaze orientation, etc. Although a lot of algorithms concerning NUI were published in the last 2-3 decades, their popularity increased dramatically in the last few years when new low-cost 3D cameras appeared. The first of such a camera was the "Microsoft Kinect", launched in November 2010 when 10 million units had been sold in 2 months. The big commercial success of MS Kinect evoked competitive models from other companies (Intel, Sony, Asus, Leap Motion, Bluetechnics, SoftKinetics, etc.).

© Springer International Publishing Switzerland 2015                                    273
P. Sinčák et al. (eds.), *Emergent Trends in Robotics and Intelligent Systems*,
Advances in Intelligent Systems and Computing 316, DOI: 10.1007/978-3-319-10783-7_30

Most of them are supplied with the corresponding Software Development Kit (SDK), libraries and practical examples simplifying the programming of applications.

In this paper we would like to present such 3D cameras as a low-cost alternative of interaction with the humanoid robots and other museum showpieces. We have developed all of them for the "Steel Park" science centre in Kosice built in the frames of "Kosice, the European Capital of Culture 2013". We used 3D cameras (Kinect, Creative Gesture Camera) and our own software based on OpenNI, OpenCV and libraries supplied with these cameras.

## 2    Principles of 3D Cameras

Typical 3D cameras offer the traditional RBG colour image plus depth map. There are 3 main ways how to obtain the depth map.

- Stereo vision based on the images from two rectified RGB cameras. Using the simple formula based on the similarity of triangles, we can obtain a "disparity map" representing the distance between x-coordinates of corresponding points from the left and right image. The disparity is proportional to the depth - closer objects have higher disparity then the farther ones.
- Kinect camera is based on a patented principle where a point on the left image is not searched but it is directly "drawn" by the laser so its position is known. Its reflected position on the right image (acquired by infra-red sensor) is compared with the known one. Their displacement (disparity) is proportional to the depth. This solution simplifies computation but it cannot be used in an outdoor environment as the IR sensor is sensitive to sunlight.
- Time of flight sensor where the time delay (light interference) is measured between transducer and receiver. This principle results in more precise detection but is more expensive.

Aligning of RGB and depth images allows calculation of (X, Y, Z) coordinates of all pixels required by most of the 3D algorithms in robotics (see, e.g., Point Cloud Library - http://pointclouds.org). However, transformation of both images into point cloud is a time consuming operation and therefore a lot of computer vision algorithms prefer their separate processing.

Application of cameras with depth sensors in robotics is a natural and quickly growing research and development area [1]. Kinect sensor can be easily integrated and tested in various robot prototypes.

## 3    Interactive Statue

Our goal was to develop a specific type of humanoid robot for the entertainment purposes with a sophisticated visual system and moving abilities limited to the upper body parts (head, arm and palm). The sitting position of the robot is static and therefore we called it "interactive statue". The detection of a human's features (distance, pose, mimics, gender, age, etc.) evokes a corresponding reaction of the statue.

### 3.1    Function and Construction

The initial status is "sleeping" (Fig. 1 left) when all servomotors are disengaged, only a sound of snoring can be heard. A camera watches the selected region of interest (ROI) on the floor. If a person (or a group of visitors) appears inside the ROI, the robot wakes up and moves its head into the straight position. At the same time, the camera detects the faces of visitors and the head motors track the face of the closest person (Fig. 1 right). The emotions of the detected human face are analysed and the statue responses by the corresponding emoticon displayed on the monitor (smile, closed eyes – Fig. 1 bottom). If a visitor comes too close to the statue, its reaction shows an angry face on the display and the "go away" gesture of the right hand.

The key hardware component of the statue is the "Creative Interactive Gesture Camera" (supplied by Intel corp.). It contains an RGB camera (640x480), the depth sensor (320x240) and the microphone array for the voice recognition system. The body of the statue is made of steel sheets as required for the concept of the Steel Park exposition. Moving parts (pan/tilt system of the head, elbows and palms of both hands) are driven by the high torque servo motors (Hitec HS-645 MG) controlled by the servo controller for 8 motors (LynxMotion Phidgets). A touch screen (Lilliput 669GL – 7") is attached to the breast of the statue to display its emotional status (smile, closed eyes, angry) or messages and instructions for the visitor.
PCSDK software controlling the camera can be downloaded from Intel's web site free of charge. It contains several individual modules (depth and colour image streams, face detection, hand gestures, voice recognition, face recognition, etc.).

We exploited the face detector, smile and eyes status detector, colour and depth map stream of images. Face detector returns bounding rectangles of all faces in real time and the depth image gives us the distances of the individual faces from the camera. Using this information, the (X,Y,Z) coordinates of the closest face is calculated in real time and sent to the independent program controlling the motors using OSC communication protocol supported by "liblo" library. The advantage of such client-server approach is that 3D coordinates of the closest detected face can be received by any program supporting UDP protocol either on the same computer or on the remote networked one. This is also a simple way how to communicate with programs created in different environments – we used it for communication with PureData and Graphics Environment for Multimedia (GEM).

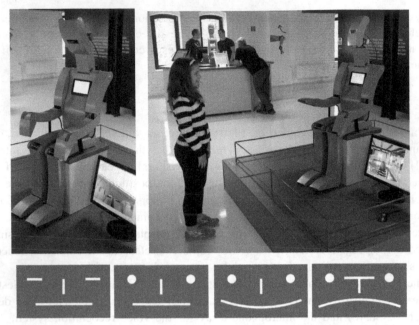

**Fig. 1.** Interactive statue and displayed emotions (designed and created by Jaroslav Tomascik, Faculty of Arts, Technical University Kosice). Left: sleeping mode, right: interacting mode (head tracking and emotions detection), bottom: displayed images of emotions (sleeping, normal, smile and angry).

## 3.2    Face and Emotion Detection

Face detection algorithm analyses the RGB image and finds a configuration of pixels representing a human face. Then algorithm returns a set of bounding rectangles around all detected faces. One of the most exploited algorithms is [2] based on Haar cascade classifier which reached high popularity thanks to its implementation in the well-known OpenCV library.

The emotions detector exploits thebounding rectangle to find the individual parts of a face (mouth, nose tip, eyes etc.) and their shapes. We exploited the implementation of "face attributes" in PC SDK face analyser module which returns the probability of the smile and closed eye attributes.

All the above mentioned algorithms are based on 2D image analysis so the information about the Z-coordinate is missing. However, the depth map acquired by the depth sensor and aligned with RGB image allows for the calculation of the camera - face distance. This is very useful in face tracking when the algorithm tracks only the closest face; the remaining ones are detected but not tracked. However, identification of the closest face is not sufficient in cases when a crowd of people stay approximately at the same distance from the robot. Then, additional information can be exploited to prefer some visitors (e.g. small children). With the help of further face attributes like gender and age detection, we could preferably track, e.g., young females (these attributes are included in the PC SDK library but were not tested yet).

Clothes recognition algorithm described in the next part represents another approach which can be used to improve tracking (e.g. the face of a visitor wearing the dress of some colour is preferred). Formally, for practical reasons, this algorithm is not included in the interactive statue exhibit but it creates a separate exposition (interactive cloakroom). However, its implementation into the robot tracking algorithm is straightforward.

# 4    Clothes Recognition

Interactive cloak room is based on the idea that a picture of a factory hall is mixed with the picture of persons in front of the Kinect camera. Following the motto "SAFETY FIRST", anybody without the proper factory cloak is not visible in the space of the virtual factory. The criterion is the similarity of the clothes colours with the predefined reference colour. The reference colour is calculated at the training stage as the average colour inside the contour of a "trainer" wearing the safety cloak. In the recognition mode, the colours inside the contours of all visible persons are calculated and compared with the saved reference colour in the RGB colour space. Clothes having significantly different colours are masked, the remaining are displayed on the factory background picture (see Fig. 2).

Hardware is based on the Kinect camera detecting the body contours of visitors and the projector displaying the resulting composed image in real time.

Our software classifying detected people into 2 groups (visible/invisible) is based only on the similarity of colours with the reference colour. As RGB colours representation is very sensitive to light change, we transformed all colours into HSV (Hue-Saturation-Value) colour space. Then all pixels having low saturation were excluded and only Hue (colour) values are compared. HSV representation showed much better resistance to the change of illumination than RGB. Anyway, homogeneous illumination is still required to achieve the same classification in all positions in front of the camera. Most of the image operations (including transformations between the colour spaces) are efficiently implemented in the OpenCV library which allows real-time processing.

**Fig. 2.** Kinect camera captures images of people and finds their contours. Visitors wearing clothes of similar colour to the colour of workers` safety cloak appear on the background image of factory hall, others are invisible.

## 5     Interactive Sandbox

Although this exhibit has no direct relation to the interactive statue, it can be used to model the environment where a robot is moving. Inspired by [3], Kinect acquires depth image exploited for the creation of the augmented reality image which is projected back to the real object (sand).

Both Kinect cameras and projectors are attached to the mount above the table with sand (see Fig. 3). Kinect evaluates the depth of sand terrain in real time and the projector illuminates the sand with different colours depending on the elevation (similarly as in the map where hills are brown, lakes blue and planes green). The change of illumination should follow the change of terrain with minimal delay.

Hardware: 3D camera Microsoft Kinect acquires the depth map and calculates the colour map using the defined look-up table. Projector BENQ MX613ST with aspect ratio 4:3 should be used in its native resolution 1024x768 without any image correction.

Vertical calibration means the detection of the reference plane. Despite the precise mounting, we cannot guarantee that both Kinect and projector are perpendicular to the sandbox. In this case the depth image doesn't correspond to the height of the terrain measured as a distance from the horizontal "sea" level. Therefore, we put a flat cover on the sandbox top and the camera captures its depth map as a reference one. This map is fitted by the analytical plane equation using the popular RANSAC algorithm. Then, we use the analytical equation to generate the virtual reference plane depth map which will be subtracted from every captured depth map.

Horizontal calibration represents a complicated problem because the alignment between Kinect, the projector and the reference plane is not perfect due to the optical distortions and inaccurate positioning. Fortunately, there are methods to eliminate hardware inaccuracies by software. It is based on the projection of a regular grid of known coordinates onto the sand surface - projector points:

$$V_p = (x_p, y_p). \tag{1}$$

Then, we place small circular objects onto the projector points, clicking their centres on the depth map captured by Kinect – we obtain a set of 3D Kinect points

$$V_k = (x_k, y_k, z_k). \tag{2}$$

The relation between Kinect and projector points at $(0, 0, 0)$ is given by transformation matrix A which contains coefficients of rotation, translation and optical distortions model

$$V_{p0} = A^*V_k \tag{3}$$

Both projector points and the corresponding Kinect points are entered into the system of 11 linear algebraic equations and solved by matrix QR decomposition. Resulting matrix contains coefficients describing the transformation between Kinect and projector coordinates (see [4] for details). Calibration should be performed only once, then the calibration data are saved as a part of configuration file. The mapping of colours to the height of terrain is defined by the user and can be also saved as XML file.

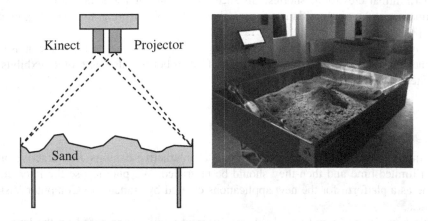

**Fig. 3.** Arrangement of hardware components (left). Height-dependent mapping of colours to the sand surface (right).

## 6    Future Work

The exhibits recently installed in "Steel Park" were continuously tested by the real visitors. Although the anonymous survey declared mostly high and very high satisfaction of visitors, practical experience revealed some problems and showed new possible improvements.

3D cameras and corresponding libraries offer much more functions than we have exploited in our interactive robotic expositions. Interactive statue is currently able to track the face, recognize smiles and closed eyes. Future extensions should include gender and age detection, gestures as well as the voice commands recognition. We currently use only very simple voice output (snoring in sleeping mode), but PCSDK allows also speech synthesis useful in communication with the robot.

Interactive statue is the ideal platform for the future supervised machine learning experiments similar to these described in [5] where we showed several different objects to the robot and entered the name of the object (by voice or text). A trained robot should be able to recognize the object and reproduce its name.

Experiments with real visitors revealed that clothes recognition based only on colour failed in some specific colours and light conditions. We plan to obtain other machine learning features (using, e.g., logo placed on the cloak). We plan image enhancement techniques to find the contour of a person more precisely (e.g. alpha matting segmentation [6]).

The calibrated Kinect/Projector/Sandbox system can be easily extended to create other augmented reality applications. One of them could be a robot navigation where Kinect acquires the depth map of terrain and projector and visualize dangerous obstacles by illuminating them with specific labels (colour, text, QR code etc.). A labelled obstacle can be easily identified by the visual system of robots. The other possibility is to exploit the 3D information about the terrain to calculate the optimal path (minimal elevation, shortest distance) from current robot position to a given destination. A lot of algorithms exist to solve such problems (e.g. live wire [7], graph-cuts [8], etc.).

Anyway, technological aspects are only one side of the story. They must be balanced not only with an overall concept of our robot but also with other exhibits in the Steel Park.

# 7    Conclusions

The quick progress in 3D cameras causes that interactive exhibits are attractive only for a limited time and then they should be upgraded. We plan to use the interactive statue as a platform for the new applications created by students in Computer Vision courses.

We have designed and tested several different applications based on the low-cost 3D cameras (ca. 150 EUR each) combined with open source libraries (OpenCV, OpenNI, Intel PCSDK). Our experience shows that they can serve not only as the cost-effective robotic visual systems but they overcome the traditional cameras and sensors. (Kinect successfully replaced the originally planned sensors that failed in two other exhibits not mentioned in this paper).

The most promising (but also the most frustrating) feature of 3D cameras is the decreasing time period between significant upgrades requiring very fast response from the developers.

**Acknowledgments.** We thank to U.S. Steel Kosice as the main contributor of Steel Park science museum and "City of Kosice" as co-founder. This work was also supported by Slovak research grant agencies APVV (grant 0526-11) and VEGA (grant 2-191-11). The design of the interactive statue was created by Jaroslav Tomascik from the Faculty of Arts of Technical University Kosice.

# References

1. Mishra, A.K., Aloimonos, Y., Cheong, L.F., Kassim, A.A.: Active Visual Segmentation. IEEE Transactions on Pattern Analysis and Machine Intelligence 34, 639–653 (2012)
2. Viola, P., Jones, M.J.: Rapid Object Detection using a Boosted Cascade of Simple Features. IEEE CVPR (2001)
3. Augmented Reality Sandbox,
   http://idav.ucdavis.edu/~okreylos/ResDev/SARndbox/
4. Hrdlicka, J.: Kinect-projector calibration, human-mapping | 3dsense interactive technologies blog,    http://blog.3dsense.org/programming/kinect-projector-calibration-human-mapping-2/
5. Tomori, Z., Gargalik, R., Hrmo, I.: Active Segmentation in 3D using Kinect Sensor. In: 20th International Conference on Computer Graphics, Visualization and Computer Vision (WSCG 2012), Pilsen, Czech Republic (2012)
6. Rother, C., Kolmogorov, V., Blake, A.: GrabCut - Interactive foreground extraction using iterated graph cuts. ACM Transactions on Graphics 23, 309–314 (2004)

7. Mortensen, E.N., Barrett, W.A.: Intelligent scissors for image composition. In: SIGGRAPH 1995: Proceedings of the 22nd Annual Conference on Computer Graphics and Interactive Techniques, pp. 191–198. ACM Press, New York (1995)
8. Boykov, Y., Kolmogorov, V.: An Experimental Comparison of Min-Cut/Max-Flow Algorithms for Energy Minimization in Vision. IEEE Trans. Pattern Anal. Mach. Intell. 26, 1124–1137 (2004)

# Artificial Intelligence and Collective Memory

Vladimír Kvasnička and Jiří Pospíchal

Institute of Applied Informatics, Faculty of Informatics and Information Technologies,
Slovak Technical University, Bratislava, Slovakia
{kvasnicka,pospichal}@fiit.stuba.sk

**Abstract.** In this chapter we study an application of artificial intelligence techniques to get a better and deeper understanding of Halbwachs concept of "collective memory" in social sciences which recently became frequently used for an interpretation of social phenomena and processes. Moreover, the concept of "collective memory" is specifically reflected in modern artificial intelligence as an effective tool for acceleration of adaptive and evolutionary problems in multiagent systems. This modern approach makes it possible to use a history of the adaptive process for a construction of its actual solution. If we compare, on one hand, the usage of the concept of collective memory in social sciences, and, on the other hand, its usage in artificial intelligence, we observe that there exist many common properties that substantially simplify a discussion of their common interaction and interpretation.

## 1    Introduction

Collective memory, as currently understood in the social sciences (where it was introduced by French sociologist M. Halbwachs [9]), is a unique type of memory which is usually distinguished from the standard types of neuroscience memory [9]. Collective memory is a distributed system of single brains/minds/memories that belong to a given social group, and is a part of its culture [2, 14]. Even though a substantial part of collective memory is located in the brain, its formation and persistence is considered a non-standard feature of the collective social group. It belongs to basic specifications of a social group which are transferred by non-genetic (i.e. cultural) methods within the lifespan of single group members. From this standpoint the concept of collective memory has many common features with generalized memetic theory initially introduced by evolutionary biologist R. Dawkins [5] (cf. also [4]). The collective memory has three well separated parts: (1) knowledge, (2) attributes, and (3) process. The first part – knowledge –, even though it may change over time, belongs to the basic constitutional elements of social group culture. The second part – attributes – forms a specific holistic view at history of social groups; it is often used as one of the basic specifications of a given social group. Finally, the third part – process – is created by a perpetual dialog among members of a group or its subgroups, and this part may change the first two parts of collective memory.

© Springer International Publishing Switzerland 2015
P. Sinčák et al. (eds.), *Emergent Trends in Robotics and Intelligent Systems*,
Advances in Intelligent Systems and Computing 316, DOI: 10.1007/978-3-319-10783-7_31

To conclude this overview of the concept of "collective memory", we turn our attention to aspects of creation, revision, continuation and termination of collective memory. In the classical (Halbwachs') theory of collective memory, these aspects are presented only cursorily without regard to artificial intelligence. Modern approaches to the revision of a knowledge base for its development and integration are now elaborated in great depth [2, 3, 15] and became an integral part of contemporary theories of logic and science.

Modern artificial intelligence developed a theoretical apparatus that is appropriate for the study of social structures in systems that contain many similar elements - individuals - agents that behave "rationally" based on certain simple rules like "if ... then ...". This approach, called "multiagent system" [17, 18], is currently among the vigorously developing fields of artificial intelligence, with a number of effective applications in different areas of both computer science as well as humanities. In multi-agent systems the course of "adaptive" process can be easily tracked using various parameters and computer graphics.

The aim of this chapter is to discuss the concept of collective memory in the multiagent approach, pointing out that this approach to the existence of memory in multi-agent systems is very close to the concept of collective memory used in social sciences. We believe that the multiagent approach to the study of collective memory provides a simple and effective interdisciplinary formalism which can be applied relatively easily in the social sciences.

## 1.1     Multiagent Systems in Artificial Intelligence

Over the last 20-30 years, a concept called "distributed intelligence" has become very popular in artificial intelligence. It is partly based on the idea of insect communities where limited cognitive abilities are combined together into a collective intelligence.

These ideas have resulted in the theory of multi-agent systems [17, 18, 16]. Such systems are widely applicable to various problems of artificial intelligence and incorporate an important class of algorithms addressing difficult problems by breaking them into much smaller and simpler sub-problems. Multiagent systems are composed of small units – agents – that have the ability (intention) to cooperate with other agents, interact with them, creating a common space-time structure.

## 2     An Attempt to DESIGN Agents with a Mind

In this section we describe the properties of agents which already have an "elemental mind" which consists of (1) memory (represented by a set of knowledge, called the theory in mathematical logic) and (2) inference apparatus by which the agents are able to deduce new knowledge by simple reasoning about an extended theory.

We provide a few remarks about memory, its symbolic and subsymbolic implementation. Symbolically represented memory can be simply identified with a knowledge (theory) represented by the set of first-order formulas

$$T = \{\varphi_1, \varphi_2, ..., \varphi_n\} \tag{1}$$

We implicitly assume that this set is consistent, i.e. from this theory, as a logical consequence, is deduced either $\varphi$ or its negation $\neg\varphi$ (but not both, exclusive disjunction), otherwise we could deduce from the theory an arbitrary formula, which is nonsense. The so-called memory can perform the revision process (see further sections) which aims to achieve a coherent theory.

## 2.1    Symbolic Specification of Agent's Minds

Let us postulate that a multiagent system $\mathcal{A} = \{A_1, A_2, ..., A_n\}$ is composed of agents, and that each agent contains a theory represented by a set - knowledge $T_A = \{\varphi_1, \varphi_2, ...\}$ (where $\varphi_1, \varphi_2, ...$ are the first-order logic formulas that represent agent's knowledge) and that an inference device is represented by a relation of logical entailment. Inference structure may be represented by the elementary laws r1, r2, r3, ... of natural deduction [6],

$$\vdash_A = \{r_1, r_2, r_3, ...\} = \left\{ \frac{p \Rightarrow q \quad p}{q}, \frac{p \Rightarrow q \quad \neg q}{\neg p}, \frac{p \Rightarrow q \quad q \Rightarrow r}{p \Rightarrow r}, \frac{(\forall x)P(x)}{P(t)} \, ... \right\} \tag{2}$$

Applying this inference device, an agent is able to deduce from an observation $\Delta = \{\psi_1, \psi_2, ...\}$ (where $\psi_1, \psi_2, ...$ are facts of the observation) a new knowledge $\chi$ which is specified as a logical entailment of the theory $T_A$ extended by an observation $\Delta$

$$T_A \cup \Delta \vdash_A \chi \tag{3}$$

Formally, an agent A has a mind $\mathcal{M}_A$ specified by an ordered couple $\mathcal{M}_A = (T_A, \vdash_A)$, i.e. the mind is specified by a memory composed of a theory $T_A$ and an inference relation $\vdash_A$. We may say that a theory formed from knowledge is representing an individual agent's mind, and, moreover, an inference relation $\vdash_A$ represents a mind which is capable of thinking.

## 2.2    Knowledge Revision

The problem of knowledge revision has a long tradition. It became an integral part of many treatises and monographs on the philosophy of knowledge (epistemology) and logic from antiquity to the present. The most-speculative phenomenological level was used to formulate the principles of our thinking, reasoning and changes in knowledge resulting from changes in underlying assumptions, their extension or partial falsification.

For example, if we specify the basic principles of epistemic development of science, the dynamics of its evolution in time, this goal can be realized at an abstract level, given that we study a consistent database of knowledge. This database is gradually modified over time by elementary operations, such as the delivery of new knowledge or removal of the original knowledge. In both cases, these changes may affect operations over other parts of knowledge, so the revision is done to remove any inconsistencies. In the initial period of the revision theory of knowledge (80s of the last century), the basic ideas were formulated by the Swedish cognitive scientist P. Gärdenfors and by a pair of American logicians C. Alchourrón and D. Makinsom who, in 1985, published major work ([1], commonly referred to by acronym AGM) that formulated the basic principles, concepts and design of the knowledge revision theory.

Note that the mind $\mathcal{M}_A$ is not a fixed entity during the existence of an agent $A$, it can be changed through a revision process which either increases or reduces some knowledge so that the resulting memory is consistent. We will distinguish the following four ways to the revision of the knowledge base $T$:

1. **Expansion**, where a consistent original theory is expanded by new knowledge taken from the consistent set $\tilde{T}$, this process creates a new expanded and consistent theory $T \cup \tilde{T}$ denoted by

$$T_{\tilde{T}}^{(+)} =_{def} T \cup \tilde{T} \tag{4}$$

2. *Contraction*, a minimal subset $\Delta \subseteq T$ is removed from the inconsistent theory T, where a maximal consistent theory is created.

$$T_{\Delta}^{(-)} =_{def} T - \Delta \tag{5}$$

3. *Revision*, this is a combination of the previous two techniques. In particular, initially a consistent theory $T$ is extended by new knowledge $\tilde{T}$, making it inconsistent, so the new theory is reduced by a minimal subset $\Delta$. A result of this whole process is a consistent theory.

$$T_{\tilde{T},\Delta}^{(+,-)} =_{def} \left(T \cup \tilde{T}\right) - \Delta \tag{6}$$

Merging, this operation can be specified as follows [10, 13]. Consider a multiagent system where each agent-explorer's task is to collect knowledge about the environment (such as positions of distributed obstacles, food and other agents). Over time, the agents return to the "headquarters" and transmit their individual knowledge $\{T_1, T_2, ..., T_n\}$ to an agent - integrator which is responsible for linking the knowledge gained from a variety of agents - explorers into a single database of knowledge about the environment. It may be that lessons learned from two different agents are mutually controversial, i.e. agent - integrator has to decide which knowledge from

$\{T_1, T_2, ..., T_n\}$ will or will not be used to create database-related knowledge. Linking knowledge to a common knowledge can take place so that all received knowledge is combined into one set $T_1 \cup T_2 \cup ... \cup T_n$. To remove inconsistencies, the integrator is looking for such a minimal set of knowledge which, when removed from the unified set, leaves the resulting set (theory) consistent

$$merge(T_1, T_2, ..., T_n) =_{def} (T_1 \cup T_2 \cup ... \cup T_n) - \Delta \qquad (7a)$$

There are many alternative options for combining different knowledge bases into a single consistent database. Thus, for example, associated database can be gradually set up so that the integrator gradually associates databases (just incoming knowledge) with a new database and the outcome of the merger eliminates the minimum subset of the previous association to make the resulting theory consistent

$$merge(T_1, T_2, T_3, T_4 ...) =_{def} \left(\left(\left(\left(\left(T_1 \cup T_2\right) - \Delta_1\right) \cup T_3\right) - \Delta_{12}\right) \cup T_4\right) - \Delta_{123}\right).... \qquad (7b)$$

In the first starting step we have a theory $T_1$, this is expanded by a next theory $T_2$, whereas from the created unification we remove a minimal subset $\Delta_1 \subseteq T_1 \cup T_2$ in such a way that resulting unified theory is consistent. In the second step a theory $T_3$ is added to the previous result and we remove a minimal subset $\Delta_{12} \subseteq ((T_1 \cup T_2) - \Delta_1) \cup T_3$, etc.

The problem of merging into a unified greater knowledge set, called the merged knowledge database, belongs in artificial intelligence to basic problems of the distributed intelligence (e.g. in multiagent systems). This problem may be, in the framework of the first-order logic, formalized in a simple way, whereas on a semantic level we usually specify "out-logical" instruments how to effectively solve the problem of inconsistency of merged theory.

There is another problem with removing a minimal subset from inconsistent knowledge to make it consistent. It can remove newly acquired true information inconsistent with old and long established supposed truth. However, to solve this problem is very difficult and out of scope of this paper.

Moreover, the inconsistencies could be considered not only on one level but on an individual agent level, on collective memory level and between agent's memory and collective memory. The problem can be solved in many different ways, none of which is known to be preferential at the moment.

We discussed here only the first order logic inferences, even though other types of logic could be used as well. However, the theme is already complicated enough without taking into account complications caused by inconsistencies allowable, e.g., in fuzzy logic.

## 2.3    Three Levels of Agent Complexity

We distinguish three levels of agent complexity; this classification is based on their properties, whether or not they contain memory, if they contain memory, then what is its type, whether it is individual or the extension of the collective memory (see Fig. 1).

1st level. Agent activities are reflexive they are not founded on an inference type of mind. In computer science, this type of memory is represented by standard evolutionary algorithms, where a genotype fitness is specified immediately by its composition - architecture. Reflexive activities are fully specified genetically, i.e. these activities are inherently instinctive.

2nd level. Agents of population are endowed by a mind $\mathcal{M} = (T_i, \vdash)$, by making use of which the agents are capable to control simple activities like an effective orientation in a given environment, assessing of closest neighbourhood, etc. Usually, a relation of logical entailment $\vdash$ is the same for all agents, i.e. we may say that an agent inference apparatus, in a particular structure of the brain, is specified by the genotype. However, an individual memory $T_i$ might be of a memetic origin [11]:

(a) By a "vertical" transfer (an education of offspring) from parents,

(b) By a "horizontal" transfer (from other agents), and

(c) By agent's own observations, i.e. an individual memory is gradually expanded by experience of the agent acquired from its activities in the environment.

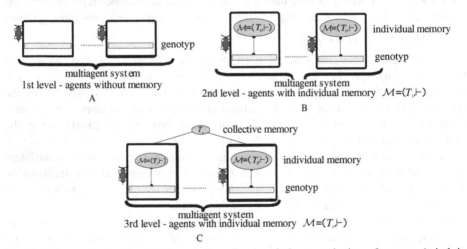

**Fig. 1.** Diagrammatic illustrations of three levels of the complexity of agents (mind is represented by a pair composed of knowledge set $T$ and an inference relation $\vdash$). (A) Diagram represents the first level agent without memory, its activities are fully specified by its genotype, i.e. they have a reflexive character; its genotype fitness is fully specified by its composition. (B) Diagram represents the second level of agent complexity, these agents are endowed by an individual mind $\mathcal{M} = (T_i, \vdash)$, where knowledge theory $T_i$ represents the individual memory of agent and the relation of entailment $\vdash$ represents an agent inference apparatus. (C) The most complex third level of mind we may get from the second level, where a collective memory is created by merging single agent memories $T_i$, this collective memory serves the whole multiagent system.

Vertical and horizontal transfer of information (not inferential apparatus!) during learning perhaps differ mainly in the fluidity of information transferred in time, where horizontal transfer should be able to spread useful information more quickly throughout the population. Even the vertical information transfer from parents is, however, considered here as teaching, not genetically based. At the end of their life cycle when agents enter into a reproduction process, each gives to its offspring only its genotype in a lightly mutated form, i.e. an agent - offspring does not inherit parental memory (the used evolution is strictly Darwinian) but inherits in a lightly mutated form of a parental inferential apparatus, $\vdash$. We do not consider here the transfer of the inference apparatus by teaching.

3rd level. Agents are endowed with a mind $\mathcal{M} = (T_i, \vdash)$ and, moreover, with a collective memory Tc which is common for the whole multiagent system. The simplest approach for an introduction of collective memory is a "postulation" of the so-called searching agents that have privileged access into single individual memories and search for proper information of a general character which will be added to collective memory. This "specialized merging" in the course of creation of the collective memory has an important role to ensure a consistency of created collective memory. It would not contain such knowledge that simultaneously logically entailed $\chi$ and its negation $\neg\chi$. (Let us note that in modern logic the problem of revision and merging of knowledge databases is very intensively studied [7, 8]).

### 2.4    Symbolic Specification of Collective Memory and Its Importance in Artificial Intelligence

In this section we postulate that the agent Ai has a specific inference relation entailment $\vdash_i$ which is different from other inference relations. A memory $T_i$ of the ith agent is divided into two parts $T_i'$ and $T_i''$

$$T_i = T_i' \cup T_i'' \tag{8}$$

where we have separated from the memory $T_i$ the knowledge $T_i''$ which is important for other agents (e.g. a device for energy recharging is localized in the right upper corner of the environment). The easiest way to ensure the emergence of collective memory is to postulate an existence of several special searching agents; these "investigators" take a copy from the agent's Ai "collective" part of memory that contains important knowledge of the agent and the properties of the environment that may be useful to other agents. Then a collective memory that is shared by the whole multiagent system can be constructed as follows:

$$T_{coll}^{(b+1)} = \left\langle T_{coll}^{(b)} \cup \left( \bigcup_i T_i'' \right) \right\rangle \tag{9}$$

where symbol $\langle T \rangle$ means that for a theory $T$ a revision was applied, i.e. from $T$ there was a minimal subset $\Delta$ removed such that $\langle T \rangle = T - \Delta$ is a consistent theory.

When creating a collective memory Tcoll, an additional condition can be used that favours only a certain type of knowledge collected from the local collective memories T 's. This can ensure that collective memory is created by using fixed criteria which are usually only very rarely changed during the evolution of multiagent system. One of the main tasks of the collective memory is to overcome information barriers between the generations, the collective memory stores essential information about the history of the adaptive process of multiagent system, which can significantly facilitate the search for an optimal forthcoming step from a set of alternative options.

The concept of collective memory implemented in evolutionary algorithm can provide a substantial acceleration of the evolutionary approach which is based on history. Instead of completely random mutations, it allows the algorithm to produce the so-called targeted mutations supported by a history of algorithm [11].

# 3    Conclusions

The concept of collective memory was introduced more than half a century ago by French sociologist M. Halbwachs [9] as a unique type of memory that is part of the culture. Collective memory is a fundamental specification of social groups and is transmitted by non-genetic (cultural) methods in the intergenerational reproductive process in multi-agent system. During the 19th century this term also became popular in an IT discipline "artificial intelligence", namely its part called "multi-agent systems". This concept is used in the theory of multi-agent systems as an accelerator of adjustment. The collective memory generalizes experience and knowledge of agents in the history of this adaptive process. This was illustrated by simple examples of genetic algorithms, where collective memory is constructed incrementally. It gradually introduces some determinism to mutations in the reproductive process based on previous experience of the adaptation process.

In sociology and humanities, the concept of "collective memory" is partly overlapping with memetics. As in the collective mind and memetics in particular, its importance lies in the fact that the information contained in the meme agent increases the chance of survival, for example, it can contain a description of its environment, its irregularities, and hidden features. Memetic approach can be reformulated into a collective memory [12], containing "historical experience" of agents from the previous history of the adaptation process.

Summarizing this chapter, we assert that the Halbwachs concept provides a multi-agent systems theory which is an effective formal tool for acceleration processes of adaptation, since it allows bridging the information barrier in the intergenerational transmission of cultural mechanisms to future generations.

Sharing and exchanging of information is used in many technological approaches from cloud computing to organization of memory in hardware or in evolutionary algorithms. Our vision is a closer cooperation between sociological theories and multiagent theories of artificial intelligence and other perhaps even technological approaches, which should mutually enrich all these scientific disciplines.

**Acknowledgments.** This chapter was supported by Grant Agency VEGA SR No. 1/0553/12 and 1/0458/13.

# References

1. Alchourròn, C.E., Gärdenfors, P., Makinson, D.: On the logic of theory change: Partial meet contraction and revision functions. Journal of Symbolic Logic 50(2), 510–530 (1985)
2. Assmann, J.: Communicative and Cultural Memory. In: Erll, A., Nünning, A., Young, S.B. (eds.) Cultural Memory Studies. An International and Interdisciplinary Handbook, pp. 109–118. Walter de Gruyter, Berlin (2008)
3. Beim, A.: The Cognitive Aspects of Collective Memory. Symbolic Interaction 30(1), 7–26 (2007)
4. Blackmore, S.: The Meme Machine. Oxford University Press, Oxford (1999)
5. Dawkins, R.: The Selfish Gene. Oxford University Press, Oxford (1976)
6. Gabbay, D.M.: Logic for Artificial Intelligence & Information Technology. College Publications, London (2007)
7. Gärdenfors, P.: Knowledge in Flux: Modelling the Dynamics of Epistemic States. MIT Press, Bradford Books, Cambridge (1988)
8. Gärdenfors, P. (ed.): Belief Revision. Cambridge University Press, Cambridge (1992)
9. Halbwachs, M.: On collective memory. The University of Chicago Press, Chicago (1992); a translation of original French edition, La mémoire collective (1950)
10. Konieczny, S., Perez, R.: Logic Based Merging. Journal of Philosophical Logic 40(2), 239–270 (2011)
11. Kvasnička, V., Pelikán, M., Pospíchal, J.: Hill Climbing with Learning. An abstraction of Genetic Algorithm. Neural Network World 5(6), 773–796 (1996)
12. Kvasnička, V., Pospíchal, J.: Artificial Chemistry and Replicator Theory of Coevolution of Genes and Memes. Collection of Czechoslovak Chemical Communication 72(2), 223–251 (2007)
13. Lin, J., Mendelzon, A.O.: Merging databases under constraints. International Journal of Cooperative Information Systems 7(1), 55–76 (1998)
14. Manier, D., Hirst, W.: A Cognitive Taxonomy of Collective Memories. In: Erll, A., Nünning, A., Young, S.B. (eds.) Cultural Memory Studies. An International and Interdisciplinary Handbook, pp. 253–262. Walter de Gruyter, Berlin (2008)
15. Roediger III, H.L., Dudai, Y., Fitzpatrick, S.: Science of Memory: Concepts. Oxford University Press, New York (2007)
16. Sugumaran, V.: Distributed Artificial Intelligence, Agent Technology, and Collaborative Applications. Information Science Reference. IGI Global, Hershey (2008)
17. Wooldridge, M.: Reasoning about Rational Agents. The MIT Press, Cambridge (2000)
18. Wooldridge, M.: An Introduction to Multiagent Systems. John Wiley & Sons, Ltd., Chichester (2005)

# On Agent-Based Virtual Economies – Pursuing Goals and Measuring Efficiency

Petr Tucnik

University of Hradec Kralove Rokitanskeho 62, 50003 Hradec Kralove, Czech Republic
petr.tucnik@uhk.cz

**Abstract.** Creating an artificial economic system also requires the definition of goals and establishing means of measuring efficiency. Our project is focused on the agent-oriented economic system where individual agents represent either production facilities or workers/consumers. In order to create an efficient system, several problems have to be addressed: goal definition, environment definition and agent specification, resulting in a well-specified embodiment of an agent. This work will be focused on the definition of goals and measuring of efficiency in such economic systems.

**Keywords:** agent, artificial economy, artificial intelligence, autonomous system, economic modelling.

## 1    Introduction

The agent-oriented technologies represent one of the major approaches for today's economic modelling, see [5], [6], [3], [4], [10] or [21] as examples of such models. The system which will be described here consists of four elementary types of agents, each representing one part of the production chain – resource processing, production facilities, customers and transportation. Some secondary elements also exist in the system but these are less important. Each of these main parts are represented by agents and the whole system is working together as a community. The goal of the proposed system is to act autonomously and to self-organize in the given environment in order to produce desired goods or services according to the given task (by the user).

It is not the purpose of this system to create a full-scale economic simulation, but it is rather focused on self-organization and decentralized problem solving. Autonomous problem solving and adaptability are features which rank such a system into the area of artificial intelligence. Economic principles are used in this context primarily to create desired and predictable behaviour. Individual agents are pursuing their own goals (maximizing their utility) and continuously try to optimize their behaviour towards maximum efficiency. In these elementary aspects, agent behaviour is consistent with standard economic principles.

© Springer International Publishing Switzerland 2015
P. Sinčák et al. (eds.), *Emergent Trends in Robotics and Intelligent Systems*,
Advances in Intelligent Systems and Computing 316, DOI: 10.1007/978-3-319-10783-7_32

## 2     Related Work

As already mentioned, the proposed system is representing a model of a community of individual agents collaborating together on a given goal. The system is different from the majority of economic models because these are usually specialized on singular economic problems. This system creates an environment with self-organizing artificial economy which is adaptable to changing conditions in a dynamic environment. This allows research of problems related to supply chains, logistics, task management and scheduling, planning, strategic decision making, etc.

There are some works worth mentioning which are focused on related topics. The problem of virtual supply chain management and its dynamic development is addressed by Golinska [8]. In this context, agents form holonic structures, consisting of cooperating nodes which task is to collect, transform, store and share information or physical objects. Expected applications are in transport services and dynamic supply chains. Similar problems of agile supply chains are addressed by Grzybowska and Kovács [9].

Another application in the industry presents Skobelev [19]. His work is focused on real time resource allocation, scheduling, optimizing, and controlling in transportation, manufacturing and related areas. Skobelev mentions interesting industrial application examples like a tankers scheduling system, a corporate taxi system, track scheduling systems, car rental scheduling systems and others. In general, use of such system results in the increase of efficiency, cost, time and risk reduction, and minimization of dependence on human factor. It also shows that such approaches are at least to some extent "battle-proofed" and used in commercial practice.

Another interesting approach of "Virtual Factory Framework" is described in the paper of Sacco et al. [17]. It is an integrated virtual environment that supports factory processes along all the phases of their lifecycle. Although it is focused mainly on factory micromanagement, it is also interesting in the context of production research and optimization.

Other interesting publications may be mentioned as well. E.g. political economy of virtual worlds by Mildenberger [14], Guo and Gong`s measuring of virtual wealth in virtual worlds [11], multi-agent organizational model for grid scheduling by Thabet et al. [20], or Hou`s personalized material flow services [12]. Also, [13], [15] and [18] may offer interesting information on the subject at hand.

In general, agent-based modelling is a quite popular modelling approach which is widely used in many disciplines. Current scientific papers indexed in the Web of Science database are classified to overall 138 research areas. Whereas the majority of research papers belong to the field of computer science or engineering (67,6 %), plethora of other application areas such as toxicology, entomology, oceanography, crystallography, or management of biological incidents [2] can be identified.

# 3    System Description

Virtual economy presented here represents the production and consumption processes similar to those in real economies. The aim is to simulate economic principles of effective price and quantity setting under specific demand and capacity constraints [16]. Hence, the focus is on trading products and services and offering work in labour market. Virtual economy simulation is similar to the work of Deguchi et al. [5], however, in that representation, the entities considered are more specific, producing a more complicated net of relations than necessary. Similarly, trust issues - discussed for example in [7] and similar concerns - are not of primary attention in the presented virtual economy.

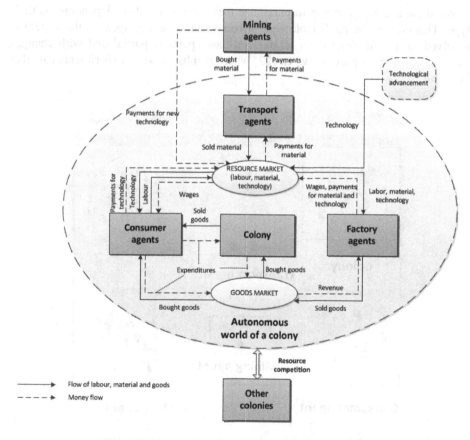

**Fig. 1.** Main components of model of virtual economy

The proposed system is already implemented in a prototype version in Netlogo platform. However, this brought to attention certain limitations related to Netlogo. First of all, Netlogo agents are supposed to operate mostly reactively and their capability to create an internal representation of the surrounding environment is limited. Also, issues with the use of external data files had to be addressed. All these issues finally resulted in an implementation of a full version of the application in another platform Anylogic. At this time, implementation is undergoing, but a working prototype of the system can be described and presented here.

In Fig. 1, there are main components of the model. There are four basic types of agents present here:

- M-agents – mining, providing resources of different types.
- F-agents – factory agents (producing goods or services).
- C-agents – representing consumers and workforces at the same time.
- T-agents – representing transportation agents (distribution of goods, etc., throughout the environment).

All these agents are trying to maximize their profit or utility, depending on their type. This creates the predictable type of behaviour, as every agent in the system is involved in it. The description of model's basic parts is partial and with changes freely adopted from previous work [22] where results of some experiments can also be found.

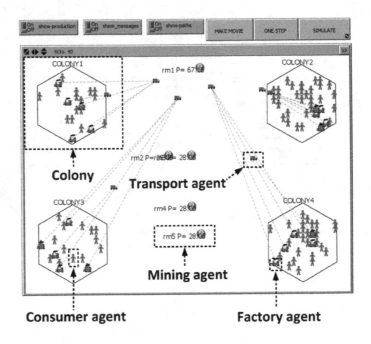

**Fig. 2.** Prototype of virtual economy environment in NetLogo

Fig. 2 is a screen capture of Netlogo environment (with an added description of its parts). Apart from that, there are charts graphically representing different types of output and variables in the model.

Consumer agent embodies the economic entity that consumes products and services (i.e. goods) and offers work (they serve as the workforce in the model). Consumer agent can buy goods based on the wealth. The wealth of a consumer agent is a product of work and qualification (the higher qualification the faster accumulation of wealth). A consumer agent makes a trade-off between investment into higher qualification (ec) and consumption. The combination of products consumed and the speed of consumption is given by the consumption function (see below). The combination of products forms a consumption pattern that can be used to divide consumers into three categories. The three categories are low income, middle income and high income consumers. The pattern determines the ratio of goods that the consumer agent is buying. There are three types of goods: necessary (basic), normal (common), and luxurious.

Factory agent corresponds with a company in a real economy. It produces either goods or services (there is a similar principle applied). Factory agent is responsible for transforming input to output, i.e. material and other products to a final product that is bought by consumer agent or a sub-product that is used by another factory agent. The consumption function determines materials and their proportions, consumed during production process. The production function determines the portfolio of goods produced. Production requires workforce, i.e. employing consumer agents. The production depends on the technological level ef and qualification of the workforce, i.e. employed consumer agents ec. The production equation is as follows:

$$\sum_{i=1}^{n} k_i^{con} x_i + WF \xrightarrow{production} e_C e_F \left( \sum_{j=1}^{m} k_j^{pro} y_j \right) \tag{1}$$

Let $k_i^{con}$ be the speed of consumption of a material $x_i$ and $WF$ is the workforce; $e_c$ is the qualification level of a consumer agent and $e_f$ technological level factory agent; $k_j^{pro}$ be the speed of production of a product $y_j$.

Mining agent is transforming resources located in the environment into raw material that is used by factory agents in the production of goods. The cost of mining is determined by the consumption function in which the energy, workforce and technology necessary for mining are reflected. The function is similar to the consumption function of a factory agent. Each mining agent supplies only one type of raw material. Possible parallel production of more than one type of raw material can be represented by several mining agents at the same location with different types of output. Raw material, as transformed from resources, is stocked in order to be later sold to transport agents and distributed to respective buyers.

Transport agents serve as intermediary between mining agents and factory agents. The cost of transportation is given by the distance. There might be barriers on the way from the mining agent to the factory agent, hence, it is the task of the transportation agent to find a route that is the most economical. The performance of a transportation agent is determined by the speed of mobility, capacity and technological level of a transportation device. Transportation agent is always buying all available material up to the capacity of transportation. Transported material that is not used directly in production is stored in factory agent´s warehouse.

The modelled virtual economy also contains a representation of a society of agents which is called "colony" in this context. The colony consists of consumer, factory and transportation agents. Mine agents do not belong to any colony, they are distributed throughout the environment, depending on the resources they process. The colony is characterized by its position in the environment and population. Colonies compete for resources that are supposed to be scarce, generally limited. Colonies exist in an environment. The environment is important in respect to transportation.

The success of a colony can be measured, for example, by accumulated wealth. Other possible means of efficiency measurement are mentioned in the following subsection. The wealth of a colony is given by the sum of wealth of all agents. Due to different colony populations, the comparison among colonies requires computing the average wealth per agent. The formula is as follows:

$$cw_{COL} = \frac{\sum_{i=1}^{n}(w_{C,i} + w_{F,i}) + w_{COL}}{\sum_{i=1}^{n} c_{COL,i}} \tag{2}$$

The model enables for various configurations and thus for conducting specific experiments.

## 4    Pursuing Goals and Measuring Efficiency

The basic prerequisite for the measuring of any form of efficiency is the specification of the goal. Goal can be described as a set of desired states of the system, a measurable value that has to be achieved, or even the state of the agent or environment which should be avoided (which is less common). All this depends on the selected approach. When defining goal, there is often a lot of abstraction involved in order to capture the complexity of the environment in some numeric value, see also [1]. It is necessary to capture all the importance and omit the unnecessary, which is not an easy task for the designer of the system.

The problem of the goal specification is even more complex in more complex environments – there is more information involved and it is difficult to distinguish important from unimportant, relevant from irrelevant. All the above mentioned leads to several recommendations which should be followed when designing goals of agents:

- Goal state has to be clearly defined and comprehensible. Formal definition should be provided in order to allow machine processing of the task.
- Goal state has to be distinguishable from other states or configurations of the environment.
- Advancement towards goal state should be measurable (the ideal way for machine processing is numeric representation).

Although some of these recommendations seem to be trivial, it is not easy to pursue goals in complex environments. Following sub-sections will focus on selected aspects of goals and efficiency measurement.

## 4.1   Individual Goals

Each agent in the environment is pursuing its own goals. This behaviour is similar to the behaviour of humans in economic systems. This is usually referred to as "maximization of utility", and artificial entity – agent – is trying to maximize it as well. At elementary level, agent needs to take care of its existential needs – food, water, housing, basic security, etc. As soon as these needs are taken care of, gradually more elaborate needs arise – healthcare, education, entertainment, etc.

It is assumed that elementary needs listed above are taken care of at the level of the whole community. These are existential imperatives - to keep populations alive, healthy and well fed. Therefore, the elementary level of needs will be mainly ignored from now on. We will focus our attention to the needs on higher levels. The behaviour of consumer agents (C-agents), representing human population in the model, will be presented here because these are basic construction blocks of the multi-agent system.

C-agents in the proposed system do have individual needs. Their behaviour is basically following elementary economic principle of descending demand (the more goods they have, the less they are willing to pay to obtain more). Demand price ($P$) is dependent on the actual level of stockpile ($s_C$). Presented version works with linear curve, but this can be easily modified if needed.

The acquisition of goods is done at "mikromarkets" representing individual commodities. In our model, these are divided into three levels reflecting technological level needed for their production. The consumption function of C-agent (con) is following:

$$con(C) = k_1^{con} x_{L1} + k_2^{con} x_{L2} + k_3^{con} x_{L3}$$

(3)

Where $k_1^{con}, k_2^{con}, k_3^{con}$ represent coefficients of consumption, $x_{Ln}$ represents production (output of F-agents from n-th layer). C-agents are divided into three groups, depending on their income – low income group (L1), middle income group (L2) and high income group (L3). For consumer categories $C_{category} = k_1^{con}, k_2^{con}, k_3^{con}$, coefficients are as follows:

- $C_{low} = 0,7; 0,2; 0,1$
- $C_{middle} = 0,4; 0,4; 0,2$
- $C_{high} = 0,2; 0,2; 0,6$

These consumption functions create consumption patterns for C-agents reflecting the combination of products and speed of their consumption.

These consumption functions reflect the consumption of goods. Similar principle is applied to consumption of services. The common principle applies that higher income groups of agents are also more demanding – requiring better quality and coverage with services.

All this results in a pressure towards efficient colony management. Agents are creating more and more demanding consumer pools for both goods and services as a colony evolves and its technological level rises through upgrades and development of production facilities.

After extensive experimentation, we found it useful to establish a special variable reflecting the satisfaction of an agent (sat(A)). The satisfaction rate is the numeric representation of an agent's ability to satisfy all of its needs – consumption of basic, common and luxury goods and services in areas of education, healthcare, security, transportation, entertainment, etc. The design of production chains with inherent dependency and succession of production of different types of products or services is considered to be extremely difficult. This is difficult to design and optimize and it still is a matter of further testing.

It is also worth mentioning that individual agents are slightly randomized in their consumption preferences, each one preferring a little different consumption basket. This reflects the fact that in the real world individual consumers are different from each other as well. This was implemented to objectify results of conducted experiments.

## 4.2     Community Goals

The situation is different on the community level. The colony management is supposed to satisfy needs of all inhabitants and pursue given goal(s), both at the same time. Goals are specified by the user as a part of problem definition and are provided at the beginning of the simulation. Community goals examples:

- Produce specified amount of selected type of goods in a given time.
- Achieve specified level of satisfaction of consumer agents in the colony.
- Become a dominant trader with selected commodity (this assumes competition present in the environment).
- Achieve specified average level of technological development in colony's production facilities.
- Maintain specified employment rate.
- Achieve specified education level of population.
- Etc.

All the examples of community goals assume parallel pursue of maintaining a good supply of food, water, energy production, and satisfaction of other possible basic needs of the population. The colony management is able to formulate rules or laws of community functioning which are mandatory for all inhabitants to follow. This results in the system where colony management creates conditions for life of its inhabitants, production efforts of factories and services providers with a common purpose in mind, given by the goal specification. The system is designed as a modular, transparent and incremental.

Although there are many things omitted in our model, we created a rather complex environment which is difficult to describe in brief. However, some areas of economy are not covered by our model at all – banking sector, capital markets, currency issues, etc. This is a matter of possible further expansion of our model in the future.

Nonetheless, our proposed model allows study and research of self-organization and different strategies formulation of agents. Individual colonies may be competing against each other in the environment, trying different strategies depending on their available resources and initial conditions and goals. This may result in conflicts or cooperation, depending on the user's specifications of scenarios.

## 5    Conclusions and Future Work

The proposed system allows agents to coordinate their activities in order to pursue their goals on both individual and community levels. However, the goal specification remains to be one of the crucial aspects in the system functioning. In general, the more complex the system, the more the complex measurement of efficiency has to be implemented. Parallel pursue of several goals at the same time is a natural result of an effort to create an economic system complex enough to capture economic behaviour of agents` collective. Appropriate levels of abstraction in the means of working knowledge of agents have to be achieved but this is mainly the task of the system designer rather than optimization of multi-agent behaviour.

At this time, an extended version of the model is being implemented in a more sophisticated environment of Anylogic platform. This allows us to pursue problems at hand from other angles since Anylogic is a simulation tool that supports all the most common methodologies in place today: system dynamics, process-centric (AKA discrete event) and agent based modelling.

Related research (see subsection 2) shows good potential for the use of obtained results in industrial or commercial applications but this would require further testing and optimization of the proposed system as well as possible expansions.

**Acknowledgments.** This research has been supported by the FIM UHK Excellence Project "Agent-based models and Social Simulation".

## References

1. Boy, G.A.: Modelling and Simulation. In: Orchestrating Human-Centered Design, pp. 139–172. Springer, London (2013)
2. Bureš, V., Otčenášková, T., Čech, P., Antoš, K.: A Proposal for a Computer-Based Framework of Support for Public Health in the Management of Biological Incidents: the Czech Republic Experience. Perspect. Public Heal. 132(6), 292–298 (2012)
3. Chavez, A., Maes, P.: Kasbah: An Agent Marketplace for Buying and Selling Goods. In: 1st International Conference on the Practical Application of Intelligent Agents and Multi-Agent Technology, pp. 75–90. Practical Application Co. Ltd., London (1996)
4. Damaceanu, R.C., Capraru, B.S.: Implementation of a Multi-Agent Computational Model of Retail Banking Market Using Netlogo. Metal. Int. 17(5), 230–236 (2012)

5. Deguchi, H., Terano, T., Kurumatani, K., Yuzawa, T., Hashimoto, S., Matsui, H., Sashima, A., Kaneda, T.: Virtual Economy Simulation and Gaming - An Agent Based Approach. In: Terano, T., Nishida, T., Namatame, A., Tsumoto, S., Ohsawa, Y., Washio, T. (eds.) JSAI-WS 2001. LNCS (LNAI), vol. 2253, pp. 218–226. Springer, Heidelberg (2001)
6. Dosi, G., Fagiolo, G., Roventini, A.: The microfoundations of business cycles: an evolutionary, multi-agent model. J. Evol. Econ. 18(3-4), 413–432 (2008)
7. Gazda, V., Gróf, M., Horváth, J., Kubák, M., Rosival, T.: Agent based model of a simple economy. J. Econ. Interact. Coor. 7(2), 209–221 (2012)
8. Golinska, P., Hajdul, M.: Multi-agent Coordination Mechanism of Virtual Supply Chain. In: O'Shea, J., Nguyen, N.T., Crockett, K., Howlett, R.J., Jain, L.C. (eds.) KES-AMSTA 2011. LNCS (LNAI), vol. 6682, pp. 620–629. Springer, Heidelberg (2011)
9. Grzybowska, K., Kovács, G.: Developing Agile Supply Chains – System Model, Algorithms, Applications. In: Jezic, G., Kusek, M., Nguyen, N.-T., Howlett, R.J., Jain, L.C. (eds.) KES-AMSTA 2012. LNCS (LNAI), vol. 7327, pp. 576–585. Springer, Heidelberg (2012)
10. Guessoum, Z., Rejeb, L., Durand, R.: Using adaptive multi-agent systems to simulate economic models. In: 3rd International Joint Conference on Autonomous Agents and Multi-agent Systems, pp. 68–75. IEEE Computer Society, Washington (2004)
11. Guo, J., Gong, Z.: Measuring Virtual Wealth in Virtual Worlds. In: Information and Technology Management 2011, pp. 121–135. Springer Science + Business Media, LLC, Heidelberg (2011)
12. Hou, H., et al.: An Enhanced Model Framework of Personalized Material Flow Services. In: Information and Technology Management, pp. 149–159. Springer Science + Business Media, LLC, Heidelberg (2011)
13. Janssen, M., de Vries, B.: The battle of perspectives: a multi-agent model with adaptive responses to climate change. Ecol. Econ. 26(1), 43–65 (1998)
14. Mildenberger, C.D.: The Constitutional Political Economy of Virtual Worlds. In: Constitutional Political Economy. Springer Science+Business Media, New York (2013)
15. Paul, C.J.M.: Productivity and Efficiency Measurement in Our "New Economy": Determinants, Interactions, and Policy Relevance. Journal of Productivity Analysis 19, 161–177 (2003)
16. Pennings, E.: Price or quantity setting under uncertain demand and capacity constraints: An examination of the profits. J. Econ. 74(2), 157–171 (2001)
17. Sacco, M., Dal Maso, G., Milella, F., Pedrazzoli, P., Rovere, D., Terkaj, W.: Virtual Factory Manager. In: Shumaker, R. (ed.) Virtual and Mixed Reality, Part II, HCII 2011. LNCS, vol. 6774, pp. 397–406. Springer, Heidelberg (2011)
18. Sinha, A.K., Aditya, H.K., Tiwari, M.K., Chan, F.T.S.: Agent oriented petroleum supply chain coordination: Co-evolutionary Particle Swarm Optimization based approach. Expert Syst. Appl. 38(5), 6132–6145 (2011)
19. Skobelev, P.: Multi-agent System for Real Time Resource Allocation, Scheduling, Optimization and Controlling: Industrial Applications. In: Mařík, V., Vrba, P., Leitão, P. (eds.) HoloMAS 2011. LNCS, vol. 6867, pp. 1–14. Springer, Heidelberg (2011)
20. Thabet, I., Bouslimi, I., Hanachi, C., Ghédira, K.: A Multi-agent Organizational Model for Grid Scheduling. In: O'Shea, J., Nguyen, N.T., Crockett, K., Howlett, R.J., Jain, L.C. (eds.) KES-AMSTA 2011. LNCS (LNAI), vol. 6682, pp. 148–158. Springer, Heidelberg (2011)
21. Tsvetovatyy, M., Gini, M., Mobasher, B., Wieckowski, Z.: MAGMA: An Agent-Based Virtual Market for Electronic Commerce. Appl. Artif. Intell. 11(6), 501–523 (1997)
22. Tučník, P., Čech, P., Bureš, V.: Self-organizational Aspects and Adaptation of Agent based Simulation Based on Economic Principles. In: Swiątek, J., Grzech, A., Swiątek, P., Tomczak, J.M. (eds.) Advances in Systems Science. AISC, vol. 240, pp. 463–472. Springer, Heidelberg (2014)

# Pedestrian Modelling in NetLogo

Jan Procházka, Richard Cimler, and Kamila Olševičová

Faculty of Informatics and Management, University of Hradec Králové, Rokitanského 62,
Hradec Králové 500 03, Czech Republic

**Abstract.** The objective of our research is to explore crowd dynamics under different circumstances, especially its optional applications in sustainable tourism. The terminology (crowd phenomena, pedestrian behaviour, local interaction, motion patterns) is explained and a brief overview of three theoretical models (cellular automata model, social force model and network model) is provided. Then our visitor flow model is suggested and the case study, the model of the crowd dynamics of visitors in the ZOO, is specified. NetLogo was used for implementation.

## 1    Introduction

The importance of pedestrian and crowd modelling can be seen in relation to a current trend of designing smart environments and ambient intelligence, where the overall behaviour, performance and success of the application strongly depends on parallel decisions and physical movements of users. It is necessary to reflect characteristics and context of individuals as well as to notice emerging effects and side-effects of coexistence of numerous users at the same time and place. This exploration can benefit from agent-based models and social simulations of transport and traffic, crowd and pedestrian models.

Agent-based models (ABM, also individual-based or bottom-up models) is an approach to modelling of complex systems composed of heterogeneous adaptive individuals (agents) interacting in the environment. The principles of ABM were explained by (Macal and North 2005, 2006, 2010, Railsback and Grimm 2012, Ulhmacher and Weyns 2009 and others). A brief explanation of the area of agent based social simulation (ABSS) from a computer scientist's perspective is presented in (Davidsson 2002). The method of developing ABM and ABSS is provided by (Grimm et al. 2010). NetLogo (Wilensky 2009) is an example of the implementation platform for ABM and ABSS. NetLogo was chosen for the implementation of our model, too.

Traffic and pedestrian models typically build on GIS data and data sources capturing characteristics and behaviour of individuals (e.g. national census). Large models often work with samples of the whole population precisely transformed to the comparable synthetic population.

In the area of our interest, which is sustainable tourism, ABM and ABSS help to develop and implement strategic policies for sustainable development (Maggi et al.

2011). (Zhang 2005) presents a GIS and agent-based social simulation of individual visitors using travel pattern data. The visitor flow modelling in tourism is closely related to the concept of the tourist carrying capacity which is defined as the maximum number of people that may visit a tourist destination at the same time without causing destruction of the physical or socio-cultural environment.

In the following part of the paper we summarize theoretical principles of pedestrian models, we present our visitor flow model that was applied in the context of visitors' moving in the ZOO and we present its implementation in NetLogo.

## 2    Pedestrian Models

The state-of-the-art article provided by (Duives et al. 2013) enumerates 8 different approaches to modelling pedestrian and crowd dynamics:

- cellular automata,
- social force models,
- activity choice models,
- velocity based models,
- continuum models,
- hybrid models,
- behavioural models,
- network models.

In general, all approaches reflect the existence of mutual interactions between individual pedestrians and between pedestrians and the environment. Ways how to cope with interactions vary from simple reasoning of the streets capacities (max. number of pedestrians) used in network models to detailed micro-simulations of pedestrian's personal space occupation used in social force models.

Pedestrian interactions are closely related to crowd phenomena. Crowd is defined by (Duives et al. 2013) as a group of more than 100 people who are located in the limited space (more than 1 person per m2) and who are moving within the time period longer than 1 minute. When observing crowd movements, it is possible to recognize different movement patterns such as:

- uni-directional flow,
- uni-directional rounding corner,
- uni-directional entering,
- uni-directional exiting,
- bi-directional flows,
- crossing two flows,
- crossing more than 2 flows,
- random flows crossing.

Our visitor *flow model* is a combination of cellular automata model, social force model and network model.

## 2.1    Cellular Automata Model

Cellular automata model of the pedestrian´s movement uses grids of cells. The state of each cell is interpreted as a presence or absence of a pedestrian. Each pedestrian is characterised by his individual physical ability called step distance. The step distance is defined in units corresponding to a distance between two neighbour cells of the environment.

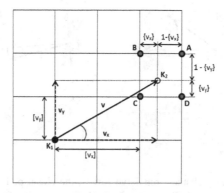

**Fig. 1.** Cellular Automata Model – one step destination cell principle

See fig. 1 for example of calculations related to one step in the grid of cells. The step begins at the discrete cell K1. The length of the pedestrian's step does not necessarily correspond to the distance of the cells. To follow the discrete rule of the cellular automata, the pedestrian cannot move to K2; one of points A, B, C or D has to be chosen according to probabilities given by distances of points A, B, C, D from K2. Movement vector splits into round and decimal parts of vx and vy:

$$v_x = [\, v_x \,] + \{\, v_x \,\} \tag{1}$$

$$v_y = [\, v_y \,] + \{\, v_y \,\} \tag{2}$$

Decimal parts determine probabilities of points A, B, C, D to be the destination:

$$P_A = \{\, v_x \,\} \cdot \{\, v_y \,\} \tag{3}$$

$$P_B = (1 - \{\, v_x \,\}) \cdot \{\, v_y \,\} \tag{4}$$

$$P_C = (1 - \{\, v_x \,\}) \cdot (1 - \{\, v_y \,\}) \tag{5}$$

$$P_D = \{\, v_x \,\} \cdot (1 - \{\, v_y \,\}) \tag{6}$$

$$P_A + P_B + P_C + P_D = 1 \tag{7}$$

## 2.2    Social Force Model

In social force model each pedestrian is influenced by one attractive and two repulsive forces (fig. 2, fig. 3). The attractive force makes pedestrian to continue towards the target and finally reach it. The repulsive forces help the pedestrian to avoid collisions with other pedestrians or obstacles in the environment. The final direction of the pedestrian movement is determined by vector composition of all three forces.

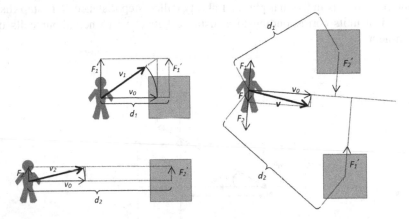

**Fig. 2.** Social force model – obstacle repulsive forces: the closer pedestrian is to the obstacle, the bigger the repulsive force (left); composition of two repulsive forces from two obstacles (right)

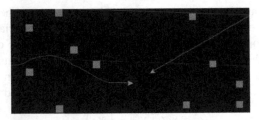

**Fig. 3.** Social force model – NetLogo implementation, trajectories influenced by repulsive forces (left) and caused by attractive force (right)

## 2.3    Network Model

In network model the environment is defined as a weighted network with nodes representing points of interest and edges representing sidewalks (pavements, streets, roads). Each pedestrian is autonomous, has got its own destination and speed and can interact with other pedestrians on the same edge. Alternatively, the edges can represent a part of the path, nodes are interpreted as intersections and the flow through the node depends on the capacities of incident edges.

## 3    Case Study: Visitor Flow Model

We propose the visitor flow model that reuses principles of cellular automata model, social force model and network model. For better explanation of our idea, we suggest the application: modelling the visitor flow in the ZOO.

The map of the ZOO can be represented as a network: nodes correspond to points of interest (entrance/exit, 19 animal houses or shops), each node has got its capacity (1-100 visitors) and attractiveness (1-10 points). Edges represent pavements between pairs of points. The length of the edge is 10-100 m (1 patch in NetLogo corresponds to 1 m2). Physical distances and the widths of pavements are taken into account in our model, while pure network models usually operate with edges' capacities only.

Visitors of the ZOO are represented by visitor-agents. Naturally, visitor agent is understood as pedestrians from previously mentioned theoretical models. In further versions of the model, visitor-agent could represent not only a single visitor but optionally also a couple or a family with relevant features. In our first version of the model, a visitor-agent has got three attributes:

- want-to-see list of animals (0-15 items),
- time limit for his visit (2-6 hours),
- walking speed (50, 80 or 100 meters per minute).

The visitor flow model consists of four components:

- control component,
- network component,
- cellular automata component,
- social force component.

## 3.1    Control Component

A control component manages visitor-agents and their movements to avoid potential collisions in case two visitor-agents want to move to the same patch (cell). It is implemented as finite state automata. See the state chart diagram of visitor-agent (fig. 4).

*Initial* and *terminal* transitions are defined as follows:

- 0.0: Visitor-agent is created and placed at the entrance/exit node. The incoming rate of visitors is an input parameter of the model. Its Poisson distribution reflects arrival rates during opening hours of the ZOO. One tick of the model clock corresponds to 1 second.
- 1.1: Transition to the terminal state: visitor-agents leave the entrance/exit node when their time limit is over or because of the closing hour.
- For *standard movement cycle*, these 4 transitions are defined:
- 1.0: All visitor-agents are ready to choose the most appropriate next step destination in *ready-to-go* state upon each simulation tick.
- 2.0: Visitor-agent is ready to negotiate with other visitor-agents who will eventually occupy the same step destination. Random choice from set of competing pedestrians solves is used.
- 3.0: Visitor-agent has won the competition for the step destination and moves there.
- 4.0: Visitor-agent gets back to *ready-to-go* state.

There are 3 visiting-a-node transitions:

- 4.1: When the last step ends in the node (point of interest) and there is free capacity, then the visitor-agent enters the node. Visitor-agents can spend 2-15 minutes at their preferred nodes and only 0-3 minutes at non-preferred nodes.
- 5.0: Time to stay in the node is over, visitor-agent gets *ready-to-go*.
- 5.1: Time to stay in the node is not over, visitor-agent stays here.

Last but not least 2 don't-move transitions are defined:

- 2.1: Visitor-agent's next step destination is not valid (e.g. other agent is occupying the destination or destination is *an obstacle* on the pavement).
- 3.1: Visitor-agent has not succeeded in the competition for step destination and gets *ready-to-go*.

## 3.2    Network Component

Nodes of the network are visitor-agents' points of interest (some of them are in their want-to-see lists). Once the visitor-agent reaches the node, the next target is chosen from the set of edge-neighbouring nodes. The selection mechanism considers visitor-agent's strategy: items from the want-to-see list have got the highest priority, not yet visited nodes are the second possible choice and random choice is the last possibility.

## 3.3    Cellular Automata Component

Visitor-agents move over the grid of cells (patches in NetLogo terminology) instead of moving directly from node to the node over edges. The calculation of the next step is based on cellular automata model algorithm.

## 3.4    Social Force Component

Social force model component is used to avoid optional obstacles placed in the environment. Visitor-agents have got their vision-distance to which they can see obstacles. Once an obstacle is seen in the collision heading, the movement is affected by the repulsive force caused by the obstacle.

# 4    Experiments

The implementation of visitor flow model in NetLogo enables experimenting and observing the overall behaviour of visitor-agents in the given environment.

## 4.1    Experiment 1 – Measurement of Intensity

It is possible to count visitor-agents at every patch to learn how intensively different parts of the environment are loaded with pedestrians. See fig. 5 for the resulting grid of the counter values. It is important to notice that these counter values can be interpreted also as a guideline for optimal design of pavements in a homogenous environment where pedestrians move freely (fig. 6).

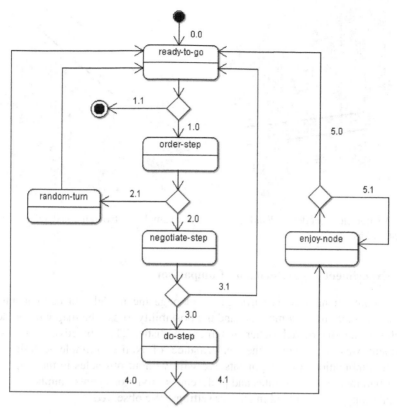

**Fig. 4.** Flow Model – state chart diagram

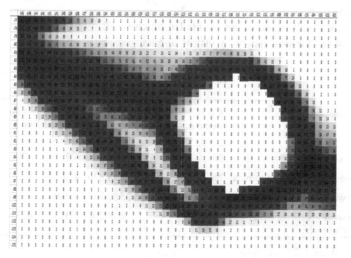

**Fig. 5.** Visitor flow model – detail of the map, darker shades of grey and higher numbers represent patches more frequently used by visitor-agents

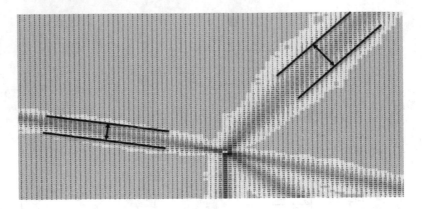

**Fig. 6.** Visitor flow model – detail of the map, example of two emerged pavements with different widths

## 4.2    Experiment 2 – Performance Comparison

Network component itself is sufficient to manage the model, but cellular automata component, social force component and the possibility to use the map with predefined obstacles make the model richer and more realistic. The objective of the second experiment was to compare the performance of NetLogo implementation under different combinations of components and with/without obstacles in the map. Values of other parameters (arrival rates and preferences of visitor-agents, number of steps of the simulation) were identical in all four settings. We observed:

- running time of simulation (in seconds),
- total number of visitor-agents,
- total distance walked by visitor-agents (in patch size),
- total number of nodes visited by visitor-agents.

**Table 1.** Experiment 2

| Component | Obstacles in the map | Run time [s] | Total walking distance | Total visited nodes | Total visitors |
|---|---|---|---|---|---|
| Network | No | 12 | 620 301 | 4 277 | 470 |
| Network + cellular automata | No | 48 | 613 826 | 4 266 | 471 |
| Network + cellular automata | Yes | 48 | 560 157 | 4 080 | 473 |
| Network + cellular automata +social force | Yes | 54 | 624 201 | 4 313 | 473 |

Results (table 1) show a big performance gap between models using only network component and models using it together with cellular automata algorithm. It is because network component operates with visitor-agents targets and the directions are given by the structure of the network only, while cellular automata model computes the movement over the grid of patches. On the other hand, cellular automata component produces data usable for estimation of widths of pavements.

The model without social force component has a better performance from the perspective of computation speed, but with the component visitor-agents, it can realistically manage obstacles in the environment.

## 5    Conclusion

The visitor flow model combines principles of cellular automata model, social force model and network model. The case study was developed for a specific situation: movement of visitors in the ZOO, but it can be adopted for analogical situations such as visitor flow in a botanical garden or a museum. Our next effort is focused on the more precise estimation of positions and widths of emerging pavements (roads) in relation to different behaviour patterns and characteristics of agents.

**Acknowledgments.** The research described in the paper was financially supported by grants GACR-402/09/0405, CZ.1.07/2.2.00/28.0327 Innovation and support of doctoral study program (INDOP) and FIM UHK specific research project no.2104 and FIM UHK project of excellence Agent-based models and Social Simulation no. 2207.

## References

1. Davidsson, P.: Agent Based Social Simulation: A Computer Science View. Journal of Artificial Societies and Social Simulation 5(1) (2002)
2. Duives, D.C., Daamen, W., Hoogendoorn, S.P.: State-of-the-art crowd motion simulation models. Transportation Research Part C (2013) (in press)
3. Grimm, V., Berger, U., DeAngelis, D.L., Polhill, J.G., Giske, J., Railsback, S.F.: The ODD protocol: A review and first update. Ecological Modelling 221(23), 2760–2768 (2010)
4. Macal, C.M., North, M.J.: Tutorial on agent-based modelling and simulation. In: Proceedings of the 37th Winter Simulation Conference, pp. 2–15 (2005)
5. Macal, C.M., North, M.J.: Tutorial on agent-based modelling and simulation part 2: how to model with agents. In: Proceedings of the 38th Winter Simulation Conference, pp. 73–83 (2006)
6. Macal, C.M., North, M.J.: Tutorial on agent-based modelling and simulation. Journal of Simulation 4, 151–162 (2010)
7. Maggi, E., Stupino, F., Fredella, F.L.: A multi-agent simulation approach to sustainability in tourism development. In: 51st European Congress of the Regional Science Association International, Barcelona (2011)
8. Railsback, S.F., Grimm, V.: Agent-Based and Individual-Based Modelling. Princeton University Press (2012)

9. Ulhrmacher, A.M., Weyns, D. (eds.): Multi-Agent Systems: Simulations and Applications. CRC Press (2009)
10. Wilensky, U.: NetLogo. Center for Connected Learning and Computer-Based Modelling. Northwestern University, Evanston (2009), http://ccl.northwestern.edu/netlogo/
11. Zhang, R.: Assessing the Carrying Capacity of Tourist Resorts: An Application of Tourists' Spatial Behaviour Simulator Based on GIS and Multi-Agent System. University Journal of Natural Sciences 10(4), 779–784 (2005)

# Communication of Mobile Robots in Temporary Disconnected MANET

Ján Papaj, Lubomír Doboš, and Anton Čižmár

Department of Electronics and Multimedia Communications, FEI,
Technical University of Košice, Park Komenského 13, 04120 Košice, Slovak Republic
jan.papaj@tuke.sk

**Abstract.** Mobile ad – hoc network (MANET) provides the new possibilities of multihop communication between mobile terminals without any fixed infrastructure. All mobile nodes are communicating with each other via wireless links and the topology of the network may change unpredictably. The MANET can be used not only for disaster events but can be used for robotic communication. The main problems of the all MANET routing protocols occur if the communication paths between mobile terminals are disconnected. In this paper we focused on the problem of temporary disconnections of the communication links. We propose the enhancement of the reactive routing protocol, also known as dynamic source routing protocol (DSR). Our enhancement provides possibilities to use opportunistic routing of the messages in the case that the routing protocol cannot find communication paths. In this case the routing protocols for MANET stop the routing process and wait for connection, and our modification provides the possibilities to send messages to mobile nodes if the routing protocols are stopped.

## 1 Introduction

A mobile ad-hoc network (MANET) is the collection of mobile nodes that are self-configuring and capable of communicating with each other, establishing and maintaining connections as needed [1]. MANETs are dynamic in the sense that each node is free to join and leave the network in a random way. It is a communications network that is capable of storing, transmitting and forwarding packets temporarily in intermediate nodes, during the time an end-to-end route is re-established or regenerated.

The Opportunistic networks (OppNets) are characterized as an evolution of the multi-hop mobile ad-hoc network (MANET). OppNets are more general than MANETs because dissemination communication is the rule rather than conversational communication. All nodes have possibilities to opportunistically exploit any pair-wise contact in order to share and forward content without any existing infrastructure [2], [3].

The main differences between OppNet and traditional mobile networks are the usability of the sources and expansion of the networks (Fig. 1). In traditional networks there are all the nodes of a single network situated together, based on the size of the

© Springer International Publishing Switzerland 2015

P. Sinčák et al. (eds.), *Emergent Trends in Robotics and Intelligent Systems*,

Advances in Intelligent Systems and Computing 316, DOI: 10.1007/978-3-319-10783-7_34

network and locations of its nodes pre-design [4]. On the other hand, in OppNet, the initial seed OppNet (which presents the basic part of OppNet) can be expanded into an expanded OppNet by considering different foreign nodes [4], [5].

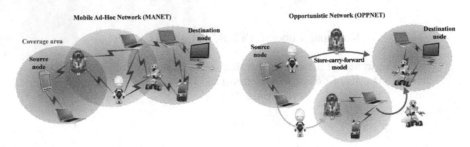

**Fig. 1.** Example of the Mobile Ad–Hoc network and Opportunistic network

In order to provide the effective communication between nodes, it is necessary to consider different aspects (disconnections, mobility, partitions, and norms instead of the exceptions). The mobility in opponent is used to provide efficient communication between unconnected groups of nodes. The mobile nodes forward data and also store data in the cache for a long time. This model is called ***store–carry–forward*** [2].

In MANET, traditional routing algorithms and protocols are based on routing schemes, which can find a path for a given node pair according to various metrics, and data packets are transmitted from one intermediate relay node to the next specified relay based on the physical condition of wireless channels.

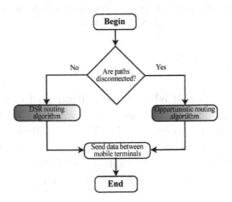

**Fig. 2.** Main idea of routing in disconnected MANET

In this paper we focused on the problem of temporary disconnections of the communication links. We propose the enhancement of the reactive routing protocol also known as Dynamic Source Routing protocol (DSR). Our enhancement (OPP_DSR) provides the possibility to use opportunistic routing of the messages in the case that the routing protocol cannot find communication paths. In this case the routing protocols for MANET stop the routing process and wait for connection. Our enhancement will provide the possibility to send messages to mobile nodes if the routing protocol cannot find the communication paths.

## 1.1   Overview of the Dynamic Source Routing Protocol for MANET

Dynamic Source Routing protocol (DSR) is a reactive protocol for MANET [6]. DSR enables to find a route between the source and destination mobile nodes. The DSR routing protocol in MANET is based on the assumption that there are end-to-end connections between the source and destination mobile nodes. The basic feature of DSR is that the routing packets in the header carry all information for routing. There are no periodic routing messages that enable the reduction of network bandwidth overhead and large routing updates throughout the ad hoc network. The basic feature of DSR is that the routing packets in the header carry all information for routing [6]. The main algorithm is based on two mechanisms, namely route discovery and route maintenance. DSR uses no periodic routing messages, thereby reducing network bandwidth overhead, conserving battery power and avoiding large routing updates throughout the MANET.

Route discovery phase deals with a process of sending the data between source and destination nodes. When a source node sends a data packet to the destination, it first checks its route cache to find existing routing paths. If no route is available, the source node initializes the route discovery process. This process of broadcasting the routing packets is called Route Request packets (RREQ). The RREQ packet has information about the source and destination node. The addresses of the source and destination nodes with a unique identification of the RREQ are stored in the routing packet RREQ. All intermediate or routing nodes read stored destination addresses, append their own address to the route record field of the RREQ packet and send to the network. This process will be repeated until the destination mobile nodes are found. If the RREQ packet reaches the destination or an intermediate node has routing information to the destination, a route reply (RREP) packet is generated. When the RREP packet is generated by the destination, it comprises addresses of nodes that have been traversed by the route request packet.

Route maintenance deals with protection of the existing connections [7], [8] and it is activated when a communication link break between two nodes along the path from the source to the destination.

## 2   Modification of the DSR for Disconnected Mobile Environment

The main problems of all MANET routing protocols occur when the communication paths between mobile terminals are disconnected. For this reason, we design a modification of the DSR's routing algorithm. Our proposal (OPP_DSR) combines the DSR routing algorithms and an opportunistic routing mechanism to find paths between disconnected MANET and the main ideas are shown in Fig. 2. The modification of the routing strategy will provide the possibilities to find paths between the source and destination nodes not only if there are existing connections, but also in the cases when the communication paths are disconnected. Our proposal of the routing protocol uses the positive characteristics of the DSR routing algorithm as well as opportunistic routing. In the event of separation, the connection activates the opportunistic routing algorithm.

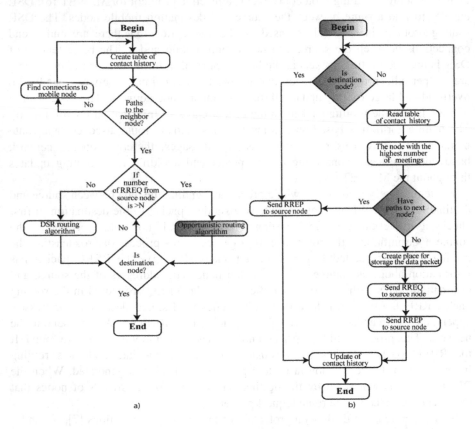

**Fig. 3.** The OPP_DSR for disconnected MANET: a) Principle of OPP_DSR routing for disconnected MANET, b) the main algorithm of the OPP_DSR

The routing algorithm of the OPP_DSR is as follows (Fig. 3a). At the beginning, the decision algorithm analyses connections between nodes. If there are paths between the source and destination nodes, the standard routing algorithm from DSR is used for the selection of the paths. In the case that the DSR routing protocol cannot find the paths to the destination or when the disconnection of the paths are occurring and also if the number of RREQ packets on node reaches number 10, the OPP_DSR is activated (Fig. 3b).

During routing, the mobile nodes create table contacts for storing the contacts. There is statistical information about all connections of the nodes which also provides useful information about how many times the mobile node was in contact with other mobile nodes. This data is necessary during the routing process and enables the enhancement of the routing properties of the OPP_DSR. The contact history table also contains the number of meetings and this parameter is then used for the selection of the candidate's nodes. Then the route cache memory allocates part of the memory for storage of the routing.

The mobile node sends route request packets (RREQ) to the mobile node selected from the contact history table and also sends the route replay packet (RREP) to the destination node. This packet will enable the transmission of the data packet from the source node to the destination mobile nodes.

## 3     Experiment Setup

In order to evaluate the proposed enhancement of the DSR, we used OPNET Modeler version 16.0 [9] to simulate the OPP_DSR routing algorithm and compared it with an existing DSR routing protocol in MANET (Fig. 4).

The 5 separated simulation scenarios, formed of 20, 40, 60, 80 and 100 nodes, were created to check the effectiveness of operation of the designed routing mechanism in MANET. The size of the simulated area was 300x300m2. Transmit power was set up to 1 mW. The random mobility model was used to simulate the mobility of nodes. The speed of the mobile nodes was in the range from 0 to 10 m/s. The simulation period was 1000 seconds in all cases. A free environment without affecting the interference was used as the simulation environment. The communication paths disconnections are created by a short transmission range of antennas and it is set up to 25 m.

The changing parameter was the initial value of movement, which gives a different initial position of individual nodes in the simulated project. The result of each simulation was a set of values that were then statistically processed and evaluated. The number of values can be chosen in the simulator environment. In our case, each sample was made up of a set of 100 values from each simulation (10 000 values were recorded). The path disconnection is simulated by the mobility of the mobile nodes and the transmission range is set up to 25 m.

Simulated parameters were used to show an average delay of MANET, the average number of RREQ and RREP packets and throughput. The average delay of MANET is assumed to be the average amount of time that is necessary for the transmission of the packet between source and destination nodes. The average number of RREQ packets provides the average number of routing packets used to find destination nodes. On the other side, the average number of the RREP presents the number of packets transmitted from the destination node to the source nodes. The parameter throughput is a parameter of the network which gives the information how the channel capacity is used for useful transmission to destination during the simulation. This channel parameter gives information about correct transmission of the data to the destination and we can say that it is a total number of data delivered to the destination during simulation.

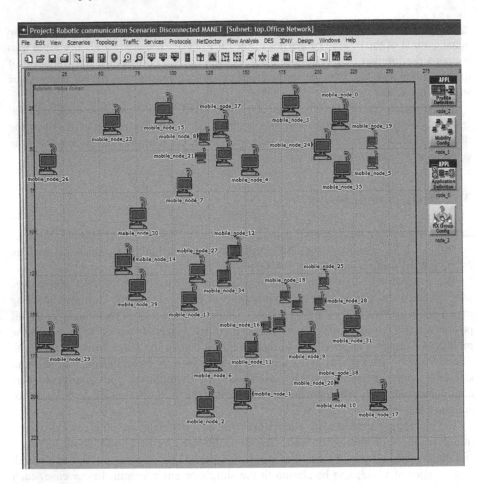

**Fig. 4.** Example of the simulation scenario with 40 mobile nodes in OPNET Modeler

In order to compare the performance of the routing protocols DSR and OPP_DSR implemented into MANET model in OPNET Modeler, 3 types of experiments were simulated.

In the first experiment, the delay of MANET was analysed. This parameter provides information about how long it takes to deliver a MANET packet from source to destination nodes when it also includes the time that is essential for processing information in disconnected environments.

Analysis in the second experiment was the average number of RREQ and RREP. We need to demonstrate how many RREQ and RREP packets are necessary to be sent until the destination node is found. During simulations, the temporary disconnection of the communication paths is taken into account.

The third experiment was focused on analysing the channel properties during routing. The parameter throughput will give information on how the modification of the routing protocol can affect the channel properties of the communication links.

This characteristic also gives quick information about how much data during the routing will be sent.

## 3.1 Simulation Results

Results of monitoring the average delay of MANET for different speed and number of the mobile nodes are shown in Table 1. It can be observed that the DSR experienced a higher average delay of MANET compared with OPP_DSR. As the speed and the number of mobile nodes increases, the DSR routing protocol experiences a huge average delay of MANET while OPP_DSR.

Tables 2 and 3 show the propagated RREQ packets and RREP for DSR with our proposed OPP_DSR and the standard DSR without possibilities to find communication paths while the communication paths will be disconnected. The values of RREQ and RREP during the OPP_DSR routing increase dramatically as the number of nodes on the network also increases and it is dependent on network mobile density.

**Table 1.** Average Delay of MANET [ms]

| Node speed | Routing algorithm | Number of mobile nodes | | | | |
|---|---|---|---|---|---|---|
| | | 20 | 40 | 60 | 80 | 100 |
| 2 m/s | DSR | 2,243 | 2,368 | 2,459 | 2,663 | 3,056 |
| | OPP_DSR | 3,728 | 3,618 | 4,299 | 6,374 | 5,532 |
| 4 m/s | DSR | 2,327 | 2,484 | 2,549 | 2,627 | 3,389 |
| | OPP_DSR | 6,721 | 5,552 | 9,213 | 3,941 | 6,497 |
| 6 m/s | DSR | 2,337 | 2,383 | 2,528 | 2,661 | 3,728 |
| | OPP_DSR | 6,897 | 3,613 | 3,964 | 5,945 | 4,9614 |
| 8 m/s | DSR | 2,337 | 2,383 | 2,383 | 2,528 | 3,374 |
| | OPP_DSR | 4,952 | 3,282 | 2,982 | 3,762 | 3,746 |
| 10 m/s | DSR | 2,275 | 2,383 | 2,522 | 2,584 | 3,192 |
| | OPP_DSR | 4,990 | 3,610 | 7,116 | 4,999 | 6,524 |

**Table 2.** Average Number of RREQ for MANET [Packets]

| Node speed | Routing algorithm | Number of mobile nodes | | | | |
|---|---|---|---|---|---|---|
| | | 20 | 40 | 60 | 80 | 100 |
| 2 m/s | DSR | 5,249 | 10,598 | 11,51167 | 12,42603 | 11,109 |
| | OPP_DSR | 11,427 | 11,845 | 10,87518 | 27,15901 | 15,521 |
| 4 m/s | DSR | 4,977 | 7,977 | 9,200119 | 8,535909 | 10,159 |
| | OPP_DSR | 18,546 | 11,289 | 19,2451 | 26,09606 | 22,565 |
| 6 m/s | DSR | 5,276 | 7,4722 | 11,6981 | 9,296944 | 12,657 |
| | OPP_DSR | 9,349 | 12,598 | 14,930 | 22,215 | 29,699 |
| 8 m/s | DSR | 4,073 | 6,168 | 7,373 | 7,373 | 6,469 |
| | OPP_DSR | 9,166 | 10,869 | 13,579 | 13,579 | 8,789 |
| 10 m/s | DSR | 4,448 | 7,208 | 7,937 | 7,558 | 9,359 |
| | OPP_DSR | 18,799 | 10,329 | 14,212 | 19,236 | 22,684 |

**Table 3.** Average Number of RREP for MANET [Packets]

| Node speed | Routing algorithm | Number of mobile nodes | | | | |
|---|---|---|---|---|---|---|
| | | 20 | 40 | 60 | 80 | 100 |
| 2 m/s | DSR | 12,495 | 12,398 | 14,565 | 11,277 | 11,709 |
| | OPP_DSR | 172,433 | 182,691 | 107,885 | 110,197 | 200,754 |
| 4 m/s | DSR | 53,249 | 43,885 | 37,466 | 30,129 | 34,592 |
| | OPP_DSR | 177,463 | 111,775 | 200,886 | 110,535 | 189,057 |
| 6 m/s | DSR | 84,017 | 71,749 | 91,214 | 59,099 | 54,356 |
| | OPP_DSR | 215,159 | 321,919 | 122,514 | 140,853 | 123,328 |
| 8 m/s | DSR | 120,325 | 82,749 | 93,7899 | 72,635 | 72,907 |
| | OPP_DSR | 377,975 | 364,567 | 209,345 | 125,037 | 136,945 |
| 10 m/s | DSR | 138,388 | 126,857 | 147,039 | 76,378 | 113,597 |
| | OPP_DSR | 334,421 | 412,566 | 512,062 | 140,923 | 393,062 |

**Table 4.** Throughput for MANET [Bits/s]

| Node speed | Routing algorithm | Number of mobile nodes | | | | |
|---|---|---|---|---|---|---|
| | | 20 | 40 | 60 | 80 | 100 |
| 2 m/s | DSR | 15417,18 | 13958,67 | 13778,91 | 12816,31 | 14263,53 |
| | OPP_DSR | 12134,45 | 15148,19 | 13372,79 | 13797,65 | 9058,92 |
| 4 m/s | DSR | 29697,07 | 30107,37 | 27053,59 | 26053,07 | 28360,87 |
| | OPP_DSR | 34913,66 | 33305,56 | 33512,07 | 29971,22 | 24964,05 |
| 6 m/s | DSR | 47242,12 | 45784,81 | 43894,51 | 42487,05 | 45625,44 |
| | OPP_DSR | 52339,31 | 53215,81 | 56859,43 | 54995,45 | 39552,44 |
| 8 m/s | DSR | 61844,65 | 60048,78 | 60923,65 | 56589,04 | 59897,11 |
| | OPP_DSR | 55310,54 | 79783,1 | 55429,99 | 52063,06 | 36941,05 |
| 10 m/s | DSR | 87477,94 | 85431,27 | 107410,2 | 80707,81 | 82919,72 |
| | OPP_DSR | 105328,8 | 92546,61 | 66955,7 | 89909,19 | 64776,05 |

Table 4 shows a comparison of the average of the throughput for the comparison of different speeds and the number of mobile nodes. The best performance is shown by DSR as it delivers data packets at a higher rate in comparison to OPP_DSR.

## 4    Conclusions

MANET and OPPNET are very interesting fields of robotic communication. These networks allow multihop communication to share the data and it enables connectivity to the internet or clouds. The main problem of the routing protocols designed for MANET is the communication in disconnected environments.

The research in the field of the temporary disconnected MANET is still at the beginning. This paper introduced DSR routing algorithm enhancements that enable the transmitting of data packets when the communication paths in MANET will be disconnected. According to the simulation results, the efficiency of the MANET and new OPP_DSR are raised in respect of some simulation metrics such as the average number of RREQ and RREP, an average delay and throughput of the disconnected MANET in OPNET Modeler.

The simulation result also shows that as the speed of nodes increase, efficiency of routing protocols decrease. In the future, our research will focus on the optimization processes on the choice of the next mobile nodes to enhance the routing.

**Acknowledgments.** This work was supported by the EU ICT Project INDECT (FP7-218086) (10%) and by the Ministry of Education of Slovak Republic under research VEGA 1/0386/12 (10%) and by Research & Development Operational Program funded by the ERDF under the ITMS projects codes 26220220141 (40%) and 26220220155 (40%).

# References

1. Čižmár, A., Doboš, Ľ., Papaj, J.: Security and QoS Integration Model for MANETs. Computing and Informatics 31(5), 1025–1044 (2012) ISSN: 1335-9150
2. Verma, A., Srivastava, A.: Integrated Routing Protocol for Opportunistic Networks. Proceeding of International Journal of Advanced Computer Science and Applications 2(3) (2011)
3. Boldrini, C., Conti, M., Passarella, A.: Modelling data dissemination in opportunistic networks. In: Proceedings of the Third ACM Workshop on Challenged Networks, Challenged Networks 2008, New York, NY, USA, vol. 6, pp. 89–96 (2008) ISBN: 978-1-60558-186-6
4. Pelusi, L., Passarella, A., Conti, M.: Opportunistic networking: data forwarding in disconnected mobile ad hoc networks. IEEE Communications Magazine 44, 134–141 (2006)
5. Sushant, J., Fall, K., Patra, R.: Routing in a delay tolerant network. In: Proceeding of Applications, Technologies, Architectures, and Protocols for Computer Communications (2004)
6. Beaubrun, R., Molo, B.: Using DSR for routing multimedia traffic in MANETs. International Journal of Computer Networks Communications 2(1), 122–124 (2010)
7. Papaj, J., Doboš, Ľ., Čižmár, A.: Routing Strategies in Opportunistic Networks. Journal of Electrical and Electronics Engineering 5(1), 167–172 (2012) ISSN 1844-6035
8. Papaj, J., Čižmár, A., Doboš, Ľ.: Implementation of the new integration model of security and QoS for MANET to the OPNET. In: Dziech, A., Czyżewski, A. (eds.) MCSS 2011. CCIS, vol. 149, pp. 310–316. Springer, Heidelberg (2011)
9. OPNET Modeler Simulation Software, http://www.opnet.com

# Application of Tracking-Learning-Detection for Object Tracking in Stereoscopic Images

Michal Puheim, Marek Bundzel, Peter Sinčák, and Ladislav Madarász

Department of Cybernetics and Artificial Intelligence, Faculty of Electrical Engineering
and Informatics, Technical University of Košice, Slovak Republic
{michal.puheim,marek.bundzel,peter.sincak,
ladislav.madarasz}@tuke.sk

**Abstract.** We use Tracking-Learning-Detection algorithm (TLD) [1]-[3] to localize and track objects in images sensed simultaneously by two parallel cameras in order to determine 3D coordinates of the tracked object. TLD method was chosen for its state-of-art performance and high robustness. TLD stores the object to be tracked as a set of 2D grayscale images that is incrementally built. We have implemented the 3D tracking system into a PC, communicating with the Nao humanoid robot [4][5] equipped with a stereo camera head. Experiments evaluating the accuracy of the 3D tracking system are presented. The robot uses feed-forward control to touch the tracked object. The controller is an artificial neural network trained using the error Back-Propagation algorithm. Experiments evaluating the success rate of the robot touching the object are presented.

**Keywords:** Tracking-Learning-Detection, TLD, Nao robot, object tracking, stereo-vision, neural network controller.

## 1 Introduction

Our goal was to develop a robust and reasonably fast (10 fps) system applicable in mobile robotics capable to track real world objects and to determine their 3D coordinates. The approach we have chosen uses two 2D tracking systems running parallel on the images sensed by a stereovision head of a Nao humanoid robot. The 3D tracking system may be used to touch or grasp objects by the robot or to plan actions based on the spatial distribution of the obstacles in the robot's world, etc.

We have applied Tracking-Learning-Detection algorithm (TLD [1]-[3]) to track objects in 2D images sensed by the stereo-vision of the Nao robot [4][5]. Using the information about object positions on the frames taken by the cameras of the stereovision system simultaneously and the parameters of the stereovision system, 3D coordinates of the tracked object are determined. The proposed system was tested in two sets of experiments. Firstly, the accuracy of the distance measurements was evaluated. Secondly, we tested the combination of the proposed system and a robot's hand controller based on the neural network with the goal to move the robot's hand in order to touch the tracked object.

© Springer International Publishing Switzerland 2015
P. Sinčák et al. (eds.), *Emergent Trends in Robotics and Intelligent Systems*,
Advances in Intelligent Systems and Computing 316, DOI: 10.1007/978-3-319-10783-7_35

## 2    Determining Depth in Stereovision

Let us assume having two identical cameras installed so that their optical axes are parallel as shown in Fig. 1. Let f be the focal distance of the cameras, c the distance between the cameras, and a the distance of cameras from the central optical axis and c = 2a. Let XR be the x-coordinate of the object as seen by the right camera and XL be the x-coordinate of the object as seen by left camera. Using the similarity of triangles we get the following equations:

$$\frac{X_L}{f} = \frac{(x-a)}{z} \tag{1}$$

$$\frac{X_R}{f} = \frac{(x+a)}{z} \tag{2}$$

By merging these equations and the elimination of x we get the formula which can be used to calculate the z coordinate:

$$z = \frac{-2af}{X_L - X_R} \tag{3}$$

This equation enables us to use the difference in projections of the same object to determine the distance of the object from the image plane (i.e. the depth information) assuming that the projections of the object are recognized and localized. This is called "the correspondence problem". We use the TLD method to solve the correspondence problem so that only the projections of the tracked object are searched for. We do not construct the depth map.

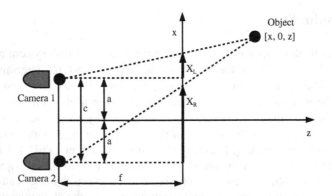

**Fig. 1.** Estimation of object $z$ coordinates using stereo cameras with parallel optical axis. Distance between cameras is $d = 2a$, where $a$ is the distance of the camera from the optical axis. $X_R$ and $X_L$ are coordinates of the object as seen by right and left camera respectively. If $f$ is focal length then $z$ coordinate can be calculated using $X_R$ and $X_L$.

# 3     Tracking-Learning-Detection

Tracking-Learning-Detection (TLD) method [1]-[3] is designed for long-term tracking of arbitrary objects in unconstrained environments. One of the advantages of the system is that it does not need to separate an offline learning stage. To initialize the tracking, the target object is delimited by a bounding box in the initial image. Further learning of the alternative appearances of the object is performed during the run-time, i.e. the longer the algorithm runs, the better it should be able to recognize the target object. TLD system is composed of three basic components:

- Tracker – a short term tracker based on the Lucas-Kanade method [6], which is used to track the given object and to generate examples for the learning of the detector.
- Detector – has the form of a randomized forest [7], enables incremental updates of its decision boundary and real-time sequential evaluations during run-time [2]. Runs independently from the tracker.
- Learning algorithm – so called "P-N Learning" [1] which uses trackers to generate positive (P) and also negative (N) examples that are further used in order to improve the model of the detector.

The object is supposed to be tracked by a tracker component and simultaneously learned in order to build a detector that is able to re-detect the object once the tracker fails. The detector is built upon the information from the first frame as well as the information provided by the tracker. Both components make errors. The stability of the system is achieved by mutual cancellation of these errors. The learned detector enables reinitialization of the tracker whenever a previously observed appearance reoccurs [2].

# 4     3D Tracking System

The proposed system applies the TLD method on the stereo images sensed by the Nao robot. The robot is able to track an arbitrary selected object and also determines the position of the target object in the three-dimensional space, see Fig. 2.

The operator delimits the target object manually by selecting its bounding box. The initialized TLD is duplicated so that the images from the stereovision camera are processed individually. We have implemented a method to synchronize the object models of the two TLD systems in order to prevent excessive difference between their object models.

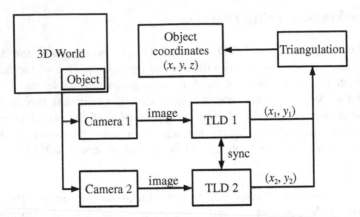

**Fig. 2.** The simplified block diagram of the proposed stereo-vision system employing the TLD method for the object tracking. The target object is simultaneously captured by two cameras. The images are used as the input to the two synchronized TLD systems producing the 2D coordinates of the target object in both images. The pair of 2D coordinates are used to calculate 3D coordinates of the target object using the triangulation method.

Each of the parallel TLD systems produces four outputs defining the bounding box of the object on the camera image, see Fig. 3.

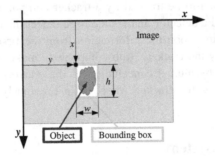

**Fig. 3.** Output of the TLD system is the bounding box of the tracked object. This bounding box is defined by four values $(x, y, w, h)$ which are horizontal and vertical coordinates of the upper left corner of the box, width and height of the box.

Using these values we can calculate the centres (xc, yc) of the bounding boxes for both TLD systems. Let (xL, yL) and (xR, yR) be the centre coordinates of the object on the image captured on the left and right camera respectively, given in pixels. The 3D coordinates of the target object are calculated as:

$$z = \frac{cf}{x_L - x_R} \tag{4}$$

$$x = \frac{c}{2} + z\frac{\left(x_L - \frac{w_i}{2}\right)}{f}$$

$$(5)$$

$$y = -z\frac{\left(y_L - \frac{h_i}{2}\right)}{f}$$

$$(6)$$

where wi is the image width (in pixels), hi is the image height (in pixels), f is the focal distance of the cameras (in pixels), and c is the distance between cameras (in meters). The distance d of the target object from the cameras is the length of the vector (x, y, z):

$$d = \sqrt{x^2 + y^2 + z^2}.$$

$$(7)$$

## 5    Robot Arm Controller

The task to test the applicability of the proposed 3D tracking system was to touch the target object with the humanoid's hand. We have implemented a neural network-based feed-forward controller of the robots arm. The polar 3D coordinates of the target object and orientation of the robots head (i.e. the neck joints angles) represent the controllers inputs, see Fig. 4.

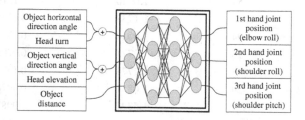

**Fig. 4.** Controller of the arm motion of the robot based on the feed-forward neural network. Inputs are the information about head turn and elevation and the information about the object position obtained from the stereo-vision TLD system. Outputs are the positions of the hand joints which can be used in order to move the hand to the object location.

The synaptic weights are optimized by the error Back-Propagation algorithm. Training data is generated by setting the 3D tracking system to follow the hand of the humanoid in order to create enough samples of corresponding input and output data. The neural network represents a transformation of the measured coordinates of the target object and the robot's arm joints angles. Feedback control is to be implemented to improve performance.

## 6     Experiments

To test the applicability of the proposed system we have performed two sets of experiments. The first evaluates the accuracy of the object distance measurement. In this experiment we place the target object 20 times in front of the robot in a variable distance ranging from 0.4 to 2.0 meters. At each measurement we compared the distance computed by the stereo-vision system with the distance measured by physical means. Results are shown in Table 1.

In the second experiment we addressed the testing of the proposed hand controller in order to determine the ability to touch the tracked object. Again, we put the target object 20 times in front of the robot in various positions but within reach of the humanoid robot. We have counted the number of times the robot successfully touched the object. Results are shown in Table 2.

**Table 1.** Results of the distance measurement experiment

|  | Real distance (m) | Measured distance (m) | Difference (m) |
|---|---|---|---|
| 1. | 0.5348 | 0.3873 | 0.1475 |
| 2. | 0.6132 | 0.4553 | 0.1475 |
| 3. | 0.6972 | 0.6403 | 0.0569 |
| 4. | 0.7849 | 0.6778 | 0.1071 |
| 5. | 0.8752 | 0.7223 | 0.1529 |
| 6. | 0.9675 | 1.0049 | −0.0374 |
| 7. | 1.0142 | 0.8145 | 0.1997 |
| 8. | 1.0611 | 1.0870 | −0.0259 |
| 9. | 1.1559 | 1.0040 | 0.1519 |
| 10. | 1.2514 | 1.1738 | 0.0776 |
| 11. | 1.2994 | 1.0040 | 0.2954 |
| 12. | 1.3476 | 1.2771 | 0.0705 |
| 13. | 1.4443 | 1.2771 | 0.1672 |
| 14. | 1.4929 | 1.0851 | 0.4078 |
| 15. | 1.5414 | 1.5327 | 0.0087 |
| 16. | 1.6389 | 1.6927 | −0.0538 |
| 17. | 1.7367 | 1.3960 | 0.3407 |
| 18. | 1.8346 | 1.5327 | 0.3019 |
| 19. | 1.9329 | 1.6927 | 0.2402 |
| 20. | 2.0313 | 1.6927 | 0.3386 |
| | | Average absolute measurement error (m): | **0.1670** |

The results of the distance measurement experiment have shown that the proposed system is not suitable for precise distance measurement. Considerable inaccuracies mainly at greater distances of the target object are caused by the nature of the sensing system (small spacing of the cameras and lower resolution used) and by inaccurate localization of the object by the TLD systems. The possible solution to this is the implementation of corrective procedures improving the localization based on epipolar geometry.

The experiment with the hand controller has shown that the closer to the robot the 3D tracking system is, the more accurate it is to enable the touching or grasping of the target object. This can be further improved by the implementation of feedback control.

As it is, the proposed object tracking system is unable to tell more about the target object besides its location. This would be a problem if manipulation with the objects of different sizes or shapes is desired. To solve this problem a different object tracking method could be used (see [8]) or we can try to track various parts of the tracked objects separately to determine their orientation additionally.

**Table 2.** Results of the hand controller testing experiment

| | NN Inputs (object coordinates) | | | NN Outputs (hand joint positions) | | | OK? |
|---|---|---|---|---|---|---|---|
| | $d$ (m) | $v$ (rad) | $h$ (rad) | $s_p$ (rad) | $s_r$ (rad) | $e_r$ (rad) | |
| 1. | 0.134146 | 0.027178 | 0.290937 | 0.057461 | 0.357181 | -0.788319 | Y |
| 2. | 0.190570 | -0.068388 | 0.589369 | -0.172792 | 0.477867 | -0.169463 | Y |
| 3. | 0.198999 | -0.073607 | 0.765424 | -0.184782 | 0.740034 | -0.160568 | Y |
| 4. | 0.164649 | 0.103553 | 0.424750 | 0.061034 | 0.431430 | -0.588861 | Y |
| 5. | 0.221017 | -0.173743 | 0.923930 | -0.495389 | 0.778589 | -0.039320 | Y |
| 6. | 0.223769 | 0.021826 | 0.735878 | -0.166607 | 0.515333 | -0.049318 | Y |
| 7. | 0.162134 | -0.435731 | 0.842525 | -0.614742 | 1.071152 | -0.477502 | N |
| 8. | 0.194244 | -0.387002 | 1.168614 | -0.553836 | 1.345710 | -0.174646 | N |
| 9. | 0.191741 | -0.055399 | 0.831764 | -0.122697 | 0.975315 | -0.333835 | Y |
| 10. | 0.185708 | -0.190491 | 0.733462 | -0.329463 | 0.733202 | -0.207705 | Y |
| 11. | 0.153799 | -0.011105 | 0.150542 | -0.038457 | 0.190594 | -0.393123 | Y |
| 12. | 0.163303 | 0.320239 | 0.386022 | 0.225971 | 0.462984 | -0.762712 | Y |
| 13. | 0.185293 | 0.017649 | 0.357128 | -0.072228 | 0.261859 | -0.185558 | Y |
| 14. | 0.217610 | -0.034607 | 0.627409 | -0.253171 | 0.383026 | -0.042686 | Y |
| 15. | 0.201207 | -0.011139 | 0.518828 | -0.137416 | 0.347757 | -0.096682 | Y |
| 16. | 0.209144 | -0.268592 | 1.175758 | -0.458409 | 1.302464 | -0.116085 | Y |
| 17. | 0.129244 | -0.330636 | 0.627409 | -0.385313 | 0.940373 | -1.098963 | N |
| 18. | 0.141164 | -0.175151 | 0.122507 | -0.166845 | 0.178159 | -0.397670 | N |
| 19. | 0.193073 | -0.401333 | 0.510929 | -0.773710 | 0.276209 | -0.030392 | Y |
| 20. | 0.193977 | 0.237270 | 0.403652 | 0.181841 | 0.320561 | -0.257261 | Y |
| | Successful touch: **16** | | | Failed touch: **4** | | Percentage: **80 %** | |

Each row represents one measurement, distance values $d$ are given in meters and all other values are given in radians. Inputs of the neural network hand controller: $d$ – object distance, $v$ – vertical directional angle of the object, $h$ – horizontal directional angle of the object. Outputs of the neural network hand controller: $s_p$ – shoulder pitch joint, $s_r$ – shoulder roll joint, $e_r$ – elbow roll joint. The last column determines if the touch was successful or not.

# 7     Conclusion

We have implemented a system processing images from two cameras of a stereovision system mounted on a Nao humanoid robot to estimate 3D coordinates of the target object. The images are captured at approximately 10 fps and individually processed by TLD tracking systems. The applicability of the 3D tracking system was tested and the ability of the humanoid to touch objects tracked in 3D by the proposed system was verified.

**Acknowledgment.** The work presented in this paper was supported by VEGA, Grant Agency of Ministry of Education and Academy Science of Slovak Republic under Grant No. 1/0298/12 – "Digital control of complex systems with two degrees of freedom." The work presented in this paper was also supported by KEGA under Grant No. 018TUKE-4/2012 – "Progressive methods of education in the area of control and modelling of complex systems object oriented on aircraft turbo-compressor engines."

# References

1. Kalal, Z., Matas, J., Mikolajczyk, K.: P-N Learning: Bootstrapping Binary Classifiers by Structural Constraints. In: Conference on Computer Vision and Pattern Recognition (2010)
2. Kalal, Z., Matas, J., Mikolajczyk, K.: Online learning of robust object detectors during unstable tracking. In: 3rd Online Learning for Computer Vision Workshop 2009. IEEE CS, Kyoto (2009)
3. Kalal, Z., Matas, J., Mikolajczyk, K.: Forward-Backward Error: Automatic Detection of Tracking Failures. In: International Conference on Pattern Recognition, Istanbul, Turkey, August 23-26 (2010)
4. Aldebaran Robotics. Nao Website, http://www.aldebaran-robotics.com/en/
5. Aldebaran Robotics. Nao Documentation v1.14.2, http://www.aldebaran-robotics.com/documentation/index.html
6. Lucas, B.D., Kanade, T.: An iterative image registration technique with an application to stereo vision. In: Proceedings of the International Joint Conference on Artificial Intelligence, pp. 674–679 (1981)
7. Breiman, L.: Random forests. ML 45(1), 5–32 (2001)
8. Yilmaz, A., Javed, O., Shah, M.: Object Tracking: A Survey. ACM Comput. Surv. 38(4), Article 13, 45 (2006)
9. Puheim, M.: Application of TLD for object tracking in stereoscopic images. Diploma thesis. Technical University of Košice. Faculty of Electrical Engineering and Informatics. Košice, 68 pages (2013) (in Slovak)

# Morphogenetic and Evolutionary Approach to Similar Image Creation

Kei Ohnishi[1] and Köppen Mario[2]

[1] Institute of Technology, 680-4 Kawazu, Iizuka, Fukuoka 820-8502, Japan
ohnishi@cse.kyutech.ac.jp
[2] Kyushu Institute of Technology, 680-4 Kawazu, Iizuka, Fukuoka 820-8502, Japan
mkoeppen@ci.kyutech.ac.jp

**Abstract.** This study presents a method for creating various images similar to an input image. The method first generates a graph which node corresponds to a pixel of an input image and which edge is made between nodes if a given condition on similarity in brightness between the corresponding pixels is met. The generated graph structure is influenced by distribution of brightness in the input image. Then, the method executes an algorithm inspired by biological development which can form various structures by changing its parameters values on the generated graph to newly determine the brightness of each pixel corresponding to a node in the graph. Next, the method uses an evolutionary algorithm to optimize parameters of the algorithm inspired by biological development to make the newly created image close to the input image. Experimental results demonstrate that the presented method can create various images similar to an input image depending on how to form the graph and design a fitness function of evolutionary algorithm.

## 1 Introduction

Recently, many proposed engineering methods have been inspired by biological behaviour and structures, such as evolutionary computation [1], neural networks [2], artificial immune systems [3], ant colony systems [4] and so on. The reason is that we can expect to obtain useful hints on creating new methods from the study of biological behaviour and structures.

Some of these methods are equivalent to the approaches used in artificial life (A-Life). These approaches create macroscopic structures or functions by only using local interaction between their elements. In these approaches, the macroscopic structures or functions are often quantitatively undefined but qualitatively defined. Consequently, we need to adjust the local interaction rules for our purpose. However, the A-Life approaches have several advantages. One advantage is that these approaches have the possibility to realize structures or functions that are barely realized by a centralized control system.

Simple systems using the A-Life approaches are feedback systems such as cellular automata [9] and L-systems [5][8]. These systems can create complex structures from a simple initial state. Complex structures are created by repeatedly applying rules of rewriting symbols corresponding to their structures. It is a simple but difficult task for us to estimate the final structure from the initial state and rules, because the rules are applied locally. Their effective use may be to discover or to create interesting structures by adjusting the rules and the initial state by trial and error. Interesting structures that we could have never imagined may be discovered by chance.

Following the idea described above, we previously proposed a new feedback system inspired by biological development [6] which will be hereinafter referred to as "BDS". Biological development mechanisms have been attracting attentions in many engineering fields [11][10]. BDS is meant to form patterns in Euclidean spaces. The advantage of BDS is the embedded rough-design structure and various possible generated structures under the given rough-design structure. Conventional feedback systems, such as the L-System or cellular automaton, apply only one rule to each element at each feedback loop, so it is difficult to design an entire final structure from the beginning and embed it into the system. Meanwhile, since BDS determines final detail from the basic structure in each module, it is easier to embed the rough-design structure before the system runs.

However, it has been shown that patterns formed in Euclidean spaces by BDS are not diverse and have some tendency. Therefore, to form more diverse patterns in Euclidean spaces, in our previous paper [7], we presented a method that extends BDS, which will be hereinafter referred to as "BDS-G". BDS-G is able to form patterns in a graph which nodes correspond to coordinates in a Euclidean space and which edges can represent arbitrary relationships among nodes corresponding to coordinates in a Euclidean space. In addition, we applied BDS-G to generate a gray-scaled image which pixels, which can be considered to be coordinates in a two-dimensional Euclidean space, correspond to nodes in a graph. Actually, in [7], a graph in which BDS-G runs is formed from a given input image. Concretely, edges among the nodes are determined, based on the distribution of brightness of the pixels in the given input image. Therefore, a pattern formed by BDS-G is influenced by the given input image. However, BDS-G has no mechanism for adjusting its parameters to create desired patterns. Thus, in this paper we propose a combination of BDS-G and a genetic algorithm (referred to as "GA" hereinafter) for adjusting parameters of BDS-G and demonstrate that this combination method can create a variety of gray-scaled images similar to an input gray-scaled image.

The present paper is organized as follows. Section 2 describes BDS which is a basis of BDS-G. In Section 3 we propose a method that combines BDS-G for pattern formation in graphs with GA and then apply the proposed method to generating gray-scale images similar to a given input image in Section 4. Section 5 describes our conclusion and future work.

## 2     Pattern Formation in Euclidean Spaces

BDS-G [7] described in Section 3 is realized based on BDS [6]. Thus, we recall BDS in this section. The hints for BDS, which is a mechanism of biological development, is described in [6]. A biological development process forms an adult body from a mother cell based only on design information of the mother cell in a rough-to-detail manner. In this process, proteins hierarchically diffuse among all or some cells and give them positional information, telling the cells what organs they will eventually become.

## 2.1    The Conceptual Framework

BDS outputs values on a space. Values of the system parameters are provided in a character code. It has an initialization module and three major modules. The module numbers like (0), (1), etc., below correspond to numbers in Figure 1. The initialization module (0) defines the area where system outputs exist. The module (1) generates the global positional information that determines what each element processes. The module (2) lets each element behave autonomously based on its positional information. Local mutual interaction among elements may take place. The module (3) exchanges the output of the module (2) into the output range of elements. The final output value of each element is determined by the feedback module (4b).

The module that generates data related with structure is the module (3). The values that the feedback process at the module (3) generates after a certain number of feedback loops determine the final structure of the system output. The modules (1) and (2) sequentially generate the structure of the final output pattern from BDS.

The modules (1) and (3) include feedback modules (4a) and (4b), respectively. The feedback module (4a) generates the positional information of the next state from that of the previous state. The feedback module (4b) gradually narrows the range of output values of each feedback process.

## 2.2    Example Algorithm

An example algorithm under the framework of BDS described in Section 2.1 generates values of y for given coordinates of x. Figure 1 shows the algorithm flow.

The task here is to draw a figure in a $(x, y)$ space by generating values of $y$ for coordinates of $x$ in a fixed closed area, $x \in [x_1, x_n]$ $(x_1 < x_n)$. Then, the modules of the algorithms are as follows. The modules are (0), (1), (2), (3), (4a) and (4b) in Figure 1. A character code presented in Figure 1 encodes values of the algorithm parameters.

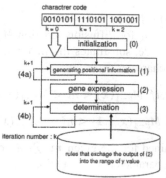

**Fig. 1.** The example of algorithm flow

## (0) initialization

The module (0) in Figure 1 inputs the range of $x$ space where $y$ values are generated to draw a figure. The initial data inputted here are $(x_1, x_n)$.

## (1) generating positional information

The module (1) in Figure 1 gives positional information of each $x$ coordinate using diffusion functions such as Equation (1) (see Figure 2(a) (1).)

$$M(x, A_k, \mu_k, \sigma_k) = M_k(x) = A exp^{-\frac{(x-\mu_k)^2}{2\sigma_k^2}} \tag{1}$$

Let us call the $M_k(x)$ in Equation (1) a diffusion function from the analogy to diffusing protein among cells in biological development. The iteration number of feedback is represented by $k$. Any function is available for the diffusing functions and we adopt a Gaussian function for $M_k$ in this paper. The positional information is determined by $M(x, A_k, \mu_k, \sigma_k)$ which parameters $A_k, \mu_k$, and $\sigma_k$ are read from the character code. Although the algorithm presented here uses only one diffusion function as in Equation (1), other algorithms may use multiple diffusion functions. That is, $\sum_i M(x, A_{ki}, \mu_{ki}, \sigma_{ki})$ is used.

## (2) gene expression

The module (2) in Figure 1 divides $x$ into several areas according to $s_i$ read from the character code and values of $M_k(x)$ (see Figure 2(a) (2)). Labels are given to the divided areas of $x$. We compared the given labels to the genetic information and named this module as a gene expression.

## (3) determination

The module (3) in Figure 1 converts the group labels outputted from the previous module (2) together with previous $y$ range into the range of values of $y$ (see Figure 2(a) (3)). The conversion rules that are different in each processing time, $k$, must be prepared.

## (4) feedback

The modules (4a) and (4b) in Figure 1 are feedback modules at the modules (1) and (3), respectively.

## (4a) feedback for the module (1)

The module (4a) calculates $\mu_k$ for diffusion function $M_k(x)$ from $M_{k-1}(x)$. It first reads $m$ from the character code, substitutes the $m$ for Equation (2) and determines $\mu_k$, where $\mu_0$ is given in the character code.

$$\mu_k = x, if \ M_{k-1}(x) < m < M_{k-1}(x + 1), or \tag{2}$$

$$if \ M_{k-1}(x + 1) < m < M_{k-1}(x).$$

When multiple $\mu_{ki}$ are obtained for the given $m$, $\sum_i M(x, A_k, \mu_{ki}, \sigma_k)$ is used in the module (1). Other necessary parameters for Equation (1), $A_k$ and $\sigma_k$, are read from the character code.

**(4b) feedback for the module (3)**

According to the feedback module (4b), the range of $y$ is gradually reduced (see Figure 2(b)). The rules that determine the new ranges of $y$ from the past ranges $y$ and group labels must be previously prepared in a rule-base at the module (3).

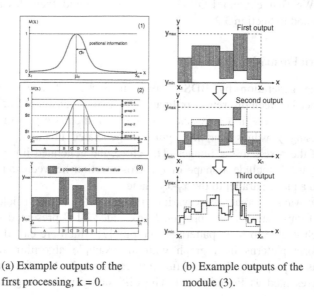

(a) Example outputs of the first processing, k = 0.

(b) Example outputs of the module (3).

**Fig. 2.** Example outputs

## 2.3   Generating Gray-Scaled Images

We demonstrate how BDS creates gray-scaled images. The algorithm used here is similar to the algorithm in Section 2.2 except for a 2-D target space, $(x_1, x_2)$ and the initial diffusion function combining 50 Gaussian functions. The algorithm used here obtains a final output just after the fifth feedback. Images are displayed by converting the generated $y$ values to the brightness of pixels. Two examples of generated images are shown in Figure 3. In Figure 3, outputs at the first and the fifth feedback are shown. Although outputs except the final output represent ranges of $y$ values on the $x_1$-$x_2$ plane, we create images from those outputs by using the smallest value in the ranges of $y$ values.

(a) First example of the outputs at $k = 1$ (left) and at $k = 5$ (right).

(b) Second example of the outputs at $k = 1$ (left) and at $k = 5$ (right).

**Fig. 3.** Two examples of gray-scaled images generated by BDS

## 3    Proposed Method

The proposed method combines BDS-G that extends BDS described in Section 2 for generating various structures in Euclidian spaces with GA that optimizes parameters of BDS-G. We first explain BDS-G in Section 3.1 and then describe the entire proposed method in Section 3.2.

### 3.1    Pattern Formation in Graphs

In the image generation by BDS shown in Section 2.3, we first considered correspondence between discrete coordinates in a two dimensional Euclidean space (a $x_1$-$x_2$ plane) and pixels in a gray-scaled image and, then, we generated gray-scaled images by using $y$ values that the algorithm generated in the $x_1$-$x_2$ plane as the brightness of the pixels in the images. Therefore, the generated images basically, as shown in Figure 3, include a shape of the cross section between multiple Gaussian functions and a plane parallel to the $x_1$-$x_2$ plane.

One of the ways to create more various images is to change a field or a space in which BDS forms patterns. We previously extended BDS presented in Section 2 to the one which is able to form patterns in graphs [7] which are called BDS-G in this paper. To form patterns in a graph with an example algorithm of BDS-G, the algorithm has to prepare a diffusion function for a graph. A diffusion function used herein is represented as Equation (3). The diffusion function is generated at some node.

$$M(h_s, A_k, \mu_k, \sigma_k) = M_k(h_s) = A \exp^{-\frac{(h_s-\mu_k)^2}{2\sigma_k^2}}, \tag{3}$$

where $h_s$ is the minimal number of hops from a node of focus to the node at which the diffusion function is generated. The diffusion function is basically the same as Equation (1) shown in Section 2.2, but $x$ in Equation 1 is changed to $h_s$ in Equation (3).

In our previous study, we applied the example algorithm of BDS-G to generating gray-scaled images. In this application, a graph is first generated from a given original image in which pixels correspond to nodes and the nodes are linked to one another in a certain way (see Figure 4), and then the algorithm is run in the generated graph. The formed pattern in the graph is a distribution of brightness of pixels which are equal to nodes.

How to generate a graph from a given image in our previous study was as follows. Each node which has a corresponding pixel makes $L$ edges to other nodes. Nodes to which each node makes edges are selected from among all of the nodes with probability inversely proportional to difference between the brightness of each node and the other node. That is, nodes with similar a brightness are likely to link to one another. The brightness of the pixels of the given image is used only to generate a graph which becomes a field to which the algorithm is applied.

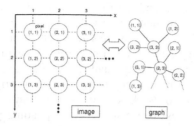

**Fig. 4.** Correspondence between nodes in a graph and pixels in a gray-scaled image

## 3.2    Optimization by a Genetic Algorithm

GA is used for optimizing parameters of BDS-G under a given fitness function. As in our previous study, we first form a graph from a given gray-scaled image and then run BDS-G in the formed graph to create new gray-scaled images in this paper. GA is used in this procedure of image creation and a fitness function is designed for BDS-G to be able to create a variety of images similar to the input image. The entire proposed method for an application of similar image creation is shown in Figure 5.

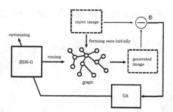

**Fig. 5.** The entire proposed method for similar image creation

## 4    Similar Image Creation

The algorithm here uses the initial diffusion function combining 50 Gaussian functions and obtains a final output just after the fifth feedback. One new image is created from one GA individual which is equivalent to a character code of 685 real-valued parameters in BDS-G. A graph in which BDS-G runs is formed by making edges between nodes with inverse proportions to difference between the brightness of each node. We set $L$, which is the number of edges that each node makes, to be 6. A fitness function of GA used here calculates the difference between the brightness of every pair of corresponding pixels in the original and the created images, and increases a fitness value by one if the absolute value of the difference between the brightness of each pair is less than five. Thus, the GA handles a maximization problem here. A population size of the GA is 50 and the stop condition of the GA is 2,000 fitness evaluations, which is 50 generations.

Figure 6 shows an example of the created images and the original image used for forming the graph. As we can see from Figure 6, the created images are not similar to

338      K. Ohnishi and K. Mario

the original image at a glance and the final fitness value was just 2,194 where the fitness value of the global optimum was 16,384. That would mean that it is hard for the combination of the BDS-G and the GA to create similar images to the original image under the use of the graph and the fitness function explained above.

Therefore, we will change the graph and the fitness function used above. A new graph is formed by considering not only the similarity of brightness of the pixels but also the closeness between pixels. A new fitness function considers not only one-to-one matching of the brightness between two corresponding pixels in the created and

(a) Original image.  (b) The best image   (c) The best image   (d) The best image
in the initial popula-   at the 20th genera-   at the 40th generation.                  tion.                 tion.

**Fig. 6.** An example of gray-scaled images created by the proposed method

(a) The best image b) The best image at c) The best image
in    the    initial the 20th generation at       the     40th
population   in   the in the first example. generation   in   the
first example.                                first example.

(d) The best image (e) The best image at (f) The best image at
in    the    initial the 20th generation the 40th generation
population   in   the in    the    second in    the    second
second example      example.           example.

**Fig. 7.** Two examples of gray-scaled images created by the proposed method when using different graphs and fitness functions

the original images but also the matching of brightness of corresponding local regions in them. Figure 7 shows an example of the created images. Compared to the created images in Figures 6 and 7, the created images using the new graph and fitness function look a little similar to the previously created ones. Thus, we saw that how similar images are created depends on how to form a graph in which BDS-G runs, as well as how to design a fitness function of GA.

## 5    Concluding Remarks

We proposed the new method for similar image creation which combines BDS-G with GA. The experiment results demonstrated that the proposed method is capable of creating various images similar to a given image depending on how to form a graph in which BDS-G runs, as well as how to design a fitness function of GA. In the proposed method, it is possible to represent fundamental desired structures or structural constraints as a form of graph topologies. Thus, the combination of the use of graph topologies, the strategy of rough-to-detail pattern formation of BDS and GA as an optimisation technique would be able to be an emergent design tool that is useable for humans. In the future work, we will examine it further.

**Acknowledgments.** This work was supported by the Japan Society for the Promotion of Science through a Grant-in-Aid for Scientific Research (C)(25330289).

## References

1. Bäck, T.: Evolutionary Algorithms in Theory and Practice: Evolution Strategies, Evolutionary Programming, Genetic Algorithms. Oxford University Press (1996)
2. Rojas, R.: Neural Networks: A Systematic Introduction. Springer, Berlin (1996)
3. Dasgupta, D. (ed.): Artificial immune systems and their applications. Springer, Heidelberg (1999)
4. Dorigo, M., Maniezzo, V., Colorni, A.: The ant system: optimization by a colony of cooperating agents. IEEE Trans. on Systems, Man, and Cybernetics-Part B 26(2), 29–41 (1996)
5. Lindenmayer, A.: Mathematical models for cellular interaction in development, PartI and PartII. Journal of Theoretical Biology 18, 280–315 (1968)
6. Ohnishi, K., Takagi, H.: Feed-Back Model Inspired by Biological Development to Hierarchically Design Complex Structure. In: IEEE International Conference on System, Man, and Cybernetics (SMC 2000), Nashville, TN, USA, October 8-11, pp. 3699–3704 (2000)
7. Ohnishi, K., Yoshida, K., Köppen, M.: Pattern Formation in Networks Inspired by Biological Development. In: Huang, F., Wang, R.-C. (eds.) ArtsIT 2009. LNICST, vol. 30, pp. 64–71. Springer, Heidelberg (2010)
8. Prusinkiewicz, P., Lindenmayer, A.: The algorithmic beauty of plants. Springer, New York (1990)
9. Wolfram, S.: Cellular automata and complexity: collected papers. Addison-Wesley, Reading (1994)

10. Doursat, R., Sayama, H., Michel, O. (eds.): Morphogenetic Engineering: Toward Programmable Complex Systems (Understanding Complex Systems). Springer, New York (2013)
11. Meng, Y., Zhang, Y., Jin, Y.: Autonomous Self-Reconfiguration of Modular Robots by Evolving a Hierarchical Mechanochemical Model. IEEE Computational Intelligence Magazine 6(1), 43–54 (2011)

# Interactive Evolutionary Computation in Modelling User Preferences

Miron Kuzma and Gabriela Andrejková

Institute of Computer Science, Faculty of Science, Pavol Jozef Safarik University in Kosice,
Jesenna 5, Kosice, Slovakia
miron.kuzma@student.upjs.sk

**Abstract.** The modelling of user preferences in many applications is very interesting and is one of the problems researched during the last year. We researched the possibilities of neural networks to predict user subjective preferences using human-machine cooperative systems that use Interactive Evolutionary Computation (IEC). In such systems a subjective preference (evaluation) is a response to a system generated proposals. We consider these preferences to present the relative discrete fitness function values. We showed that searching for a preferred solution can be accelerated and evaluation characteristics can be obtained quicker if the target fitness values are converted from relative values to absolute values. We described a formula for a conversion of relative fitness function values to absolute values in IEC algorithms. We used a recurrent neural network to predict user preferences on a problem of the most attractive font face. Our experiments showed a substantial improvement of the error of the neural network in testing phase when using absolute fitness function values.

## 1 Introduction

The human fatigue is a serious problem in every human-machine system that is used to find optimal solutions according to subjective preferences. The acceleration of methods used in such systems is one of the possible solutions to overcome this problem. The faster we predict user subjective preferences, the more we reduce the human's fatigue.

Interactive evolutionary computation (IEC) is widely used in human-machine systems in various application fields. The IEC has some boundaries which are still a subject of research mainly by Takagi [24] and his laboratory. The human's fatigue limits the search space that is explored (e.g. the number of generations and the number of individuals displayed in one generation). Therefore, the research [19] aims at approaches that try to reduce the dimension of the problem and explore the reduced dimension [20], model the fitness landscape [21], recalculate the fitness values given by the user [25] or include some background information about the user including some psychological aspects which describe each human as an individual combination of basic human values [11, 12].

In order to not bring noise to the modelling user preferences, we consider the fitness values given as an evaluation by the user in each generation to be relative. If we don't consider them to be relative, we introduce the noise into the learning process which further results in averaging all evaluations of the same individuals. We need to find an approach that performs transformation of these values to absolute scale. There exists a proposal for performing such recalculation of relative values introduced by Wang and Takagi [25] but it expects the evaluation of the same individuals in generations following one after another. The difference between the individuals in compared generations is added to or subtracted from the relative fitness value depending on minimizing or maximizing the fitness function. After the shift from relative values the neural networks are used to learn the user evaluation characteristics. These experiments were performed on simulated data generated from Hartmann's and Schwefel's functions. Wang and Takagi showed that convergence of the interactive evolutionary computation is much faster with the transformation of relative fitness function to absolute fitness function.

In our research we tried to design a method that avoids the limitation of differential comparisons of individuals between generations and proposed a method that will transform the relative fitness function values to absolute values without comparing the same individuals between generations. The method also gives a variable amount of weight of data from the history driven by parameter. In our experiments we used the recurrent neural network to predict the user preferences comparing various amounts of weight of data from the history. In comparison to Wang and Takagi, we performed these experiments on data obtained from real users. Collecting data from real users often results in smaller data set compared to the generated data set.

In this paper we briefly describe the basic types of optimization problems. Then, we describe the Interactive Evolutionary Computation methods in Sec. 2. Next, we explain the problem of the most attractive font face we use to collect the data for our experiments in Sec. 3. We describe our formula in Sec. 4. We describe the experiments we performed in Sec. 5 and we conclude in Sec. 6 the results we obtained from the experimental part.

## 2    Optimization Problems with Interactive Evolutionary Computation

There are commonly known two types of optimization problems with the following characteristics:

1. Problems that have explicitly declared optimization function $f$ by some mathematical formula [13]. There exist many methods for solving such optimization problems, e.g.: gradient, non-gradient methods and also evolutionary optimization algorithms [1-3][14][15] which find acceptable solutions or solutions identical to global minimum.

2. The second type of optimization problems are those that do not have a mathematical expression of optimization function $f$. Some methods for solving such optimization problems were introduced by Takagi [23]. These methods

replace the optimization function $f$ which is not directly observable by embedding human preference, intuition, emotion, knowledge and psychological aspects, also called KANSEI.

In this paper we focus on the second type of optimization problems. The already published survey [24] gives a summary of recent knowledge about Interactive Evolutionary Computation (IEC) which is used to solve these types of problems. There already exists a large variety of systems that use IEC, e.g.: [4][5], and other systems such as [6-8][17][18]. IEC is a set of methods used for optimization problems with unknown optimization function. It uses evolutionary computation (EC) for an optimization based on human subjective preferences. It is an EC algorithm which fitness function is replaced by a human user choice input from a set of discrete values, e.g.: {1; 2; 3; 4; 5}. We consider these values to be relative in each generation of the evolutionary algorithm. One of the hidden side effects of this relative evaluation is that one individual can obtain different relative fitness function values in different generations (What fitness value is the right one?) or more individuals can obtain the same relative fitness function values in different generations (Which individual is the better one?) - it means that there is a contradiction in data.

**Fig. 1.** User evaluates what he sees or hears. Illustrative image according to Takagi [24].

Fig. 1 shows a general IEC system where the user evaluates what he sees or hears and the EC optimizes target parameters of a particular solution proposal for the given problem in order to obtain the preferred output based on the user subjective preferences. Fig. 1 also illustrates one loop (or iteration, generation of the evolutionary algorithm) of a system that is based on IEC. The global optimum of the IEC is rough because every system output that a human user cannot distinguish is considered to be psychologically the same (e.g. it is difficult to distinguish 256 tones of blue colour). The global optimum of the IEC is not a point but rather a small area. The population size of the evolutionary algorithm is limited by the number of phenotypes (generally 3 to 16) that are simultaneously spatially displayed on a computer monitor or by the human capacity to remember sounds, images, movies, etc. The number of search generations (generally 10 to 20) is limited by human fatigue as well [24]. That is the main reason why future research can be focused on methods that reduce user's fatigue in IEC systems. The acceleration of evolutionary computation is one of the possible solutions to overcome this problem.

# 3     The Problem of the Most Attractive Font Face

One of the problems that can be solved by IEC is the problem of the most attractive font face [8]. We have given one type of font which face is described by a finite set of parameters, where a domain of a parameter could be a finite set of values or an interval. The task is to set the values of parameters. By applying values of these parameters to the given font user, it should get the most attractive font for him. We use Computer Modern Font's Metafont configuration file [16]. By changing the value of one of its parameters at one location in the Metafont file, one can produce a consistent change throughout the entire font. Computer Modern Roman illustrates many uses of this feature (Roman Font uses essentially the same Metafont file but with different global parameters). The font configuration file contains 62 parameters. We have experimentally determined 21 parameters [8] that have major impact on the final font look. The list of parameters is in Table 1 with a corresponding description, lower and upper thresholds. The original font definition file contains a parameter vector $\vec{x}$ of dimension 62. Each element in vector presents one parameter from those 62 parameters. Each element in vector $\vec{x}$ is equal to zero if it presents one of the 21 estimated parameters of our modification. The modification vector $\Delta\vec{x}$ is a vector of dimension 62 and its elements present parameters we use for modification. Those elements which are equal to zero are not used for the modification of the font face. We obtain a new parameter vector $\vec{y}$ by addition of the vector $\Delta\vec{x}$ to the vector $\vec{x}$:

$$\vec{y} = \Delta\vec{x} + \vec{x} \tag{1}$$

The genotype of a candidate solution of this problem has 21 parameters and the phenotype is a proposed font. We developed an IEC application for testing purposes [8][9] where each user has to find his "most attractive font face". The application spatially displays whole population - 12 fonts in each generation of the EC. The user has to evaluate (i.e.: assign relative fitness function values from {1; 2; 3; 4; 5}) displayed fonts (that is why we consider the user's response to be a relative discrete fitness function) in each generation according to his subjective preferences. The session of each user was recorded for a future user preference prediction with neural networks.

**Table 1.** Parameter domain of the problem of the most attractive font face

| $D_i$ | Lower limit | Upper limit | $D_i$ | Lower limit | Upper limit |
|-------|-------------|-------------|-------|-------------|-------------|
| P1 | 0 | 100 | P12 | 0 | 100 |
| P2 | 0 | 100 | P13 | 0 | 100 |
| P3 | 0 | 100 | P14 | 0 | 0,5 |
| P4 | 0 | 80 | P15 | 0 | 1,5 |
| P5 | 1 | 10 | P16 | 0 | 1,0 |
| P6 | 0 | 80 | P17 | 0 | 0,7 |
| P7 | 0 | 60 | P18 | False | True |
| P8 | 0 | 60 | P19 | False | True |
| P9 | 0 | 100 | P20 | False | True |
| P10 | 0 | 30 | P21 | False | True |
| P11 | 0 | 100 | Time | 0 | Individual |

# 4    Converting Relative Evaluations to Absolute Fitness Function

We would like to contribute to overcome the human fatigue problem in IEC systems by proposing a method that recalculates the absolute fitness values from the relative fitness values but it does not expect that $l$ individuals (system generated proposals) will appear in two or more generations of an evolutionary algorithm to determine the difference in their evaluation. We consider the user evaluation of proposed solutions in one loop of the IEC system to be our relative discrete fitness function. Next, we consider the result of the formula we propose to be the absolute fitness function for each generated proposed solution. In our recent work [10] we performed an experimental conversion of relative to absolute fitness inspired by Pei, Takagi and Wang [19][23][25]. We tried to model the optimization function (the prediction of user preference) with a feed forward neural network with one hidden layer. We concluded that using a relative fitness function as a desired output from the model slowed down the convergence and caused the learning process (the neural network error) to oscillate. When we used absolute fitness function values as a target value, this oscillation disappeared, e.g., the relative values act as a noise in the prediction of the user preferences.

The first proposal [25] shows how to perform such calculation considering that the same individual or $l$ individuals intentionally appear in more generations during the user session. Comparing the difference of fitness of $l$ individuals among generations, one can calculate the absolute fitness values from the relative.

Our method is also inspired by Pei [19]. The author stated it would be interesting to research the policy named as "Dynamic Fitness Threshold", i.e. only individuals with a higher fitness than a certain threshold (which also increases after each generation) are accepted in the next generation. According to Takagi [24], after every loop of the IEC we are closer to the user's desired preferences, i.e. we explore the areas with higher fitness function values. Considering these ideas, we reconstructed the formula from Wang [25], we included the idea of the dynamic threshold of fitness from Pei [19] which is expressed as a weighted amount of historical data in Eqn. 2:

$$v_j^{'i} = v_j^i + \gamma * \frac{\sum_{j=1}^n v_j^{'i-1}}{n} \tag{2}$$

where $v_j^{'i}$ is the absolute fitness of individual $j$ in a generation $i$, $v_j^{'i+1}$ is its relative fitness, $n$ is the number of individuals in a generation. Notice that for the generation $i = 1$ the expression is $\sum_{j=1}^n v_j^{'i-1} = 0$. Our formula does not expect $l$ individuals to appear in more generations, it evaluates the whole previous generation. We apply this formula to our relative fitness values and recalculate the target values for the neural network. In addition to our previous work [10], we updated the neural network structure we are going to use and we also updated the transformation formula by adding a variable that specifies the amount of weight of historical data. The formula can produce a continuous function. So, as a result, our absolute fitness function (the formula result) calculated from the relative fitness values (which are discrete) can also be continuous depending on $\gamma$. The proposed formula produces a non-decreasing function.

## 5     Experiments

### 5.1     Data Description

We have a collection of dataset from our IEC font design application [8]. The set was also used in previous experiments in [9][10]. It contains 2490 samples from 4 subjects. Each sample is one font from one generation of some of the user's session: a vector of 21 + 1 parameters of font and also time from Table 1 and it also includes the relative evaluation according to the user's subjective preference.

### 5.2     Experiment Description

We try to predict the user's subjective preference with a recurrent neural network. We expect a decrease of the error of the neural network after transformation of relative fitness values to absolute fitness values. We used the cross-validation technique to split the data set into 4 folds and ran experiments 4 times, e.g.: 75% training set and 25% testing set. We use the proposed formula from Eqn. 2 for transformation of relative fitness values to absolute fitness values. We also tested $a$ values in $\langle 0; 1 \rangle$ interval by step of 0.1 - a 10% increase of historical data from 0% (that actually means no calculation is made and the relative fitness is used) to 100%.

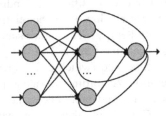

**Fig. 2.** The structure of the neural network used by our experiment showing the recurrent connections

From the left: input layer, hidden layer and output layer. We updated the topology of the neural network from [10] which now uses recurrent connections. The neural network we use has a topology of $22 - 6 - 1$ neurons with recurrent connections from output to the neurons in the hidden layer, as we can see in Fig. 2. Each sample from dataset is a 22 dimensional vector, so the size of the input layer is 22. We expect the absolute fitness function value as the desired output from the neural network. We use $1 \times 10^4$ and $2 \times 10^4$ epochs to adapt the neural network, considering the second number of cycles to be sub-optimal. We used the backpropagation learning algorithm in time with learning parameter $\eta = 0.05$ to perform the neural network learning phase. The model of the neural network was created and tested in the PyBrain machine learning suite for Python programming language [22].

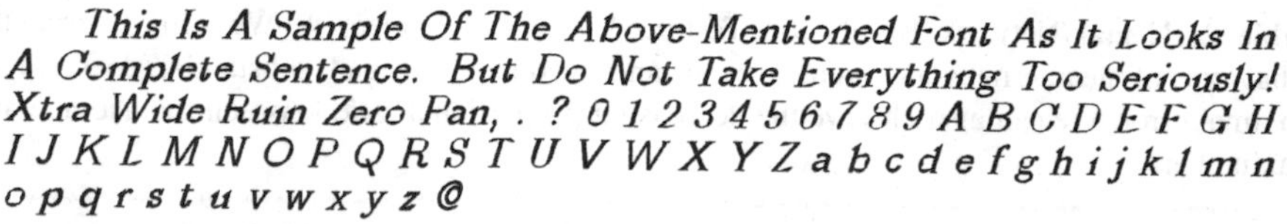

This Is A Sample Of The Above-Mentioned Font As It Looks In
A Complete Sentence. But Do Not Take Everything Too Seriously!
Xtra Wide Ruin Zero Pan, . ? 0 1 2 3 4 5 6 7 8 9 A B C D E F G H
I J K L M N O P Q R S T U V W X Y Z a b c d e f g h i j k l m n
o p q r s t u v w x y z @

**Fig. 3.** Illustrative figure of a final font face with the highest absolute fitness from subject 4

### 5.3 Experiment Results

We obtained the results that have similar characteristics for every subject from the data set but are different in some small aspects. The neural network test using data from subject 1 and subject 2 has the lowest average and maximum error at $\gamma = 0.7$ and $\gamma = 0.6$, using data from subject 3 and subject 4 at about $\gamma = 0.8$. Next, we can observe similar characteristics: the decrease of average error and also maximum error of the network with increasing value of parameter $\gamma$ (see Fig. 4 and Table 2). This behaviour was expected. We also attached an illustrative image of the most attractive font face for subject 4 on Fig. 3. This font face got the highest absolute fitness value according to our experiments on data from subject 4.

Neural Network Results on Testing Data from Subject 4.
Error
Weight of Historical Data - Parameter γ
0.4
0.35
0.3
0.25
0.2
0.15
0.1
0.05
0
0
0.2
0.4
0.6
0.8
1
Average10
Maximum10
Median10
Average20
Maximum20
Median20

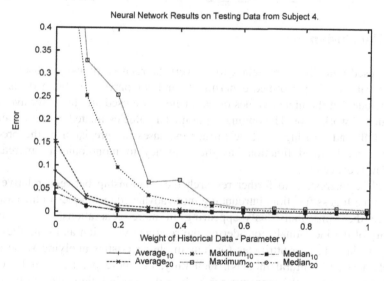

**Fig. 4.** The testing phase of neural network on data from subject 4 shows that with the increase of parameter $\gamma$ we get higher precision. Notice that indexes accompanying error variables indicate $1 \times 10^4$ and $2 \times 10^4$ epochs of the neural network training.

We were able to observe that the neural network average, maximum and median error in 1×104 epochs is higher than in 2×104 epochs on data from subject 1 and 3 and slightly varies on data from subjects 2 and 4. This can be caused by the random splitting of the dataset into training and testing sets. This side effect could be eliminated repeating the experiments using more than 4 folds in cross-validation.

348    M. Kuzma and G. Andrejková

**Table 2.** Neural Network Results on Testing Data from Subject 4. We can observe the behaviour of neural network average error and median error and substantial decrease of maximal error. We can also observe the decreasing proportional ratio between the average and maximal error.

| Epochs γ | Average Error | | Max. Error | | Median Error | |
|---|---|---|---|---|---|---|
| | $1\times10^4$ | $2\times10^4$ | $1\times10^4$ | $2\times10^4$ | $1\times10^4$ | $2\times10^4$ |
| 0.0 | 0.088418 | 0.152117 | 0.748708 | 1.100739 | 0.039442 | 0.058237 |
| 0.1 | 0.028654 | 0.035244 | 0.252271 | 0.328782 | 0.013008 | 0.011233 |
| 0.2 | 0.007879 | 0.014055 | 0.096204 | 0.254284 | 0.002787 | 0.004285 |
| 0.3 | 0.004090 | 0.009872 | 0.036370 | 0.064711 | 0.001750 | 0.002695 |
| 0.4 | 0.003119 | 0.006090 | 0.022898 | 0.069772 | 0.001467 | 0.001260 |
| 0.5 | 0.002076 | 0.001925 | 0.014247 | 0.019059 | 0.000804 | 0.001073 |
| 0.6 | 0.001306 | 0.001354 | 0.006889 | 0.008999 | 0.000787 | 0.000620 |
| 0.7 | 0.000771 | 0.001031 | 0.004044 | 0.009427 | 0.000322 | 0.000536 |
| 0.8 | 0.000741 | 0.000672 | 0.003942 | 0.008611 | 0.000388 | 0.000220 |
| 0.9 | 0.000471 | 0.000924 | 0.002899 | 0.009570 | 0.000176 | 0.000460 |
| 1.0 | 0.000746 | 0.000668 | 0.004358 | 0.007072 | 0.000461 | 0.000252 |

## 6    Conclusion

We proposed a modified formula 2 to convert the relative fitness to absolute fitness function values. We performed experiments and compared the obtained results. We can conclude that the higher values of parameter γ we used are the lower average and maximum network error. The optimal γ parameter value estimated by our experiments in ⟨0.6; 0.8⟩ and also higher value further increases the precision of the prediction. Since the recalculated function has the tendency to monotonically increase, this behaviour is expected.

It will be interesting to further research the relationship between relative fitness and absolute fitness function, introducing more advanced techniques in the case we do not have the same individuals appearing in more generations, e.g. to measure similarity of the individuals considering the relative evaluation and time they appear but also to include user attributes (or human characteristics applying some existing psychological and cultural models). Generally, the more precise model we obtain using neural network, the more we can reduce the user's fatigue caused by the long time spent with IEC based system. We also proposed the solution of the problem of contradiction in data samples in datasets in IEC applications as a consequence of the application of the proposed formula, e.g. the cases when the user changes his mind over time about the same generated proposals.

**Acknowledgements.** This work was partially supported by the grant – 1/0479/12 VEGA – Combinatorial structures and complexity of algorithms by the Slovak Research and Development Agency.

# References

1. Bäck, T.: Evolutionary Algorithms in Theory and Practice. Evolution Strategies, Evolutionary Programming, Genetic Algorithms. Oxford University Press, New York (1996)
2. Brunovská, A.: Malá optimalizácia, Metódy, programy, príklady. Alfa, Bratislava (1990)
3. Fogel, D.B.: Evolutionary Computation. Toward a New Philosophy of Machine Intelligence. IEEE Press, New York (1995)
4. Gajdoš, M.: Reduction of Human Fatigue in IEC with Neural Networks for Graphic Banner Design. Master's Thesis, Košice, Technical University of Košice, Faculty of Electrical Engineering and Informatics, Department of Cybernetics and Artificial Intelligence (2006)
5. Jakša, R., Takagi, H., Nakano, S.: Image Filter Design with Interactive Evolutionary Computation. In: Proc. of the IEEE International Conference on Computational Cybernetics (ICCC 2003), Siofok, Hungary, August 29-31 (2003) ISBN 963 7154 175
6. Jakša, R., Takagi, H.: Tuning of Image Parameters by Interactive Evolutionary Computation. In: Proc. of 2003 IEEE International Conference on Systems, Man & Cybernetics (SMC 2003), Washington DC, October 5-8, pp. 492–497 (2003)
7. Kováč, J.: Image Database Search Using Self-Organizing Maps and Multi-scale Representation Master's Thesis, Košice, Technical University of Košice, Faculty of Electrical Engineering and Informatics, Department of Cybernetics and Artificial Intelligence (2007)
8. Kuzma, M.: Interactive Evolution of Fonts. Master thesis, Technical University of Košice (2008)
9. Kuzma, M., Jakša, R., Sinčák, P.: Computational Intelligence in Font Design. In: Computational Intelligence and Informatics: Proceedings of the 9th International Symposium of Hungarian Researchers, Budapest, pp.193–203 (November 2008)
10. Kuzma, M., Andrejková, G.: Interactive Evolutionary Computation in Optimization Problem Solving. In: Cognition and Artificial Life XII, Praha, vol. XII, pp. 120–125 (2012)
11. Kuzma, M.: Estimating Font Face Attributes According to User Preferences. In: Cognition and Artificial Life XIII, Star Lesn, vol. XIII, pp. 148–152.
12. Kuzma, M.: Improving the Estimation of a Font Face Attributes According to User Preferences. In: ITAT 2013: Information Technologies - Applications and Theory, vol. XII (2013)
13. Kvasnička, V., Pospíchal, J., Tiňo, P.: Evolučné algoritmy, STU Bratislava (2000)
14. Lukšan, L.: Metody s proměnou metrikou. Academia, Praha (1990)
15. Maňas, M.: Optimalizační metody. SNTL, Praha (1979)
16. Metafont Tutorial,
   http://metafont.tutorial.free.fr/downloads/mftut.pdf
   (cited May 8, 2008)
17. Neupauer, M.: Analysis of Medical Data using Interactive Evolutionary Computation. Master's Thesis, Košice, Technical University of Košice, Faculty of Electrical Engineering and Informatics, Department of Cybernetics and Artificial Intelligence (2006)
18. Pangráč, L.: Interactive Evolutionary Computation for Satellite Image Processing. Master's Thesis, Košice, Technical University of Košice, Faculty of Electrical Engineering and Informatics, Department of Cybernetics and Artificial Intelligence (2007)

19. Pei, Y., Takagi, H.: A Survey on Accelerating Evolutionary Computation Approaches. In: 2nd International Conference of Soft Computing and Pattern Recognition (SoCPaR 2011), Dalian, China, October 14-16, pp. 201–206 (2011)
20. Pei, Y., Takagi, H.: Accelerating IEC and EC Searches with Elite Obtained by Dimensionality Reduction in Regression Spaces. Journal of Evolutionary Intelligence 6(1), 27–40 (2013), doi:10.1007/s12065-013-0088-9
21. Pei, Y., Takagi, H.: Fourier analysis of the fitness landscape for evolutionary search acceleration. In: IEEE Congress on Evolutionary Computation (CEC), Brisbane, Australia, June 10-15, pp. 1–7 (2012)
22. Schaul, T., Bayer, J., Wierstra, D., Sun, Y., Felder, M., Sehnke, F., Rückstiess, T., Schmidhuber, J.: PyBrain. Journal of Machine Learning Research 11, 743–746 (2010)
23. Takagi, H.: Interactive Evolutionary Computation: System Optimization Based on Human Subjective Evaluation. In: IEEE International Conference on Intelligent Engineering Systems (INES 1998), Vienna, Austria, September 17-19, pp. 1–6 (1998)
24. Takagi, H.: Interactive Evolutionary Computing: Fusion of the Capacities of EC Optimization and Human Evaluation. In: Proc. of 7th Workshop on Evaluation of Heart and Mind, Kita Kyushu, Fukuoka, November 8-9, pp. 37–58 (2002)
25. Wang, S., Takagi, H.: Improving the Performance of Predicting Users' Subjective Evaluation Characteristics to Reduce Their Fatigue in IEC. Journal of Physiological Anthropology and Applied Human Science 24(1), 81–85 (2005)

# Author Index